DYNAMIC FUZZY PATTERN RECOGNITION WITH APPLICATIONS TO FINANCE AND ENGINEERING

INTERNATIONAL SERIES IN INTELLIGENT TECHNOLOGIES

Prof. Dr. Dr. h.c. Hans-Jürgen Zimmermann, Editor
European Laboratory for Intelligent
Techniques Engineering
Aachen, Germany

Other books in the series:

Distributed Fuzzy Control of Multivariable Systems
by Alexander Gegov

Fuzzy Modelling: Paradigms and Practices
by Witold Pedrycz

Fuzzy Logic Foundations and Industrial Applications
by Da Ruan

Fuzzy Sets in Engineering Design and Configuration
by Hans-Juergen Sebastian and Erik K. Antonsson

Consensus Under Fuzziness
by Mario Fedrizzi, Janusz Kacprzyk, and Hannu Nurmi

Uncertainty Analysis in Enginerring Sciences:
Fuzzy Logic, Statistices, and Neural Network Approach
by Bilal M. Ayyub and Madan M. Gupta

Fuzzy Modeling for Control
by Robert Babuška

Traffic Control and Transport Planning:
A Fuzzy Sets and Neural Networks Approach
by Dušan Teodorović and Katarina Vukadinović

Fuzzy Algorithms for Control
by H.B. Verbruggen, H.-J.Zimmermann. and R. Babŭska

Intelligent Systems and Interfaces
by Horia-Nicolai Teodorescu, Daniel Mlynek, Abraham Kandel
and H.J. Zimmermann

Practical Applications of Computational Intelligence Techniques
by Lakhmi Jain and Philippe De Wilde

DYNAMIC FUZZY PATTERN RECOGNITION WITH APPLICATIONS TO FINANCE AND ENGINEERING

LARISA ANGSTENBERGER

Kluwer Academic Publishers

Boston/Dordrecht/London

Distributors for North, Central and South America:
Kluwer Academic Publishers
101 Philip Drive
Assinippi Park
Norwell, Massachusetts 02061 USA
Telephone (781) 871-6600
Fax (781) 871-6528
E-Mail <kluwer@wkap.com>

Distributors for all other countries:
Kluwer Academic Publishers Group
Distribution Centre
Post Office Box 322
3300 AH Dordrecht, THE NETHERLANDS
Telephone 31 78 6392 392
Fax 31 78 6546 474
E-Mail <orderdept@wkap.nl>

Electronic Services <http://www.wkap.nl>

Library of Congress Cataloging-in-Publication Data

Angstenberger, Larisa.
Dynamic fuzzy pattern recognition with applications to finance and engineering / Larisa Angstenberger.
p. cm. -- (International series in intelligent technologies ; 17)
Includes bibliographical references and index.

1. Pattern perception. 2. Fuzzy systems. I. Title. II. Series.

Q327 .A57 2001
006.4--dc21

2001038476

ISBN 978-90-481-5775-4

Printed on acid-free paper.

Printed in the United States of America

'Omnia mutantur, nihil interit'.
(Everything is changing, nothing is getting lost.)
Phylosoph Ovid, methamorphoses 15, 165

'All the real knowledge which we possess, depends on methods by which we distinguish the similar from the dissimilar'.
Swedish naturalist Linnaeus, 1737

TABLE OF CONTENTS

LIST OF FIGURES

LIST OF TABLES

FOREWORD

Data mining, knowledge discovery and pattern recognition, which are very often used synonymously, have become very large and important areas in many disciplines, such as in management, engineering, medicine etc. The main reason is, that we have moved from a situation characterised by a lack of computer readable data into a situation with an abundance of data in data banks and data warehouses. Very often masses of data have been stored, but the human observer is not able to see or find the information that is relevant to him and which is hidden somewhere in the data. That is exactly when computer based data mining becomes important.

Originally a multitude of statistical methods have been and are still being used in data mining. These methods, such as regression analysis, cluster methods etc., have the disadvantage, that generally they are based on dual logic, i.e., they are dichotomous in character and do not always fit to the problems they are applied to. In the recent past methods of Computational Intelligence (Fuzzy Set Theory, Neural Nets) are used, which circumvent this dichotomy and are sometimes also more efficient than classical approaches.

This book focuses on fuzzy methods and, in particular, on fuzzy clustering methods, which have proven to be very powerful in pattern recognition.

So far these methods, whether crisp or fuzzy, have been static. This means, that the "elements" or "objects" that are clustered are points (or vectors) in the state or feature space. Distances or similarities, on the bases of which cluster algorithms work, are relations between points. When dynamic systems behaviour is considered, then normally the analysis is comparative static, i.e., it consists of a sequence of static observations (like snap shoots) and analyses, generally, with a fixed feature space. This is obviously a type of analysis in which often important information can be lost. In a really dynamic analysis the "elements" are trajectories or (in discrete cases) vectors. These, however, cannot be clustered by normal cluster methods. One of the reasons is, that the similarity between functions is generally not defined. Also for a system, the states of which have to be considered over time, the feature space might change over time and so may the clusters.

Dr. Angstenberger has intensively researched this problem area in the past and contributed very interesting research results. She has published numerous results of her research under her maiden name Mikenina. In this book she considers the entire process of dynamic pattern recognition. After setting a general framework for Dynamic Pattern Recognition, she describes in detail the monitoring process using fuzzy tools and the adaptation process in which the classifiers have to be adapted, using the observations of the dynamic process. She then focuses on the problem of a changing cluster structure (new clusters, merging of clusters, splitting of clusters and the detection of gradual changes in the cluster structure). She integrates these parts into a complete algorithm for dynamic fuzzy classifier design and classification.

One of the core chapters of this book is certainly chapter 5 in which similarity concepts for dynamic objects are investigated. It was already mentioned, that the usual similarity concepts are defined for points and not for functions. There exist general mathematical definitions of similarity which have certain mathematical properties. Unluckily, in practice the usefulness or appropriateness of a similarity concepts depends very much on the context. Hence the goal is, to find similarity measures which are both as general and as adaptable to different contexts as possible. Dr. Angstenberger distinguishes in this book between two different kinds of similarity: pointwise similarity and structural similarity, a concept that has proven quite useful in practice. For both types she points to existing problems and also to ways of using those measures operationally.

So far the book could be considered as a lucid, clear and well to read theoretical treatise about fuzzy dynamic pattern recognition. It is particularly commendable, that Dr. Angstenberger does not stop here, but that she shows very clearly, how her suggestions really work in applications. As mentioned above, data mining problems occur in management as well as in engineering and in other areas. Dr. Angstenberger presents one case each of a management and an engineering application. The management problem that she solves is the customer segmentation problem for banks. This is a problem that very recently has gained enormous visibility and importance. As engineering problem she has chosen the optimisation of a computer network based on dynamic network load classification. This is, obviously, a very timely problem of growing importance. For both problem solutions real data were used and the results that she achieved look in deed very promising.

In my view this book really presents a breakthrough in a new area of pattern recognition. It will certainly be the basis for many other research projects and real applications. I can only congratulate her to this excellent scientific and application oriented work and hope that it may be of benefit to many scientists and practitioners.

Aachen, January 2001 H.-J. Zimmermann

ACKNOWLEDGEMENTS

In 1995 I was awarded a one-year grant from the German Academic Exchange Service (DAAD) for research in Germany during my PhD studies. I was very interested in investigating the field of Fuzzy Logic and in fuzzy data analysis and very happy when I received the invitation from Prof. Dr. Dr. h.c. H.-J. Zimmermann to study at his chair of Operations Research at the RWTH Aachen and to write my PhD thesis under his scientific supervision. Thanks to a four year financial support by DAAD I managed to graduate in the master course of Operations Research and to perform the research in the area of fuzzy pattern recognition which has resulted in this book. I am very grateful to Prof. Zimmermann for allowing me to share his knowledge and scientific experience, for his support and encouragement as well as his attention and interest in my research and his scientific suggestions concerning the research direction, possible applications, and book structure. It was a great pleasure to work with Prof. Zimmermann and to be a part of the team of young researchers at the institute of Operations Research where, thanks to his particular personality and humour, a very exhilarating and creative atmosphere had been developed over the years. I am also much obliged to Prof. Bastian for his critical remarks and suggestions for improvements to this manuscript.

I would like to thank all my colleagues at the chair of Operations Research for their mutual understanding, for knowledge and experience exchange and for a lot of fun. Especially the competent advice from Dr. Uwe Bath and Dr. Tore Grünert and the technical support of Gilberto von Spar during my work on this book were of great importance for me. I appreciate the numerous discussions about fuzzy clustering and the philosophical discussions about life with my colleague Dr. Peter Hofmeister. His stable optimism and energy have helped me to stay always in high spirits despite of hard work.

During my studies at the RWTH Aachen I had the opportunity to work at MIT - Management Intelligenter Technologien GmbH - where I participated in different projects in the area of intelligent data analysis. From 1997 I worked on a new scientific project on dynamic fuzzy data analysis sponsored by DFG (German Research Society) for three years. This project motivated me to investigate this new area in greater detail and has resulted in a number of publications which served as a basis for my further research. Helpful discussions with my project partner Arno Joentgen have positively influenced my research and brought me many new ideas.

Due to the technical support of Sebastian Greguletz, who provided me with a special software for network monitoring, it was possible to gather enough data for the technical application of dynamic pattern recognition methods introduced in this book. I also want to thank one of my best friends, Jens Junker, who helped me to better understand the process of data transmission in computer networks, to identify practical goals in computer network optimisation and to interpret the results of classification.

I am very grateful to Thomas MacFarlane for his very intensive proof-reading of this manuscript, which considerably improved the language and the style of this book. I am also obliged to Kluwer Academic Publishers for the opportunity to publish this book and for the good co-operation in preparing it.

I would like to express my great gratitude to my husband Joachim Angstenberger for his unwavering support and numerous fruitful discussions during my almost two years as a PhD student. I appreciate a lot that he has dedicated so much time to reading the first and all subsequent versions of this manuscript, since his valuable suggestions and corrections have significantly improved my book. I am very happy that he showed so much patience and understanding for my work whereas I did not have enough time on weekends.

This book is a result of many years of studies and research, in which I was constantly supported and motivated by my parents. They showed me the path to knowledge and always helped me to find the most efficient means to reach my goals.

I hope that the results of my work meet the expectations of my parents, my friends and my teachers.

Hamburg, January 2001 Larisa Angstenberger
(maiden name Mikenina)

1 INTRODUCTION

The phenomenal improvements in data collection due to the automation and computerisation of many operational systems and processes in business, technical and scientific environments, as well as advances in data storage technologies, over the last decade have lead to large amounts of data being stored in databases. Analysing and extracting valuable information from these data has become an important issue in recent research and attracted the attention of all kinds of companies in a big way. The use of data mining and data analysis techniques was recognised as necessary to maintain competitiveness in today's business world, to increase business opportunities and to improve service. A data mining endeavour can be defined as the process of discovering meaningful new correlations, patterns and trends by examining large amounts of data stored in repositories and by using pattern recognition technologies as well as statistical and mathematical techniques. Pattern recognition is the research area which provides the majority of methods for data mining and aims at supporting humans in analysing complex data structures automatically.

Pattern recognition systems can act as substitutes when human experts are scarce in specialised areas such as medical diagnosis or in dangerous situations such as fault detection and automatic error discovery in nuclear power plants. Automated pattern recognition can provide a valuable support in process and quality control and function continuously with consistent performance. Finally, automated perceptual tasks such as speech and image recognition enable the development of more natural and convenient human-computer interfaces. Important additional benefits for a wide field of highly complex applications lie in the use of intelligent techniques such as fuzzy logic and neural networks in pattern recognition methods. These techniques permitted the development of methods and algorithms that can perform tasks normally associated with intelligent human behaviour. For instance, the primary advantage of fuzzy pattern recognition compared to the classical methods is the ability of a system to classify patterns in a non-dichotomous way, as humans do, and to handle vague information. Methods of fuzzy pattern recognition gain constantly increasing ground in practice. Some of the fields where intelligent pattern recognition has obtained the greatest level of endorsement and success in recent time are database marketing, risk management and credit-card fraud detection.

The advent of data warehousing technology has provided companies with the possibility to gather vast amounts of historical data describing the temporal behaviour of a system under study and allows a new type of analysis for improved decision support. Amongst other applications in which objects must be analysed in the process of their motion or temporal development, the following are worth mentioning:

- Monitoring of patients in medicine, e.g. during narcosis when the development rather than the status of the patient's condition is essential;
- The analysis of data concerning buyers of new cars or other articles in order to determine customer portfolio;
- The analysis of monthly unemployment rates;
- The analysis of the development of share prices and other characteristics to predict stock markets;
- The analysis of payment behaviour of bank customers to distinguish between good and bad customers and detect fraud;
- Technical diagnosis and state-dependent machine maintenance.

The common characteristic of all these applications is that in the course of time objects under study change their states from one to another. The order of state changes, or just a collection of states an object has taken, determines the membership of an object to a certain pattern or class. In other words, the history of temporal development of an object has a strong effect on the result of the recognition process. Such objects representing observations of a dynamic system/process and containing a history of their temporal development are called dynamic. In contrast to static, they are represented by a sequence of numerical vectors collected over time.

Conventional methods of statistical and intelligent pattern recognition are, however, of limited benefit in problems in which a dynamic viewpoint is desirable since they consider objects at a fixed moment of time and do not take into account their temporal behaviour. Therefore there is an urgent need for a new generation of computational techniques and tools to assist humans in extracting knowledge from temporal historical data. The development of such techniques and methods constitutes the focus of this book.

1.1 Goals and Tasks of the Book

Dynamic pattern recognition is concerned with the recognition of clusters of dynamic objects, i.e. recognition of typical states in the dynamic behaviour of a system under consideration. The goal of this book is to investigate this new field of dynamic pattern recognition and to develop new methods for clustering and classification in a dynamic environment.

Due to the changing properties of dynamic objects, the partitioning of objects, i.e. the cluster structure, is not obviously constant over time. The appearance of new observations can lead to gradual or abrupt changes in the cluster structure such as, for instance, the formation of new clusters, or the merging or splitting of existing clusters. In order to follow temporal changes of the cluster structure and to preserve the desired performance, a classifier must posses adaptive capabilities, i.e. a classifier must be automatically adjusted over time according to detected changes in the data structure. Therefore, the development of methods for dynamic pattern recognition consists of two tasks:

- Development of a method for dynamic classifier design enabling a design of an adaptive classifier that can automatically recognise new cluster structures as time passes;
- Development of new similarity measures for trajectories of dynamic objects.

The procedure of dynamic classifier design must incorporate, in addition to the usual static steps, special procedures that would allow the application of a classifier in a dynamic environment. These procedures representing dynamic steps are concerned with detecting temporal changes in the cluster structure and updating the classifier according to these changes. In order to carry out these steps the design procedure must have a monitoring procedure at its disposal, which must supervise the classifier performance and the mechanism for updating a classifier dependent on the results of the monitoring procedure.

Different methods suggested in the literature for establishing the monitoring process are based on the observation and analysis of some characteristic values describing the performance of a classifier. If the classifier performance deteriorates they are able to detect changes but cannot recognise what kind of temporal changes have taken place. Most of the updating procedures proposed for dynamic classifiers are based either on recursive updating of classifier parameters or re-learning from scratch using new objects, whereas the old objects are discarded as being irrelevant.

In this book a new algorithm for dynamic fuzzy classifier design is proposed, which is based partly on the ideas presented in the literature but also uses a number of novel criteria to establish the monitoring procedure. The proposed algorithm is introduced in the framework of unsupervised learning and allows the design of an adaptive classifier capable of recognising automatically gradual and abrupt changes in the cluster structure as time passes and adjusting its structure to detected changes. The adaptation laws for updating the classifier and the template data set are coupled with the results of the monitoring procedure and characterised by additional features that should guarantee a more reliable and efficient classifier.

Another important problem arising in the context of dynamic pattern recognition is the choice of a relevant similarity measure for dynamic objects, which is used, for instance, for the definition of the clustering criterion. Most of the pattern recognition methods use the pairwise distance between objects as a dissimilarity measure used to calculate the degree of membership of objects to cluster prototypes. As was mentioned above, dynamic objects are represented by a temporal sequence of observations and described by multidimensional trajectories in the feature space, or vector-valued functions. Since the distance between vector-valued functions is not defined, classical clustering and in general pattern recognition methods are not suited for processing dynamic objects.

One approach used in most applications for handling dynamic objects in pattern recognition is to transform trajectories into conventional feature vectors during pre-processing. The alternative approach addressed in this book requires a definition of a similarity measure for trajectories that should take into account the dynamic behaviour of trajectories. For this approach it is important to determine a specific criterion for similarity. Depending on the application this may require either the best match of trajectories by minimising the pointwise distance or a similar form of trajectories independent of their relative location to each other. The similarity measure for trajectories can be applied instead of the distance measure to modify classical pattern recognition methods.

The combination of a new method for dynamic classifier design, which can be applied and modified for different types of classifiers, with a set of similarity measures for trajectories leads to a new class of methods for dynamic pattern recognition.

1.2 Structure of the Book

This book is organised as follows:

In Chapter 2, a general view on the pattern recognition problem is given. Starting with the main principles of the knowledge discovery process, the role of pattern recognition in this process will be discussed. This will be followed by the formulation of the classical (static) problem of pattern recognition and the classification of methods in this area with respect to different criteria. Particular attention will be given to fuzzy techniques, which will constitute the focus of this book. Since dynamic pattern recognition represents a relative new research area and a standard terminology does not yet exist, the main notions and definitions used in this book will be introduced and the main problems and tasks of dynamic pattern recognition will be considered.

Chapter 3 provides an overview of techniques that can be used as components for designing dynamic pattern recognition systems. The advantages and drawbacks of different statistical and fuzzy approaches for the monitoring procedure will be discussed, followed by the analysis of updating strategies for a dynamic classifier. It will be shown that the classifier design cannot be separated temporally from the phase of its application to the classification of new objects and is carried out in a closed learning-and-working-cycle. The adaptive capacity of a classifier depends crucially on the chosen updating strategy and on the ability of the monitoring procedure to detect temporal changes.

Based on the results of Chapters 2 and 3, a new method for dynamic fuzzy classifier design will be developed in Chapter 4. The design procedure consists of three main components: the monitoring procedure, the adaptation procedure for the classifier and the adaptation procedure for the training data set used to learn a classifier. New heuristic algorithms proposed for the monitoring procedure in the framework of unsupervised learning facilitate the recognition of gradual and abrupt temporal changes in the cluster structure based on the analysis of membership functions of fuzzy clusters and density of objects within clusters. The adaptation law of the classifier is a flexible combination of two updating strategies, each depending on the result of the monitoring procedure, and provides a mechanism to adjust parameters of the classifier to detected changes in the course of time. The efficiency of the dynamic classifier is guaranteed by a set of validity measures controlling the adaptation procedure.

The problem of handling dynamic objects in pattern recognition is addressed in Chapter 5. After considering different types of similarity and introducing different similarity models for trajectories, a number of definitions of specific similarity measures will be proposed. They are based on the set of characteristics describing the temporal behaviour of trajectories and representing different context dependent meanings of similarity.

The efficiency of the proposed method for dynamic classifier design, combined with new similarity measures for trajectories, is examined in Chapter 6 using two application examples based on real data. The first application is concerned with the load optimisation in a computer network based on on-line monitoring and dynamic recognition of current load states. The second application from the credit industry regards the problem of segmentation of bank customers based on their behavioural data. The analysis allows one to recognise tendencies and temporal changes in the customer structure and to follow transitions of single customers between segments.

Finally, Chapter 7 summarises the results and their practical implications and outlines new directions for future research.

2 GENERAL FRAMEWORK OF DYNAMIC PATTERN RECOGNITION

Most methods of pattern recognition consider objects at a fixed moment in time without taking into account their temporal development. However, there are a lot of applications in which the order of state changes of an object over time determines its membership to a certain pattern, or class. In these cases, for the correct recognition of objects it is very important not only to consider properties of objects at a certain moment in time but also to analyse properties characterising their temporal development. This means that the history of temporal development of an object has a strong effect on the result of the recognition process. Classical methods of pattern recognition are not suitable for processing objects described by temporal sequences of observations. In order to deal with problems in which a dynamic viewpoint is desirable, methods of dynamic pattern recognition must be applied.

The field of pattern recognition is a rapidly growing research area within the broader field of machine intelligence. The increasing scientific interest in this area and the numerous efforts at solving pattern recognition problems are motivated by the challenge of this problem and its potential applications. The primary intention of pattern recognition is to automatically assist humans in analysing the vast amount of available data and extracting useful knowledge from it. In order to understand the mechanism of extracting knowledge from data and the role of pattern recognition in fulfilling this task, the main principles of the knowledge discovery process will be described in Section 2.1. Then the problem of classical (static) pattern recognition will be formulated in Section 2.2, which should provide a general framework for the investigations in this book. Since the main topic of this book centres on the relative new area of dynamic pattern recognition, the main notions and definitions used in this area will be presented in Section 2.3. Finally, the goal and basic steps of the dynamic pattern recognition process will be summarised.

2.1 The Knowledge Discovery Process

The task of finding useful patterns in raw data is known in the literature under various names including knowledge discovery in databases, data mining, knowledge extraction, information discovery, information harvesting, data archaeology, and data pattern processing. The term *knowledge discovery in databases* (KDD), which appeared in 1989, refers to the process of finding knowledge in data by applying particular data mining methods at a high level. In the literature KDD is often used as a synonym of data mining since the goal of both processes is to mine for pieces of knowledge in data. In [Fayyad et al., 1996, p. 2] it is, however, argued that KDD is related to the overall process of discovering useful knowledge from data while data mining is concerned with the application of algorithms for extracting patterns from data without the additional steps of the KDD process, such as considering relevant prior knowledge and a proper interpretation of the results.

The burgeoning interest in this research area is due to a rapid growth of many scientific, business and industrial databases in the last decade. Advances in data collection in science and industry, the widespread usage of bar codes for almost all commercial products, and the computerisation of many business and government transactions, have produced a flood of data which has been transformed into 'mountains' of stored data using modern data storage technologies. According to [Fraway et al., 1992], 'the amount of information in the world doubles every 20 months'. The desire to discover important knowledge in the vast amount of existing data makes it necessary to look for a new generation of techniques and tools with the ability 'to intelligently and automatically assist humans in analysing the mountains of data for nuggets of useful knowledge' [Fayyad et al., 1996, p. 2]. These techniques and tools are the object of the field of knowledge discovery in databases.

In [Frawley et al., 1991] the following definition of the KDD process is proposed:

'Knowledge discovery in databases is the non-trivial process of identifying valid, novel, potentially useful, and ultimately understandable patterns in data'.

The interpretation of the different terms in this definition is as follows.

- *Process*: KDD represents a multi-step process, which involves data preparation, search for patterns, knowledge evaluation and allows return to previous steps for refinement of results. It is assumed that this process has some degree of search autonomy, i.e. it can investigate by itself complex dependencies in data and present only interesting results to the user. Thus, the process is considered to be non-trivial.
- *Validity:* The discovered patterns should be valid not only on the given data set, but on new data with some degree of certainty. For the evaluation of certainty, a certainty measure function can be applied.
- *Novelty:* The discovered patterns should be new (at least to the system). A degree of novelty can be measured based on changes in data or knowledge by comparing current values or a new finding to previous ones.

- *Potentially useful:* The discovered patterns should be potentially usable and relevant to a concrete application problem and should lead to some useful actions. A degree of usefulness can be measured by some utility function.
- *Ultimately understandable:* The patterns should be easily understandable to humans. For this purpose, they should be formulated in an understandable language or represented graphically. In order to estimate this property, different simplicity measures can be used which take into account either a size of patterns (syntactic measure) or the meaning of patterns (semantic measures).

In order to evaluate an overall measure of pattern value combining validity, novelty, usefulness and simplicity, a function of significance is usually defined. If the value of significance for a discovered pattern exceeds a user-specified threshold, then this pattern can be considered as knowledge by the KDD process.

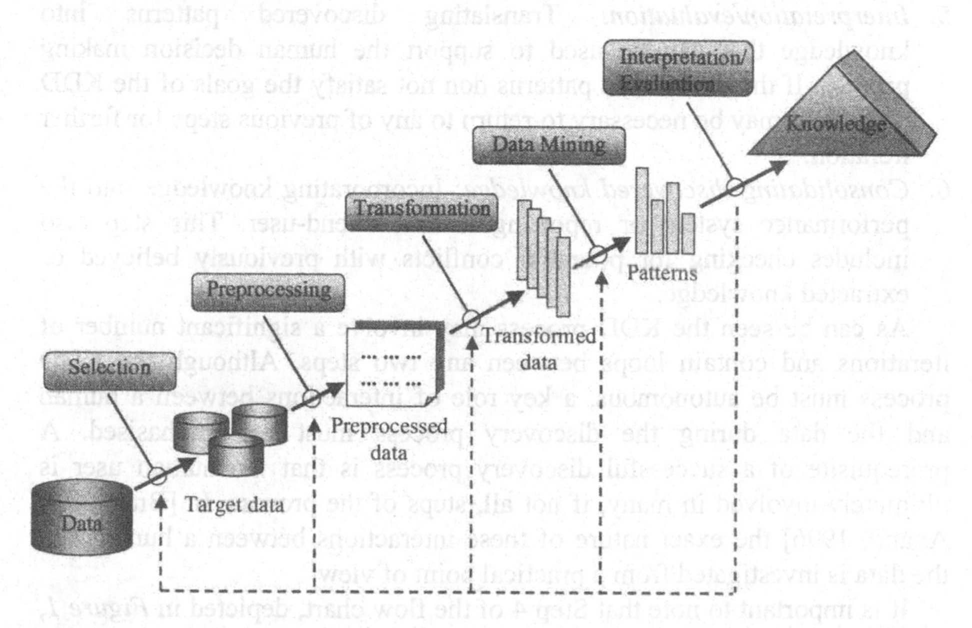

Figure 1. An overview of the steps of the KDD process

Figure 1 provides an overview of the main steps/stages of the KDD process emphasising its iterative and interactive nature [adapted from Fayyad et al., 1996, p. 10]. The prerequisite of the KDD process is an understanding of the application domain, the relevant prior knowledge, and the goals of the user. The process starts with the raw data and finishes with the extracted knowledge acquired during the following six stages:

1. *Selection:* Selecting a target data set (according to some criteria) on which discovery will be performed.

2. *Pre-processing:* Applying basic operations of data cleaning including the removal of noise, outliers and irrelevant information, deciding on strategies for handling missing data values, and analysing information contained in time series. If the data has been drawn from different sources and has inconsistent formats, the data is reconfigured to a consistent format.
3. *Transformation:* Selecting relevant features to represent data using dimensionality reduction, projection or transformation techniques. The data are reduced to a smaller set of representative usable data.
4. *Data Mining:* Extracting patterns from data in a particular representational form relating to the chosen data mining algorithm. At first, the task of data mining such as classification, clustering, regression etc. is defined according to the goal of the KDD process. Then, the appropriate data mining algorithm(s) is (are) selected. Finally, the chosen algorithm(s) is (are) applied to the data to find patterns of interest.
5. *Interpretation/evaluation:* Translating discovered patterns into knowledge that can be used to support the human decision making process. If the discovered patterns don not satisfy the goals of the KDD process, it may be necessary to return to any of previous steps for further iteration.
6. *Consolidating discovered knowledge:* Incorporating knowledge into the performance system or reporting it to the end-user. This step also includes checking for potential conflicts with previously believed or extracted knowledge.

As can be seen the KDD process may involve a significant number of iterations and contain loops between any two steps. Although the KDD process must be autonomous, a key role of interactions between a human and the data during the discovery process must be emphasised. A prerequisite of a successful discovery process is that the human user is ultimately involved in many, if not all, steps of the process. In [Brachman, Anand, 1996] the exact nature of these interactions between a human and the data is investigated from a practical point of view.

It is important to note that Step 4 of the flow chart, depicted in *Figure 1*, is concerned with the application of data mining algorithms to a concrete problem. On the subject of data mining or data analysis as a process, however, this usually involves the same seven steps described above emphasising the need for pre-processing and transformation procedures in order to obtain successful results from the analysis. This understanding explains the fact that data mining, data analysis and knowledge discovery processes are often used as equivalent notions in the literature ([Angstenberger, 1997], [Dilly, 1995], [Petrak, 1997]).

KDD overlaps with various research areas including machine learning, pattern recognition, databases, statistics, artificial intelligence, knowledge acquisition for expert systems, and data visualisation and uses methods and techniques from these diverse fields [Fayyad et al., 1996, p. 4-5]. For instance, the common goals of machine learning, pattern recognition and KDD lie in the development of algorithms for extracting patterns and models from data (data mining methods). But KDD is additionally concerned with the extension of these algorithms to problems with very large real-world databases while machine learning typically works with smaller data sets. Thus, KDD can be viewed as part of the broader fields of machine learning and pattern recognition, which include not only learning from examples but also reinforcement learning, learning with teacher, etc. [Dilly, 1995]. KDD often makes use of statistics, particularly exploratory data analysis, for modelling data and handling noisy data.

In this book the attention will be focussed on pattern recognition which represents one of the largest fields in KDD and provides the large majority of methods and techniques for data mining.

2.2 The Problem of Pattern Recognition

The concept of patterns has a universal importance in intelligence and discovery. Most instances of the world are represented as patterns containing knowledge, if only one could discover it. Pattern recognition theory involves learning similarities and differences of patterns that are abstractions of objects in a population of non-identical objects. The associations and relationships between objects make it possible to discover patterns in data and to build up knowledge. The most frequently observed types of patterns are as follows: (rule-based) relationships between objects, temporal sequences, spatial patterns, groups of similar objects, mathematical laws, deviations from statistical distributions, exceptions and striking objects [Petrak, 1997]. A human's perceptive power seems to be well adapted to the pattern-processing task. Humans are able to recognise printed characters and words as well as hand-written characters, speech utterances, favourite melodies, the faces of friends in a crowd, different types of weave, scenes in images, contextual meanings of word phrases, and so forth. Humans are also capable of retrieving information on the basis of associated clues including only a part of the pattern. Humans learn from experience by accumulating rules in various forms such as associations, tables, relationships, inequalities, equations, data structures, logical implications, and so on. A desire to understand the basis for these powers in humans is the reason for the growing interest in investigating the pattern recognition process.

The subject area of pattern recognition belongs to the broader field of machine learning, whose primary task is the study of how to make machines learn and reason as humans do in order to make decisions [Looney, 1997, p. 4]. In this context learning refers to the construction of rules based on observations of environmental states and transitions. Machine learning algorithms examine the input data set with its accompanying information and the results of the learning process given in form of statements, and learn to reproduce these and to make generalisations about new observations. Ever since computers were first designed the intention of researchers has been that of making them 'intelligent' and giving them the same information-processing capabilities that humans possess. This would make computers more efficient in handling real world tasks and would make them more compatible with the way in which humans behave.

In the following section the pattern recognition process will be considered and its main steps examined. This will be followed by a classification of pattern recognition methods regarding different criteria. Finally, the characteristics of a special class of fuzzy pattern recognition methods will be discussed along with the advantages that they afford dynamic pattern recognition.

2.2.1 The process of pattern recognition

Pattern recognition is one of the research areas that tries to explore mathematical and technical aspects of perception – a human's ability to receive, evaluate, and interpret the information as regards his/her environment - and to support humans in carrying out this task automatically. 'The goal of pattern recognition is to classify objects of interest into one of a number of categories or classes' [Therrien, 1989, p. 1]. This process can be viewed as a mapping of an object from the observation space into the class-membership space [Zadeh, 1977] or 'a search for structure in data' [Bezdek, 1981, p. 1]. Objects of interest may be any physical process or phenomenon. The basic scheme of the pattern recognition process is illustrated in *Figure 2* [adapted from Therrien, 1989, p. 2].

Information about the object coming from different measurement devices is summarised in the observation vector. The observation space is usually of a high dimension and transformed into a feature space. The purpose of this transformation is to extract the smallest possible set of distinguishing features that lead to the best possible classification results. In other words it is advantageous to select features in such a way that feature vectors, or patterns, belonging to different classes occupy different regions of the feature space [Jain, 1986, p. 2]. The resulting feature space is of a much lower dimension than the observation space. A procedure of selecting a set of sufficient features from a set of available features is called *feature extraction*. It may be based on intuition or knowledge of the physical characteristics of the problem or it may be a mathematical technique that reduces the dimensionality of the observation space in a prescribed way.

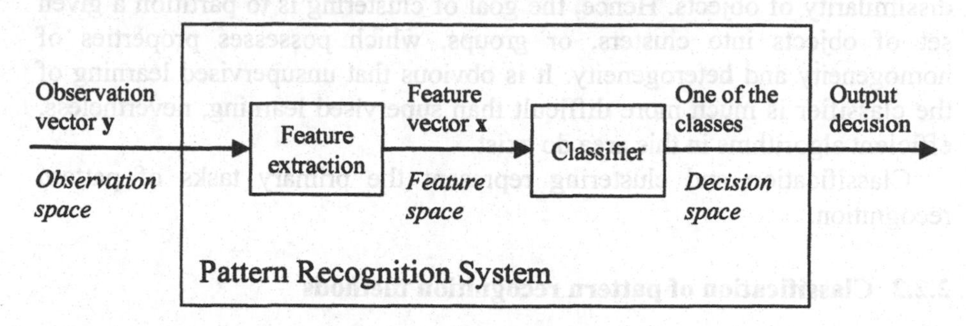

Figure 2. Basic scheme of the pattern recognition process

The next step is a transformation of the feature space into a decision space, which is defined by a (finite) set of classes. A *classifier*, which is a device or algorithm, generates a partitioning of the feature space into a number of decision regions. The classifier is designed either using some set of objects, the individual classes of which are already known, or by learning classes based on similarities between objects. Once the classifier is designed and a desired level of performance is achieved, it can be used to classify new objects. This means that the classifier assigns every feature vector in the feature space to a class in the decision space.

Depending on the information available for classifier design, one can distinguish between *supervised* and *unsupervised* pattern recognition ([Therrien, 1989, p. 2], [Jain, 1986, p. 6-7]). In the first case there exists a set of labelled objects with a known class membership. A part of this set is extracted and used to derive a classifier. These objects build the training set. The remaining objects, whose correct class assignments are also known, are referred to as the test set and used to validate the classifier's performance. Based on the test results, suitable modifications of the classifier's parameters can be carried out. Thus, the goal of supervised learning, also called *classification*, is to find the underlying structure in the training set and to learn a set of rules that allows the classification of new objects into one of the existing classes.

The problem of unsupervised pattern recognition, also called *clustering*, arises if cluster memberships of available objects, and perhaps even the number of clusters, are unknown. In such cases, a classifier is designed based on similar properties of objects: objects belonging to the same cluster should be as similar as possible (homogeneity within clusters) and objects belonging to different clusters should be clearly distinguishable (heterogeneity between clusters). The notion of similarity is either prescribed by a classification algorithm, or has to be defined depending on the application. If objects are real-valued feature vectors, then the Euclidean distance between feature vectors is usually used as a measure of dissimilarity of objects. Hence, the goal of clustering is to partition a given set of objects into clusters, or groups, which possesses properties of homogeneity and heterogeneity. It is obvious that unsupervised learning of the classifier is much more difficult than supervised learning, nevertheless, efficient algorithms in this area do exist.

Classification and clustering represent the primary tasks of pattern recognition.

2.2.2 Classification of pattern recognition methods

Over the past two decades a lot of methods have been developed to solve pattern recognition problems. These methods can be grouped into two approaches ([Fu, 1982b, p. 2], [Bunke, 1986, p. 367]): the decision-theoretic and the syntactic approach. The decision-theoretic approach, described in the previous section, is the most common one. The origin of this approach is related to the development of statistical pattern recognition ([Duda, Hart, 1973], [Devijer, Kittler, 1982], [Therrien, 1989]). The goal of statistical methods is to derive class boundaries from statistical properties of feature vectors through procedures known in statistics as hypothesis testing. The hypotheses in this case are that a given object belongs to one of the possible classes. The validity measure used in the decision rule is the probability of making an incorrect decision, or the probability of error. The decision rule is optimal if it minimises the probability of error or another quantity closely related to it.

Decision-theoretic methods can be classified depending on the representation form of information describing objects and on their application area into three groups: algorithmic, neural networks-based and rule-based methods [Zimmermann, 1996, p. 244]. Among algorithmic methods one can distinguish between statistical (described above), clustering and fuzzy pattern matching methods. Clustering methods (crisp or fuzzy) represent a big class of algorithms for unsupervised learning of structure in data ([Bezdek, 1981], [Gustafson, Kessel, 1979], [Gath, Geva, 1989], [Krishnapuram et al., 1993], [Krishnapuram, Keller, 1993]). They aim at grouping of objects into homogeneous clusters that are defined so that the intra-cluster distances are minimised while the inter-cluster distances are maximised.

A fuzzy pattern matching technique proposed by [Cayrol et al., 1980, 1982] is based on possibility and necessity measures and aims to estimate the compatibility between an object and prototype values of a class. The fuzzy pattern matching approach was extended by [Dubois et al., 1988] and its general framework summarised by [Grabisch et al., 1997]. The idea of this group of methods is to build fuzzy prototypes of classes in the form of fuzzy sets and, during classification, to match a new object with all class prototypes and select the class with the highest matching degree.

Rule-based classification methods ([Zimmermann, 1993, p. 78-83], [Kastner, Hong, 1984], [Boose, 1989]) are based on principles of expert systems. The first step of these methods consists of the knowledge acquisition corresponding to a classifier design. Knowledge about causal dependencies between feature vectors and decisions (classes) is formulated in the form of if-then-rules, which are obtained either using experts' decisions or generated automatically from a training data set. The resulting rule base allows a mapping of a set of objects into a set of classes imitating human decision behaviour. Classification of a new object is performed in the inference engine of the expert system. The result of inference (a membership function or a singleton) is matched with class descriptions by determining the similarity of the result with class descriptions, and an object is assigned to the class with the highest degree of similarity.

Pattern recognition methods based on artificial neural networks have proven to be a very powerful and efficient tool for pattern recognition. They can be categorised into two general groups: feed-forward neural networks ([Rosenblatt, 1958], [Minsky, Papert, 1988], [Hornik et al., 1989]) used for supervised classification and self-organising networks ([Kohonen, 1988]) enabling unsupervised clustering of data. Neural networks are able to learn a highly non-linear mapping of feature vectors in the input space into class vectors in the output decision space without applying any mathematical model. The desired mapping is achieved by adjusting appropriately the weights of neurones during supervised training or self-organisation.

Hybrid neuro-fuzzy methods combine the advantages of neural networks and fuzzy sets and compensate their disadvantages. Neural networks techniques enable neuro-fuzzy systems to learn new information from a given training data set in a supervised or unsupervised mode, but the behaviour of neural networks cannot be interpreted easily. On the other hand, fuzzy systems are interpretable, plausible and transparent rule-based systems, which in general cannot learn. The knowledge has first to be acquired and provided to the system in the form of if-then-rules. Furthermore, fuzzy set theory enables neuro-fuzzy systems to present the learned information in a humanly understandable form. The most important works on this subject were carried out by ([Lee, Lee, 1975], [Huntsberger, Ajjimarangsee, 1990], [Kosko, 1992], [Nauck et al., 1994]).

With the syntactic pattern recognition approach ([Fu, 1974], [Fu, 1982a]), objects are represented by sentences in contrast to feature vectors in decision-theoretic pattern recognition. Each object is described by structural features corresponding to the letters of an alphabet. Recognition of a class assignment of an object is usually done by hierarchically parsing the object structure according to a given set of syntax rules. One way to parse the sentences is to use a finite state machine or finite automaton [Kohavi, 1978]. In this approach an analogy between the object structure and the syntax of a language can be seen.

A taxonomy of pattern recognition methods is summarised on *Figure 3*.

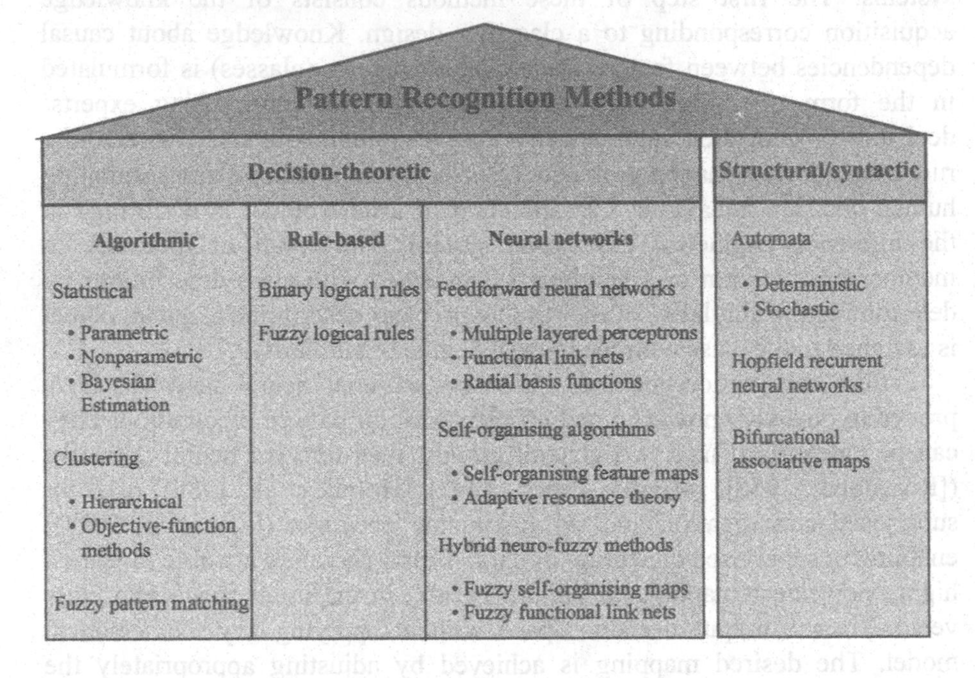

Pattern Recognition Methods
Decision-theoretic
Structural/syntactic
Algorithmic
Statistical
• Parametric
• Nonparametric
• Bayesian Estimation
Clustering
• Hierarchical
• Objective-function methods
Fuzzy pattern matching
Rule-based
Binary logical rules
Fuzzy logical rules
Neural networks
Feedforward neural networks
• Multiple layered perceptrons
• Functional link nets
• Radial basis functions
Self-organising algorithms
• Self-organising feature maps
• Adaptive resonance theory
Hybrid neuro-fuzzy methods
• Fuzzy self-organising maps
• Fuzzy functional link nets
Automata
• Deterministic
• Stochastic
Hopfield recurrent neural networks
Bifurcational associative maps

Figure 3. A taxonomy of pattern recognition methods

Considering the taxonomy of pattern recognition methods an additional general distinction of all different techniques of the decision-theoretic approach into two classes can be drawn: classical (crisp) and fuzzy methods. Crisp algorithms for pattern recognition generate partitions such that an object belongs to exactly one cluster. In many cases, however, there is no distinct cluster structure allowing the clear assignment of an object to strictly one cluster. For instance, there may be a situation where objects are located between clusters building bridges, or where clusters overlap to some degree, so that objects belong to several clusters. Another example is given by outliers belonging to none of clusters or to all of them to the same degree. Such situations can be handled well by a human who has an ability to classify in a non-dichotomous way [Zimmermann, 1996, p.242]. However, traditional pattern recognition methods can not provide an adequate solution to such problems, since they do not take into consideration similarity of objects to cluster representatives. Fuzzy methods can avoid this information loss by generating a degree of membership of objects to clusters.

In this book the attention will be mainly focussed on clustering methods which can be categorised according to the underlying techniques and the type of clustering criterion used in the following main (non-distinct) groups [Höppner et al., 1996, p. 8-9]:

- *Incomplete clustering methods*: This group of methods is represented by geometric methods and projection techniques. The goal of these methods is to reduce the dimensionality of a given set of multidimensional objects (using e.g. principal component analysis) in order to represent objects graphically in two or three-dimensional space. Clustering of objects is performed manually by visual examination of the data structure.
- *Deterministic crisp clustering methods*: Using these methods, each object is assigned to exactly one cluster so that the resulting cluster structure defines a clear partition of objects.
- *Overlapping clustering methods*: Each object is assigned to at least one cluster, but it may as well belong to several clusters at the same time.
- *Probabilistic clustering methods*: For each object, a probability distribution over all clusters is determined, which gives a probability of an object belonging to a certain cluster.
- *Fuzzy clustering methods*: This group of methods generates degrees of membership of objects to clusters, which indicate to what extent an object belongs to clusters. Degrees of membership of an object to all clusters are usually normalised to one.
- *Possibilistic clustering methods*: This group represents a sub-class of fuzzy clustering methods that produce possibilistic memberships of objects to clusters. The probabilistic constraint that the membership of an object across clusters must sum to one is dropped in possibilistic methods. The resulting degrees of membership are interpreted as degrees of compatibility, or typicality, of objects to clusters.
- *Hierarchical clustering methods:* These methods generate a hierarchy of partitions by means of successive merging (agglomerative algorithms) or splitting (divisive algorithms) of clusters. This hierarchy is usually represented by a dendrogram, which might be used to estimate an appropriate number of clusters. These methods are not iterative, i.e. the assignment of objects to clusters made on preceding levels can not be changed.
- *Objective-function based clustering methods:* These methods use the formulation of the clustering criterion in the form of an objective function to be optimised. The objective function provides an estimate of the partition quality for each cluster partition. In order to find a partition with the best value of the objective function the optimisation problem must be solved.

The considerations in this work will be primarily limited to fuzzy and possibilistic clustering methods with an objective function.

2.2.3 Fuzzy pattern recognition

The desire to fill the gap between traditional pattern recognition methods and human behaviour has lead to the development of fuzzy set theory, introduced by L.A. Zadeh in 1965. The fundamental role of fuzzy sets in pattern recognition, as it was stated by [Zadeh, 1977], is to make the opaque classification schemes, usually used by a human, transparent by developing a formal, computer-realisable framework. In other words, fuzzy sets help to transfer a qualitative knowledge regarding a classification task into the relevant algorithmic structure. As a basic tool used for this interface serves a membership function. Its meaning can be interpreted differently depending on the application area of fuzzy sets. Three semantics of a membership grade can be generalised, according to [Dubois, Prade, 1997] in terms of similarity, uncertainty, or preference, respectively. A view as a degree of uncertainty is usually used in expert systems and artificial intelligence, and interpretation as a degree of preference is concerned with fuzzy optimisation and decision analysis. Pattern recognition works with the first semantic, which can be formulated as follows:

Consider a fuzzy set $\widetilde{A}$, defined on the universe of discourse X, and the degree of membership $u_{\widetilde{A}}(x)$ of an element x in the fuzzy set $\widetilde{A}$. Then $u_{\widetilde{A}}(x)$ is the degree of proximity of x to prototype elements of $\widetilde{A}$ and is interpreted as a degree of similarity.

This view, besides a meaning of the semantic, shows distinctions between a membership grade and different interpretations of a probability value. Consider a pattern $\mathbf{x}$ and a class A. On observing $\mathbf{x}$ the prior probability $P(\mathbf{x} \in A) = 0.95$, expressing that the pattern $\mathbf{x}$ belongs to the class A, becomes a posterior probability: either $P(\mathbf{x} \in A \mid \mathbf{x}) = 1$ or $P(\mathbf{x} \in A \mid \mathbf{x}) = 0$. However, a degree $u_{\widetilde{A}}(x) = 0.95$ to which the pattern $\mathbf{x}$ is similar to patterns of the class A remains unchanged after observation [Bezdek, 1981, p. 12].

Fuzzy set theory provides a suitable framework for pattern recognition due to its ability to deal with uncertainties of the non-probabilistic type. In pattern recognition uncertainty may arise from a lack of information, imprecise measurements, random occurrences, vague descriptions, or conflicting or ambiguous information ([Zimmermann, 1997], [Bezdek, 1981]) and can appear in different circumstances, for instance, in definitions of features and, accordingly, objects, or in definitions of classes. Different methods process uncertainty in various ways. Statistical methods based on probability theory assume features of objects to be random variables and require numerical information. Feature vectors having imprecise, or incomplete, representation are usually ignored or discarded from the classification process. In contrast, fuzzy set theory can be applied for handling non-statistical uncertainty, or fuzziness, at various levels. Together with possibility theory, introduced by Dubois and Prade [Dubois, Prade, 1988], it can be used to represent fuzzy objects and fuzzy classes. Objects are considered to be fuzzy if at least one feature is described fuzzily, i.e. feature values are imprecise or represented as linguistic information. Classes are considered to be fuzzy, if their decision boundaries are fuzzy with gradual class membership. The combination of two representation forms of information such as crisp and fuzzy with two basic elements of pattern recognition such as object and class induces four categories of problems in pattern recognition [Zimmermann, 1995]:

- Crisp objects and crisp classes;
- Crisp objects and fuzzy classes;
- Fuzzy objects and crisp classes;
- Fuzzy objects and fuzzy classes.

The first category involves the problem of classical pattern recognition, whereas the latter three categories are concerned with fuzzy pattern recognition. In this book the attention is focused on methods dealing with the second category of problems - crisp objects and fuzzy classes.

It is obvious that the concept of fuzzy sets enriches the basic ideas of pattern recognition and gives rise to completely new concepts. The main reasons for the application of fuzzy set theory in pattern recognition can be summarised in the following way [Pedrycz, 1997]:

1. Fuzzy sets design an interface between linguistically formulated features and quantitative measurements. Features are represented as an array of membership values denoting the degree of possession of certain properties. Classifiers designed in such a way are often logic-oriented and reflect the conceptual layout of classification problems.
2. Class memberships of an object take their values in the interval [0, 1] and can be regarded as a fuzzy set defined on a set of classes. Thus, it is possible that an object belongs to more than one class, and a degree of membership of an object to a class expresses a similarity of this object with typical objects belonging to this specific class. Using a gradual degree of membership the most 'unclear' objects can be identified.
3. Membership functions provide an estimate of missing or incomplete knowledge.
4. A traditional distinction between supervised and unsupervised pattern recognition is enriched by admitting implicit rather than explicit object labelling or allowing for a portion of objects to be labelled. In the case of

implicitly supervised classification objects are arranged in pairs according to their similarity levels.

Fuzzy set theory has given rise to a lot of new methods of pattern recognition, some of which are extensions of classical algorithms and others completely original techniques. The major groups of fuzzy methods are represented by fuzzy clustering, fuzzy-rule-based, fuzzy pattern matching methods and methods based on fuzzy relations. Since the focus of this work is on algorithmic methods, only fuzzy clustering methods will be discussed in the following sections.

Fuzzy techniques seem to be particularly suitable for dynamic pattern recognition when it is necessary to recognise gradual temporal changes in an object's states. Considering the temporal development of objects, it is often difficult to assign objects to classes crisply and precisely. One can imagine, for instance, a system with two possible states as classes: proper operation and faulty operation. When the state of the system is changing measurement values express that the system operation is not more proper, but there is no error in operation yet. This means that the observed objects do not belong to any of these classes, or belong to a small degree to both classes. The use of fuzzy set theory provides a possibility to produce fuzzy decision boundaries between classes and allows a gradual (temporally changing) membership of objects to classes. This primary advantage of fuzzy set theory is crucial for pattern recognition in general and for dynamic approach in particular, because of the possible temporal transition of objects between classes and changes of classes themselves. Therefore, the development of methods for dynamic pattern recognition in this book will be based on fuzzy techniques.

2.3 The Problem of Dynamic Pattern Recognition

Since a standard terminology in the area of dynamic pattern recognition does not yet exist, the main notions employed in this book must first be defined. And as the topic of this book is concerned with the development of pattern recognition methods for dynamic systems, the basic principles of analysis and modelling of dynamic systems will first be presented. This will be followed by the formulation of the main definitions used in the area of dynamic pattern recognition. Finally, specific problems arising in dynamic pattern recognition and the ways to handle them will be discussed, which will then lead to the formulation of the goal and basic steps of the dynamic pattern recognition process.

2.3.1 Mathematical description and modelling of dynamic systems

This section is intended to provide a general overview about the description of dynamic systems used in different research areas and about tools used for their analysis.

In economics, medicine, biology, and control theory and a lot of other areas, a system or a process is called *dynamic*, if one of the variables is time-dependent ([Stöppler, 1980], [Rosenberg, 1986]). The dynamic behaviour of the system demonstrates how it performs with respect to time.

Time-domain analysis and the design of dynamic systems use the concept of the states of a system. A system's dynamics is usually modelled by a system of differential equations. These equations are normally formulated based on physical laws describing the dynamic behaviour of a system.

A dynamic system, or process, is defined by input variables $u_1(t), ..., u_p(t)$, which are used to influence the system, and output variables $y_1(t), ..., y_q(t)$, whose dynamic behaviour is of major interest [Föllinger, Franke, 1982]. In order to obtain a system of normal differential equations representing the relationship between input and output variables, intermediate variables, which are called *state variables* $x_1(t), ..., x_n(t)$, are used. A set of state variables $x_1(t_0), ..., x_n(t_0)$ at any time t_0 determines the state of the system at this time. If the present state of a system and values of the input variables for $t > t_0$ are given, the behaviour of a system for $t > t_0$ can be described clearly.

Hence, the *state of a system* is a set of real numbers such that the knowledge of these numbers and the input variables will provide the future state and the output of the system, with the equations describing the system's dynamics. The *state variables* determine the future behaviour of a system when the present state of a system and the values of input variables are known [Dorf, 1992].

The multidimensional space induced by the state variables is called the *state space*. The solution to a system of differential equations can be represented by a vector $\mathbf{x}(t)$. It corresponds to a point in the state space at a given moment in time. This point moves in the state space as time passes. The trace, or path, of this point in the state space is called *a trajectory* of the system. For a given initial state $\mathbf{x}_0 = \mathbf{x}(t_0)$ and a given end state $\mathbf{x}_e = \mathbf{x}(t_e)$, there is an infinite number of input vectors and corresponding trajectories with the same start and end points. On the other side, considering any point of the state space there is exactly one trajectory containing this point [Föllinger, Franke, 1982].

Considering dynamic systems in control theory, great attention is dedicated to adaptive control. The primary reason for introducing this research area was to obtain controllers that could adapt their parameters to changes in process dynamics and disturbance characteristics. [Åström, Wittenmark, 1995] propose the following definition:

'An *adaptive controller* is a controller with adjustable parameters and a mechanism for adjusting the parameters'.

During extensive research carried out in the last two decades it was found that adaptive control is strongly related to ideas of learning that are emerging in the field of computer science.

In this book some definitions of control theory are reformulated and applied to the area of pattern recognition in dynamic systems.

2.3.2 Terminology of dynamic pattern recognition

Consider a dynamic complex system that can assume different states in the course of time. Each state of the system at a moment in time represents an object for classification. As stated in Section 2.3.1, a dynamic system is described by a set of state variables characterising its dynamic behaviour. These variables can be voltages, currents, pressures, temperatures, unemployment rates, share prices, etc. depending on the type of system. They are called features in pattern recognition. If a dynamic system is observed over time, its feature values vary constituting time-dependent functions. Therefore, each object is described not only by a feature vector at the current moment but also by a history of the feature values' temporal development.

Objects are called *dynamic* if they represent measurements or observations of a dynamic system and contain a history of their temporal development. In other words, each dynamic object is a temporal sequence of observations and is described by a discrete function of time. This time-dependent function is called *a trajectory* of an object.

Thus, in contrast to static objects, dynamic objects are represented not by points but by multi-dimensional trajectories in the feature space extended by an additional dimension 'time'. The components of the feature vector representing a dynamic object are not real numbers but vectors describing the development of the corresponding feature over time.

Dynamic pattern recognition is concerned with the recognition of classes, or clusters, of dynamic objects, i.e. recognition of typical states in the dynamic behaviour of a system under consideration.

To illustrate the difference between static and dynamic pattern recognition consider a set of dynamic objects described by two features X_1 and X_2 that were observed over time interval [0, 100]. From a static viewpoint, the current states of objects are represented by the momentary snapshot of objects at current moment t = 100 and can be seen in the cut of the three-dimensional feature space at this moment (*Figure 4*). Using the spatial closeness of points as a criterion of object similarity (a typical criterion in static pattern recognition), two clusters of objects can easily be distinguished in this plane (squares and circles). The way in which objects arrive at the current state is however irrelevant to the recognition process.

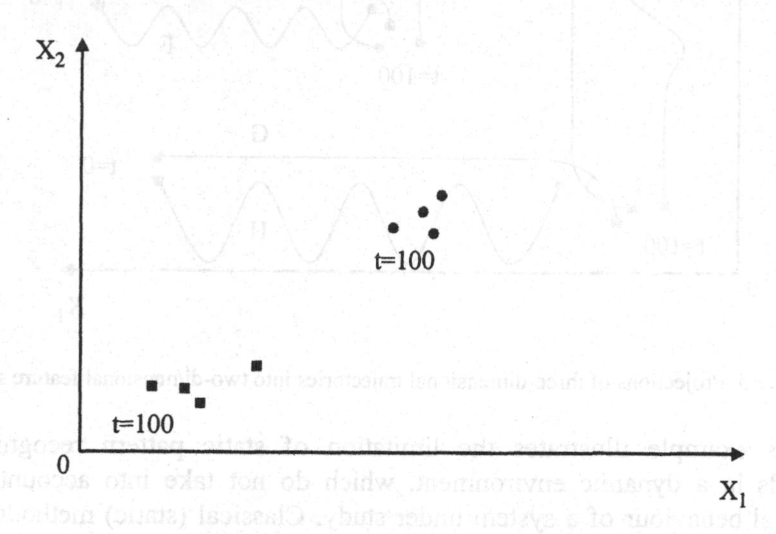

Figure 4. Current states of dynamic objects from a static viewpoint

From a dynamic viewpoint the states of objects are characterised not only by their momentary location but also by the history of their temporal development, which is represented by a trace, or trajectory, of each object from its initial state to its current state in the three-dimensional feature space. *Figure 5* shows the projections of three-dimensional trajectories of objects into the two-dimensional feature space (without dimension 'time'). If the form of trajectories is chosen as a criterion of similarity between trajectories, then three clusters of dynamic objects can be distinguished: {A, C}, {B, D, E, G}, and {F, H}. Obviously, these clusters are different from the ones recognised for static objects at the current moment. If the form and the orientation of trajectories in the feature space is chosen as the criterion for similarity, then dynamic objects B, D, E and G can not be considered as similar any more and they are separated into two clusters: {B, D} and {E, G}. Thus, based on such a similarity criterion four clusters of trajectories are obtained. In the third case, if the form and orientation of the trajectories are irrelevant but their spatial pointwise closeness is a base for a similarity definition, then another four clusters can be recognised: {A, B}, {C, D}, {E, F}, and {G, H}.

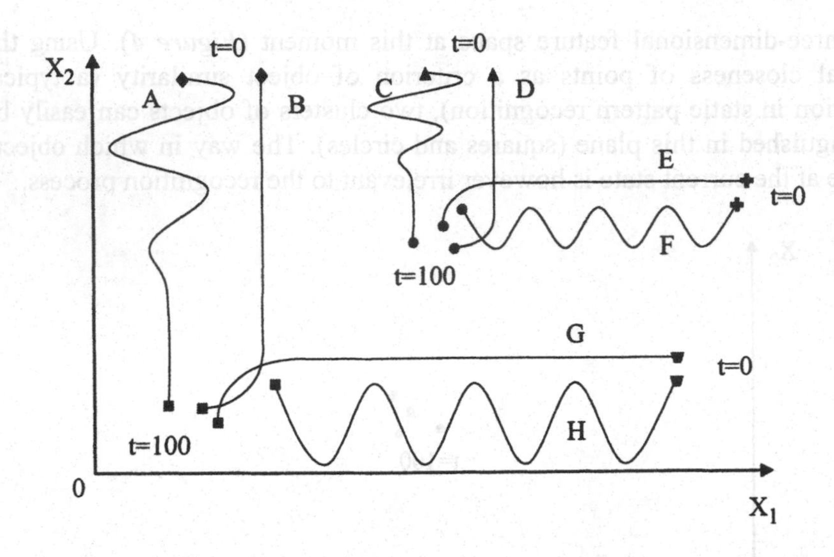

Figure 5. Projections of three-dimensional trajectories into two-dimensional feature space

This example illustrates the limitation of static pattern recognition methods in a dynamic environment, which do not take into account the temporal behaviour of a system under study. Classical (static) methods are usually applied in order to recognise a (predefined) number of a system's states. However, there is a number of applications, in particular diagnosis problems, in which gradual transitions from one state to another are observed over time. In order to follow such slow changes of a system's state and to be able to anticipate the occurrence of new states, it is important to consider explicitly the temporal development of a system. In general, there can be two possibilities to deal with dynamic objects in pattern recognition:

1. To pre-process trajectories of dynamic objects by extracting some characteristic values (temporal features, trends) that can represent components of conventional feature vectors. The later can be used as valid inputs for static methods of pattern recognition.
2. To modify classical (static) methods or to develop new dynamic methods that can process trajectories directly.

In the first case, methods for extraction of relevant temporal features from trajectories are required. This way to deal with dynamics is often chosen in practice. The second case is of primary interest because it is concerned with a new research area. Since most methods of pattern recognition use a distance, or dissimilarity, measure to classify objects, dealing with trajectories in pattern recognition requires a definition of similarity measure for trajectories. As the example above has shown, different definitions of similarity, leading to different classification results, are possible. However, two general viewpoints on similarity between trajectories can be distinguished from among all these possible definitions [Joentgen, Mikenina et al., 1999b, p. 83]:

1. *Structural similarity*: Two trajectories are the more similar, the better they match in form/ evolution/characteristics,
2. *Pointwise similarity*: Two trajectories are the more similar, the smaller their pointwise distance in feature space is.

Structural similarity relates to the similar behaviour of trajectories over time. It is concerned with a variety of aspects of the trajectories (in general functions) under consideration such as the form, evolution, size or orientation of trajectories in the feature space. The choice of relevant aspects for a description of similar trajectories is related to a concrete application. Depending on the chosen aspect, different criteria can be used to define a similarity, for instance, slope, curvature, position and values of inflection points, smoothness or some other characteristics of trajectories representing trends in trajectories. In order to detect similar behaviour of trajectories it is important that a measure of structural similarity be invariant to such changes as scaling, translation, addition or removal of some values and some incorrect values due to measurement errors called outliers.

Pointwise similarity relates to the closeness of trajectories in the feature space. This type of similarity can be defined based directly on functions values without taking into consideration special characteristics of the functions. In this case the behaviour of trajectories is irrelevant and allows some variations in form such as fluctuations or outliers.

The example in *Figure 5* illustrates the difference between structural and pointwise similarity. In terms of structural similarity, especially if the form of trajectories is relevant, three clusters can be recognised: {A, C}, {B, D, E, G}, and {F, H}. In terms of pointwise similarity, four clusters {A, B}, {C, D}, {E, F}, and {G, H} seem to be more natural.

Thus, the definition of similarity measure is a crucial point in pattern recognition and a non-trivial task in the case of dynamic objects because of its strong dependence on the application at hand.

Due to the non-stationary character of objects in dynamic pattern recognition, the partitioning of objects is not necessarily fixed over time. The number of clusters and the location of cluster centres at a moment in time constitute the *cluster structure*. It can be either static or dynamic. If the number of clusters is fixed and only the location of cluster centres is changing slightly as time passes, that cluster structure can be considered static. If in the course of time the number of clusters and the location of cluster centres vary, then one has to deal with the dynamic cluster structure. Its temporal development is represented by trajectories of the cluster centres in the feature space. Changes in the cluster structure correspond to changes of a state, or behaviour, of a system under study and will also be referred to as *structural changes*. This term comes from econometrics where it is used to describe changes in a regression model either at an unknown time point or at a possible change point.

Due to the arrival of new objects and the discarding of old data being irrelevant, the following changes in the dynamic cluster structure can appear [Mann, 1983]:

1. *Formation of new clusters*: If new data can not be clearly assigned to existing clusters, one or more new clusters may be formed either subsequently (one after another) or in parallel (from the first cluster to many different clusters) (*Figure 6*). This situation can appear if the degrees of membership of new objects to all fuzzy clusters are approximately equal or very low (e.g. in the case of possibilistic c-means they must not sum to one as in probabilistic c-means, and can be very low). If new data can not be assigned to existing clusters and their number is not large enough to form a new cluster, these new data are recognised as being stray. It should be noted that the projections of discrete trajectories into the feature space without time axis are shown in *Figure 6*, where points correspond to observations of dynamic objects at different moments in time.
2. *Merging of clusters*: Two or more clusters may be merged into one cluster *Figure 7, a*). If a large number of new data has equally high degrees of membership (> 0.5) to two clusters, for example, these two clusters cannot be considered as heterogeneous any more and in contrast are considered as similar and must be merged.
3. *Splitting of clusters*: One cluster may appear as two or more clusters (*Figure 7, b*). If a large number of new data has been absorbed, distinct groups of objects with high density within a cluster may be formed, whereas a cluster centre may be located in the area of very low density. Alternatively it may happen that there is no new data being absorbed into a cluster and old data are discarded as irrelevant. Due to the discarding of old data, an area that has a very low density of data, or is even empty, may appear around the cluster centre and along some direction separating the data into two or more groups. A cluster cannot be considered as homogeneous any more and must be split to find the better partition.
4. *Destruction of clusters*: One or more clusters may disappear if there are no new data being assigned to these clusters and old data are discarded. The cluster centre must, however, be saved in order to preserve the discovered knowledge. It may happen that this cluster appears once again in the future. In this case a new cluster can be recognised and identified faster if the knowledge about this cluster already exists and does not need to be learned anew.
5. *Drift of clusters*: The location of clusters (cluster centres) in the feature space may be slightly changed over time.

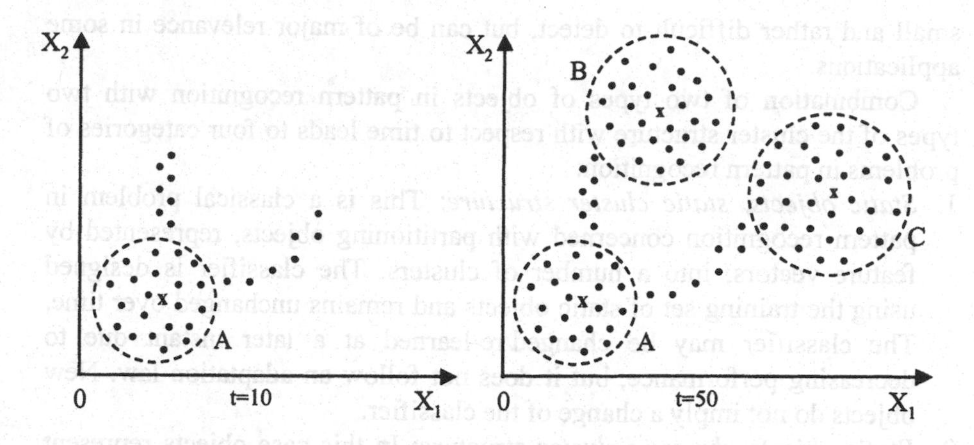

Figure 6. Formation of new clusters

a) At time t=10 there is one cluster A; b) At time t=50 two new clusters, B and C, are formed

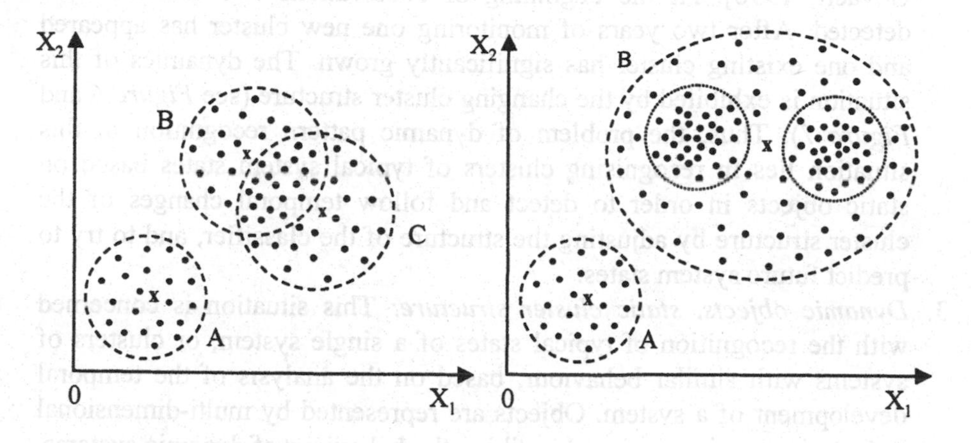

Figure 7. Changes of the dynamic cluster structure

a) Merging clusters B and C; b) Splitting cluster B

The first four types of structural changes represent abrupt (serious) changes in the cluster structure. In many applications abrupt changes are associated with faults in the operational behaviour of a system, which have to be detected as early as possible to avoid dangerous consequences (e.g. fatal damage to equipment or faults in chemical or power plants). The fifth type of structural changes is referred to as gradual changes, which can be useful to predict the occurrence of abrupt changes. They are usually very small and rather difficult to detect, but can be of major relevance in some applications.

Combination of two types of objects in pattern recognition with two types of the cluster structure with respect to time leads to four categories of problems in pattern recognition:

1. *Static objects, static cluster structure*: This is a classical problem in pattern recognition concerned with partitioning objects, represented by feature vectors, into a number of clusters. The classifier is designed using the training set of static objects and remains unchanged over time. The classifier may be changed/re-learned at a later instant due to decreasing performance, but it does not follow an adaptation law. New objects do not imply a change of the classifier.
2. *Static objects, dynamic cluster structure*: In this case objects represent measurements of a dynamic system, which are sufficient for a description of the system's behaviour. For instance, the data can come from an environmental monitoring application and consist of daily single-point measurements of water quality parameters [Denoeux, Govaert, 1996]. At the beginning of observations two clusters were detected. After two years of monitoring one new cluster has appeared and one existing cluster has significantly grown. The dynamics of this situation is exhibited by the changing cluster structure (see *Figure 6* and *Figure 7*). Thus, the problem of dynamic pattern recognition in this situation lies in recognising clusters of typical system states based on static objects in order to detect and follow temporal changes of the cluster structure by adjusting the structure of the classifier, and to try to predict future system states.
3. *Dynamic objects, static cluster structure:* This situation is concerned with the recognition of typical states of a single system, or clusters of systems with similar behaviour, based on the analysis of the temporal development of a system. Objects are represented by multi-dimensional trajectories, or time series, describing the behaviour of dynamic systems. The problem of dynamic pattern recognition in this situation is to recognise clusters based on similar behaviour of trajectories. There is a large number of applications in which clustering and classification of dynamic objects is of primary importance whereas the cluster structure remains unchanged over time, e.g. classification of share prices or other market characteristics, analysis of customer behaviour or recognition of typical scenarios in scenario analysis. For instance, stock market buy/sell decisions are usually made based on past values of share prices showing specific (predefined) patterns. If the behaviour of share prices over a period of time corresponds to a certain pattern, then shares are sold or bought. It is not sufficient to consider only current share values to make a correct decision. Clustering of dynamic objects can also be applied in

scenario analysis for complexity reduction. In order to find typical scenarios of the future development of economic characteristics for strategic planning it is reasonable to consider the characteristics' temporal behaviour instead of their final values only. *Figure 8* shows three typical scenarios found after clustering 150 different scenarios [Hofmeister, 2000, p. 280-287]. Thus, due to the clustering of trajectories (scenarios) it is possible to reduce a large set of raw scenarios to a few typical scenarios, which can be used to make strategic decisions and to avoid an important loss of information in the case of considering the final values of scenarios.

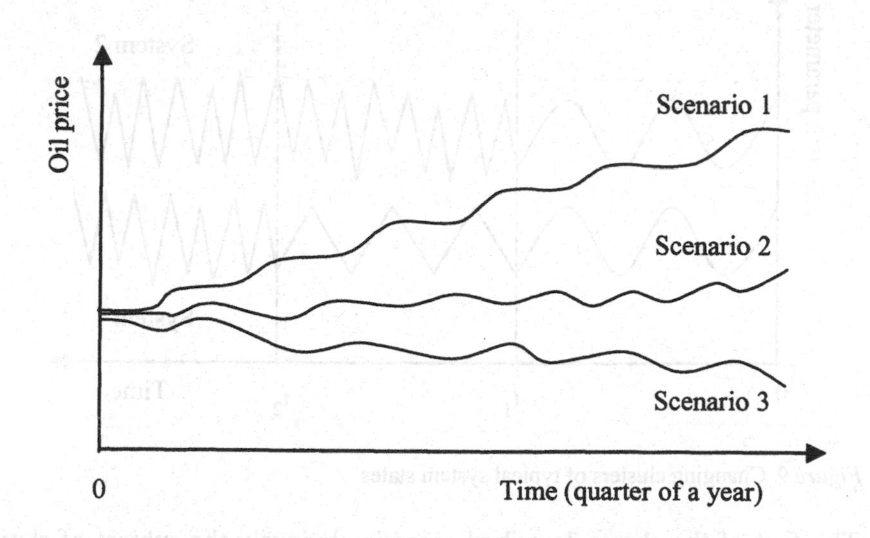

Figure 8. Typical scenarios of future temporal development of oil price

4. *Dynamic objects, dynamic cluster structure*: This situation is a combination of cases 2 and 3. The task of dynamic pattern recognition in this case is to recognise typical system states based on the analysis of the temporal behaviour of a system, to detect and follow changes in the cluster structure, and to try to predict future states. In other words, in the case of a single system the problem is to recognise whether the current trajectory contains changes in temporal behaviour indicating a new system state or remains unchanged in its behaviour. In the case of several systems the aim is to recognise different types of systems and to detect whether clusters and the assignment of systems to these clusters change over time. This problem can appear in real-time diagnosis in different application areas such as medicine, biology, chemical and industrial engineering, etc. A typical example is preventive machine maintenance where several systems are monitored simultaneously (*Figure 9*).

Comparing system dynamics during different periods of operation, different clusters of systems can be distinguished: for example, in the 1st period Systems 1 and 2 are assigned to cluster A; in the 2nd period System 1 is assigned to a new cluster B and System 2 is assigned to another new cluster C; after the 2nd period both systems are assigned to cluster C. Hence, the dynamics of this situation shows itself in changing cluster structure and in the transition of dynamic objects (represented by trajectories) between clusters.

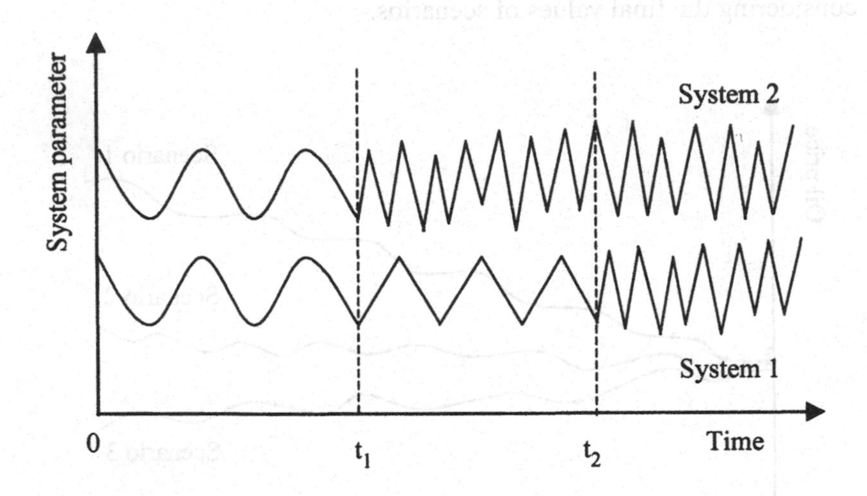

Figure 9. Changing clusters of typical system states

The first of the above described cases is obviously the subject of static pattern recognition, whereas the last three cases represent problems of dynamic pattern recognition and, therefore, will be treated in this book.

A classifier which is able to deal with dynamic objects can be called dynamic with respect to objects. A classifier which is able to deal with dynamic cluster structure can be called dynamic according to its design principle.

A *dynamic fuzzy classifier* is defined by time-dependent cluster centres and by time-dependent degrees of membership of objects to clusters. Since, in the course of time, cluster centres and membership functions of clusters can be changed due to the arrival of new objects, the dynamic classifier must be updated to preserve its performance. There are two strategies for updating the classifier with respect to a set of objects: one involves using the complete history given by the whole set of objects obtained from the beginning of observation until the current moment, and the other involves using only the most recently obtained objects. Since old data may become irrelevant over time for representing the current situation and may have negative effects on the clustering procedure, it is not reasonable to preserve the constantly growing set of objects for clustering. A better strategy is to update the classifier within a rolling horizon defined as a *moving time window*. The idea of using time windows is that only objects of the current time window are considered for clustering. Preceding ('old') objects are not taken into consideration because they are deemed irrelevant. Time windows are defined as subsequent (overlapping) time intervals of constant or variable length, which are shifted along the time axis (*Figure 10*).

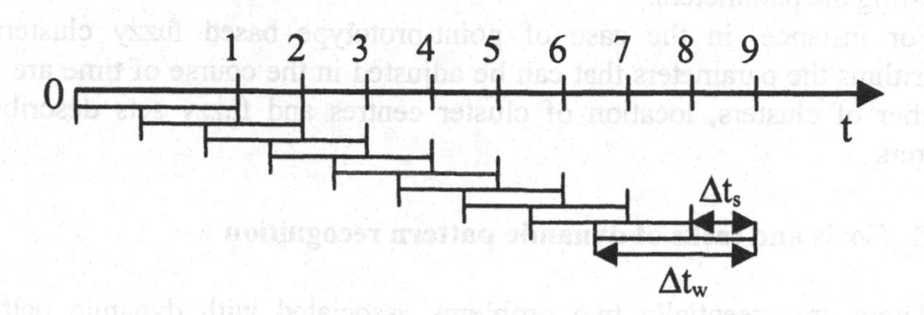

Figure 10. Moving time windows of constant length

Moving time windows are characterised by two parameters: the length of the window Δt_w and the length of the shift Δt_s. The length of the time window determines the number of objects considered for clustering. If the window is too large, then the changes in a system's state may be recognised too late and updating of the classifier may be too slow. If the window is too small, then it may not contain enough information to design a reliable classifier. Thus the 'optimal' length of the window must be chosen carefully depending on the application.

The length of the shift of the time window determines the number of new objects that is taken into consideration for the update of the classifier. In other words, the length of the shift defines the frequency of the update of the classifier. If the length of the shift is too large, then the updating of the classifier to the changes of a system's state may be too slow. If the length of the shift is too small, then the classifier may be updated much more frequently than changes appear in the state of a system. For instance, the shift of the time window by one time unit corresponds to the update of the classifier at each moment in time. Such a frequent adaptation of a classifier may be very costly and time-consuming. While choosing the 'optimal' length of the shift it should be taken into account that only a 'sufficient' number of new objects can cause such changes in the cluster structure as merging or splitting of clusters. An exception to this rule may be provided by applications where significant changes, such as the formation of new clusters, must be recognised as early as possible based on just a few objects (e.g. early fault detection).

A dynamic classifier must have an adaptive capacity in order to be able to follow temporal changes in the data and to preserve its performance.

Based on the notion of an adaptive controller, an *adaptive classifier* can be defined as a classifier that can modify its structure in response to changes in the dynamics of the system/process under consideration. An adaptive classifier is characterised by adjustable parameters and a mechanism for adjusting the parameters.

For instance, in the case of point-prototype based fuzzy clustering algorithms the parameters that can be adjusted in the course of time are the number of clusters, location of cluster centres and fuzzy sets describing clusters.

2.3.3 Goals and tasks of dynamic pattern recognition

There are essentially two problems associated with dynamic pattern recognition (in addition to those arising in static pattern recognition [Taylor et al., 1997]):

1. To detect any change in the cluster structure,
2. To react to any detected change.

The first problem can be solved during the *monitoring process*. Its task is to monitor the performance of the classifier by observing and evaluating some calculated measures based on the results of classification of new objects. By doing this, the monitoring process tries to check how well new objects arriving at each moment fit into the existing cluster structure. In particular, if new objects do not fit the existing cluster structure and the performance of the classifier is rather poor, it is assumed that changes have taken place. There can be two reasons for the change of the cluster structure in the course of time [Kunisch, 1996, pp. 36-37]:

1. The number of features changes. The number of relevant features describing objects can decrease or increase over time. The reasons for the decrease of the number of features may be that some features have to be dropped since they become irrelevant for representing objects due to fundamental changes in properties of objects (e.g. changes in the operating conditions of a system), or it is no longer possible for some reasons to observe some feature. On the other hand, there might be more features available for describing objects as time passes. If they contain a relevant information for the recognition process they have to be added to the feature set to improve the performance of the classifier. In both cases of a quantitative change of the feature set it seems reasonable to re-learn the classifier from scratch.

2. The values of feature change. The number of features remains the same but the values of some feature change their range or interpretation. The following types of changes can be distinguished:
   - Drift of the feature values: The range of feature values is changed slowly and gradually.
   - Shift in the feature values: An abrupt change of the range of feature values is observed. The interpretation of features may though remain the same.
   - Semantic change: Although the feature values are not changed, their meaning may become completely different. This case is very hard to deal with during the classifier design.

The problem of a variable feature set can present real difficulties for the recognition process, if new relevant features are not known a priori. This problem is related to feature selection and feature generation rather than classifier design. The selection of relevant features is a non-trivial task anyway, but the detection of features whose relevance to pattern recognition is time-dependent can be viewed as a challenging new research area and requires the development of new methods for time-dependent feature selection/generation. For solving this problem two cases can be distinguished: 1) an initial set of features is given, and the goal is to select a subset of features relevant at each moment in time (or within a period of time); 2) relevant features at any given moment are combinations of available features such as a product, a ratio, linear combination, etc. In the first case, one possible approach can be a repeated application of some feature selection procedure that is indeed computationally expensive. In the second case, approaches for feature generation depending on the current feature values are needed.

The problem of changing feature values is related to the detection of changes in the cluster structure. These changes depend to some degree on the number of changed features and their importance for the recognition process. The importance of features is determined by the contribution of features to the recognition process. As stated above, the following types of changes can take place in the cluster structure due to the arrival of new objects: drift of clusters, formation of new clusters, merging of clusters, and splitting of clusters. In order to detect these structural changes new sophisticated procedures must be developed. The following chapters will focus on this problem.

The monitoring process can either be comparative static and based on the analysis of changes caused by new static objects (without the history) at the current moment or dynamic and based on the consideration of the temporal development of objects during a certain period of time (trajectories). The first approach can employ classical (static) methods for classifier design and classification, whereas the second approach requires the development of new methods for dealing with trajectories of dynamic objects.

After the temporal changes in the cluster structure are recognised by the monitoring process, the problem of reacting to these changes arises. It means that the pattern recognition process must contain a mechanism to update the existing classifier according to current changes in order to preserve its validity over time. One extreme strategy for solving this problem is to re-learn a classifier from scratch. It does not require any adaptive mechanism, therefore, classical methods for classifier design can be used. The other extreme solution is to ignore changes and to improve the current classifier using, for example, incremental updating based on new objects. The optimal approach lies between these two extremes and is concerned with the adaptation of the existing classifier using the results of the monitoring process to fit the changed cluster structure [Nakhaeizadeh et al., 1997]. A strategy chosen for the updating process is a crucial point for the performance of the whole pattern recognition system with respect to the accuracy of results and time consumption.

To be able to make a diagnosis about a current system's state, the information concerning detected structural changes must be interpreted correctly. In contrast to static pattern recognition the interpretation of results includes not only the description of the current situation but also a description of the temporal development of the situation. The diagnostic results can be used to make a short-term prognosis about a tendency of the future development of the current situation or to formulate a control action to modify the dynamics of a system. For instance, one possible diagnosis may be: 'A system's state has been moved away from cluster A and is approaching cluster B.'

Thus, compared with static pattern recognition, the process of dynamic pattern recognition must incorporate three additional steps:

1. Monitoring process: comparative static or dynamic.
2. Updating process: re-learning, incremental updating, or adaptation.
3. Diagnostics: description and interpretation of the current cluster structure.

The general scheme of the process of dynamic pattern recognition is presented on *Figure 11*.

The *goal* of dynamic pattern recognition is to detect and follow changes in the cluster structure and to adapt the classifier to these changes in the course of time. Since dynamic objects are represented by multidimensional trajectories, methods of dynamic pattern recognition must be based on a similarity measure for trajectories.

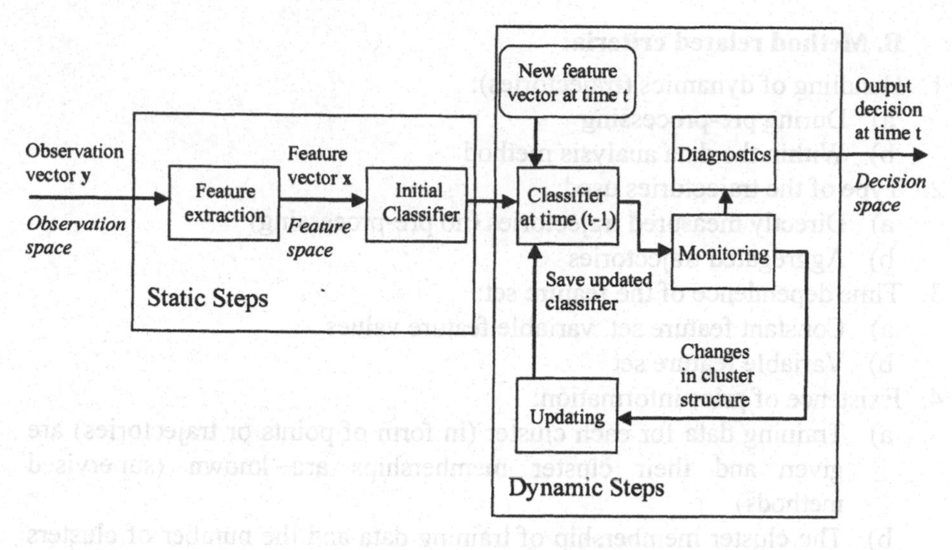

Figure 11. The process of dynamic pattern recognition

All the aforementioned considerations about dynamic pattern recognition can be summarised in a *taxonomy* of this field. The taxonomy is an extended version of the one presented in ([Joentgen, Mikenina et al., 1998], [Joentgen, Mikenina et al., 1999a]) and contains different criteria which are suitable for structuring this field. The criteria themselves are divided into two groups: problem related criteria and method related criteria. Problem related criteria are concerned with the structure and type of system observed and the character of observations. Method related criteria provide a basis for the distinction of methods and techniques used in classifier design and classification. Essentially they include criteria for selecting and processing relevant features.

The field of dynamic pattern recognition can be structured using the following taxonomy:

A. Problem related criteria:

1. Classification object:
   a) Classification of several systems
   b) Classification of the states of one system
2. Type of cluster structure:
   a) Static cluster structure
   b) Dynamic cluster structure
3. Observation period:
   a) Moment in time
   b) Rolling horizon
   c) Complete history

B. Method related criteria:

1. Handling of dynamics (trajectories):
   a) During pre-processing
   b) Within the data analysis method
2. Type of the trajectories used:
   a) Directly measured trajectories (no pre-processing)
   b) Aggregated trajectories
3. Time dependence of the feature set:
   a) Constant feature set, variable feature values
   b) Variable feature set
4. Existence of prior information:
   a) Training data for each cluster (in form of points or trajectories) are given and their cluster memberships are known (supervised methods)
   b) The cluster membership of training data and the number of clusters are unknown (unsupervised methods)
5. Detection of temporal changes in cluster structure:
   a) Comparative static
   b) Dynamic
6. Type of design of dynamic classifier:
   a) Re-learning
   b) Incremental updating
   c) Adaptation.

In the next chapter the analysis of dynamic steps in the pattern recognition process is carried out and different approaches for the realisation of these steps in a dynamic classifier are considered.

3 STAGES OF THE DYNAMIC PATTERN RECOGNITION PROCESS

As stated in the previous chapter, the design procedure of a dynamic classifier must couple the results of the monitoring process with a mechanism for updating the classifier according to detected changes in the cluster structure. Different approaches used for establishing the monitoring process are usually based on the observation and the analysis of some characteristic values describing the performance of a classifier or the cluster structure. The temporal change of these characteristics points to some structural changes in the underlying cluster structure and to the need for adapting a classifier. According to the nature of the monitored characteristics, one can distinguish between statistical and fuzzy techniques for the monitoring process, the most important of which will be discussed in Section 3.1. In order to preserve the performance of a dynamic classifier over time, it must be adapted to temporal changes detected by the monitoring process. The updating strategies of a dynamic classifier presented in Section 3.2 depend on the type of temporal changes in the cluster structure (gradual or abrupt) and can require either the adjustment of classifier parameters or complete re-learning of a classifier. As will be shown, the most flexible adaptation law of a dynamic classifier must combine both these techniques and include additional mechanisms supporting the intelligent design of a dynamic classifier. An updating strategy represents a crucial component of a dynamic patter recognition system since it determines an adaptive capacity of a dynamic classifier and its ability to follow temporal changes in the cluster structure.

3.1 The Monitoring Process

In many application areas it has become increasingly important to monitor the behaviour of dynamic systems based on multiple measurements. According to [Denoeux et al., 1997] the task of a monitoring system is 'to detect the departure of a process from normal conditions, to characterise the new process state, and to prescribe appropriate actions.' The purpose of this book is to solve this task by means of a dynamic pattern recognition system. It must consist of different components, one of which is the monitoring process. The aim of the monitoring process in this context, as stated in section 2.3.3, consists in detecting gradual and abrupt changes in the cluster structure where clusters represent typical states, or the typical behaviour, of a system under consideration.

In order to detect changes in the cluster structure different characteristic values for the evaluation of the performance of the classifier have to be monitored regularly. The following statistical measures are usually used for monitoring ([Kunisch, 1996], [Lanquillon, 1997]):

1. *Accuracy of the approximation*: Many classifiers respond with approximated output values instead of class labels. If the true output values for all clusters become known the accuracy of the approximation in the current time window, defined as the difference between the output values and the expected values corresponding to each cluster, is compared to the accuracy of the approximation in previous time windows. If the difference increases this means a deterioration of the approximation and the classifier performance.
2. *The number of misclassified objects*: If the true class labels of previously (crisply) classified objects become known the error rate of classification in the current time window can be compared to the average error rate of previous windows. This is one of the most common characteristics used to evaluate the performance of a classifier. Error rates, which are less or approximately equal to the average, provide evidence about the stable cluster structure and satisfactory performance of the classifier. If error rates are greater than the average, changes in the cluster structure can be assumed.
3. *Unambiguity of the classification:* If, in a classifier, a binary 1-out-of-c coding has been selected to represent the existing c clusters the approximated output with the highest value usually determines the class label (the so called 'winner takes all' principle). Under certain conditions, the output values can be considered as probabilities (or proportional to probabilities) for each cluster. Then if the difference between two maximum output values is very small the classifier is somewhat indifferent between the corresponding clusters, i.e. its discriminating ability is rather pure. In the most uncertain case where each cluster is equally likely all outputs take value 1/c.
4. *Class distributions*: The relative number of objects (the percentage in the total number) assigned to each cluster during the current time window can be compared to the average relative number of objects assigned to each cluster in previous time windows. The true class labels of previously classified objects are only required for the previous time windows. The number of objects in each cluster represents the class distribution describing a certain structure of the data. If the class distribution obtained after classification of new objects in the current time window differs from the average this can indicate a change in the underlying cluster structure.
5. *Means and variances of features*: For each feature and class, the mean and the variance of its values in the current time window can be

compared to the corresponding statistics in the previous time windows. Based on these measures, a drift or shift in the feature values can easily be detected. These measures require true class labels in order to be able to calculate the measures separately for each class.

In the case of fuzzy clusters an additional measure can be formulated:

6. *Unambiguity of fuzzy classification:* Monitoring the performance of the fuzzy classifier can be based on the analysis of membership functions representing fuzzy clusters. In order to evaluate the quality of fuzzy cluster assignment, the difference between the two maximum degrees of membership of each object to the clusters can be considered. If the value of this difference is large an object can be assigned clearly to one of the clusters. If the value of this difference is small an object belongs to both clusters to the same degree and its assignment is very ambiguous. The measure of unambiguity can be estimated for each time window using, for instance, the maximum and average values of the difference between the two largest membership degrees over all objects. If the maximum value of the difference over all objects in the current time window is much smaller compared to the one in previous time windows, then the assignment of objects to clusters becomes very ambiguous. Depending on the absolute degrees of membership, two situations of ambiguous assignment can be distinguished. Equally low degrees of membership of new objects to all clusters indicate that objects do not belong to any of the existing clusters. Equally high degrees of membership indicate that objects belong to more than one cluster to the same degree that signalises the overlapping of clusters. These cases correspond to different kinds of changes in the cluster structure and lead to a decrease in classifier performance.

Measures 1, 2 and 5, which require knowledge of the true class memberships of new objects, can only be used in methods of supervised learning. Since the true class labels can become known after classification within a certain period of time, changes can only be detected with a certain delay. Moreover, the true class labels are needed for updating the classifier by means of supervised learning, thus a delay can not be avoided [Lanquillon, 1997]. In contrast, measures 3, 4 and 6, which do not require any information concerning true class labels, can be used in methods of unsupervised learning. They rely solely on the values provided by new objects and on new classification results. Using these measures it is possible to avoid delays and to recognise changes in the cluster structure immediately. The true class labels are only required for the interpretation of the changes.

In the following section some methods for the monitoring process are presented, which use some of the described measures for the evaluation of the performance of the classifier.

3.1.1 Shewhart quality control charts

From a statistical point of view, the sequence of observed characteristic values represents a discrete random process that is characterised by some inherent variations. They are caused by stochastic fluctuations and are usually considered to be noise. Thus, the key problem of the monitoring process is to distinguish between variations due to stochastic perturbations and variations caused by unexpected changes in a system's state. If the sequence of observations is noisy it may contain some inconsistent observations or measurements errors (outliers) that are random and may never appear again. Therefore, it is reasonable to monitor a system and to process observations within time windows in order to average and reduce noise. Moreover, the information about possible structural changes within time windows can be interpreted and processed more easily. As a result, a more reliable classifier update can be achieved using monitoring within time windows.

One of the most popular statistic methods for monitoring the quality of a product, a process or a system's state is the Shewhart quality control chart introduced in [Shewhart, 1931] as a change detection method. The idea of this method is to observe over time some characteristic values describing a process and to check whether they remain within predefined limits. If they do, then the process is stable. Otherwise, it can be assumed that some structural change in the process has occurred.

A control chart is defined in statistics [Hogg, Ledolter, 1992] as a plot of a certain characteristic of a process obtained from samples (observations) sequentially in time and the characteristic's corresponding values. These values can be defined, for example, as the mean value of a characteristic and the measure of variability such as the standard deviation. Besides the characteristic's values plotted as time series, control charts also include bounds, called control limits, which are used to determine whether an observation is within acceptable limits of random variations. Control limits are employed in control charts to distinguish between variations caused by expected stochastic fluctuations and those caused by unexpected changes in a process or a system's state. If the plotted characteristic moves outside the control limits this can indicate that something has happened to the process leading to a new behaviour and to a new possible state. The use of control charts is illustrated in *Figure 12*.

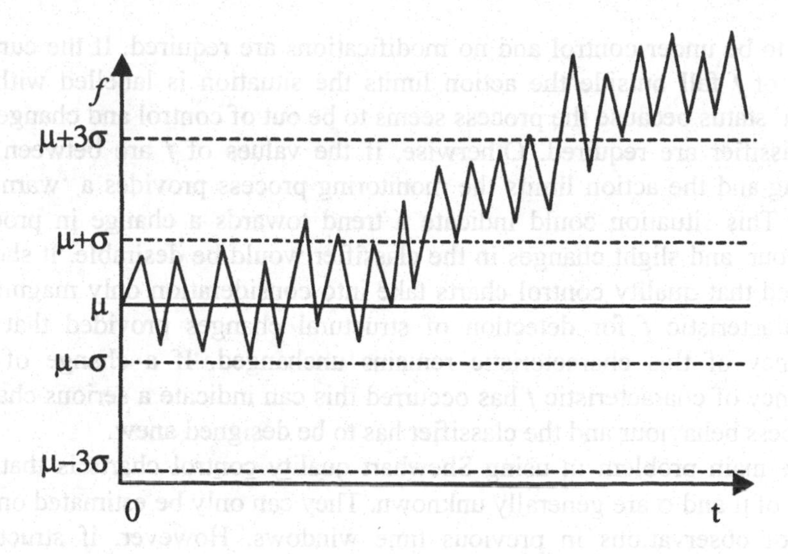

Figure 12. Shewhart quality control chart for the characteristic f

Suppose that characteristic f of the process derived from observations within the current time window is monitored over time. Let mean value μ, calculated during previous time windows, be the expected value for characteristic f with a standard deviation of σ. The values of f are plotted in *Figure 12* in time-series fashion versus time. So-called warning limits are used to determine slight departures of the process from its stable state and are usually defined in the literature at $\mu\pm2\sigma$. In [Lanquillon, 1997] the warning limits are chosen as $\mu\pm\sigma$ to provide a wider range in which the classifier must be updated. Action limits are equivalent to control limits and are usually set at $\mu\pm3\sigma$. They are suitable for detecting serious changes in process behaviour. The reasons for this choice of action limits can be found in statistics. If it is supposed that the values of f are normally distributed with mean μ and variance σ^2, then the probability that any observed value of f falls between the limits $\mu+3\sigma$ and $\mu-3\sigma$ is 0.9974. Normality assumptions should not be considered as an obstacle, since in the case of the time window having a long length (i.e. a large number of observations) the distribution of values of f can be considered approximately normal according to the central limit theorem. Therefore, if the process is stable it is rather unlikely that any values of f would be outside the action limits. Consequently, if values of f fall outside the action limits this indicates instability of the process, which is most likely caused by a serious change of the process state.

The monitoring process based on the quality control chart is carried out as follows. If the current values of characteristic f are within the warning limits the monitoring process responds with the status 'okay'. The process seems to be under control and no modifications are required. If the current values of f fall outside the action limits the situation is labelled with an 'action' status because the process seems to be out of control and changes in the classifier are required. Otherwise, if the values of f are between the warning and the action limits the monitoring process provides a 'warning' status. This situation could indicate a trend towards a change in process behaviour, and slight changes in the classifier would be desirable. It should be noted that quality control charts take into consideration only magnitude of characteristic f for detection of structural changes provided that the frequency of this characteristic remains unchanged. If a change of the frequency of characteristic f has occurred this can indicate a serious change in process behaviour and the classifier has to be designed anew.

The main problem in using Shewhart quality control charts is that the values of μ and σ are generally unknown. They can only be estimated on the basis of observations in previous time windows. However, if structural changes have occurred in the past the estimation of μ and σ based on all previous observations may not provide appropriate values [Lanquillon, 1997]. This problem can be solved in two ways. The first possibility is to exclude from future calculations values of μ obtained in time windows where changes were detected. This approach is suitable for situations in which values of the considered characteristic do not change in magnitude due to structural changes. Another possibility is to ignore values of μ obtained in all previous time windows up until the time window where the last change was detected. This strategy can be applied to characteristics whose values may change in magnitude after structural changes.

In [Taylor et al., 1997] the idea of quality control charts was used with a slight modification of warning limits to monitor the error rates of a classifier with supervised learning. In order to evaluate the current classification error rate, the expected error rate $\bar{e}$ (a weighted mean of the error rates of the previous time windows) and its standard deviation σ are estimated. If an error rate falls below the expected error rate the classifier performance is satisfactory. Thus, it is sufficient to consider only upper limits during the monitoring process, i.e. the warning and the action limits are set at $\bar{e}+\sigma$ and $\bar{e}+3\sigma$, respectively. Since it is not possible to estimate the expected error rate depending on the previous error rates in the first time window or after the classifier is completely re-learned based on the most recent observations, cross-validation can be applied for estimating the expected error rate $\bar{e}$ based only on the training data set. The idea of the cross-validation test, which is also called k-fold cross-validation, ([Stone, 1974], [Geisser, 1975]) is to divide the training set into k distinct subsets and to use (k-1) subsets for training a classifier and one subset for classifier validation.

This procedure is repeated for all k combinations of (k-1) subsets leading to k different classifiers and k error rates. The overall error rate is estimated as the mean of these error rates and represents a rather unbiased estimate.

In the literature there are some other methods to interpret the results of Shewhart quality control charts. For instance, in [Smith, 1994] quality control charts are used in a neural network setting to detect changes.

Shewhart quality control charts have the following drawbacks:

1. Small and persistent changes remain undetected. However, this is not crucial as long as the error rate remains within the warning limits.
2. They are able to detect changes of the characteristic's values but not those of the behaviour of a process or system.
3. Due to crisp control limits they are not able to detect gradual changes, i.e. to detect the degree to which a change in a system's state has occurred.
4. They are not able to detect the kind of changes that have occurred.

Drawbacks 2 and 4 can be avoided by the proper choice of characteristic *f*, which can represent a derived feature of a quality of a process or a measure of compactness or separability of clusters, respectively. Two other drawbacks can be avoided by applying fuzzy techniques to monitor the performance of a classifier and to detect gradual and abrupt changes in its cluster structure, as will be described in the next section.

3.1.2 Fuzzy techniques for the monitoring process

The potential of fuzzy modelling in system diagnosis was recognised very soon after the introduction of fuzzy set theory in pattern recognition. Since then, methods of fuzzy clustering and classification have been successfully applied to a number of real-world problems such as, for example, tool wear monitoring [Zieba, Dubuisson, 1994], human car driver performance monitoring [Peltier, Dubuisson, 1993], or the supervision of state evolution of telephone networks [Boutlex, Dubuisson, 1996]. The success of the fuzzy approach is due to the possibility of modelling clusters by fuzzy sets in the feature space. The membership function of a fuzzy set gives the degree to which an arbitrary object may be considered as a representative of a cluster. Therefore, each object is assumed to belong to each cluster with a certain degree of membership. Due to this interpretation and the continuous nature of membership functions, the fuzzy framework seems to be well-adapted to dynamic pattern recognition. Using the concept of membership functions it is possible to model the temporal development of a system and a transition of objects between clusters in the course of time.

3.1.2.1 Fuzzy quality control charts

As stated in the previous section, one of the drawbacks of statistical quality control charts is that it is not possible to recognise to what degree a system state has been changed. Monitoring characteristic *f* of a system, or process, it is only possible to detect whether or not the behaviour of a process is stable, has a tendency to a change or has become poor. Control limits provide a crisp discrimination between good and bad behaviour or between good and bad quality of the process under consideration. As long as the values of characteristic *f* are within the so-called range of tolerance defined by the control limits, the behaviour or the quality of the process can be considered as good. Otherwise, the quality is characterised as bad. However, the transition from good to bad quality is often continuous and gradual. In order to be able to represent a gradual change in the quality of a process, a fuzzy set 'good quality' with membership function u(f) over the domain of the monitored characteristic *f* can be defined (*Figure 13*). The further the actual value of characteristic *f* deviates from its expected value μ, the smaller its degree of membership to the fuzzy set 'good quality'. Such a representation of the range of tolerance corresponds to a definition of fuzzy control limits [Schleicher, 1994, p. 18].

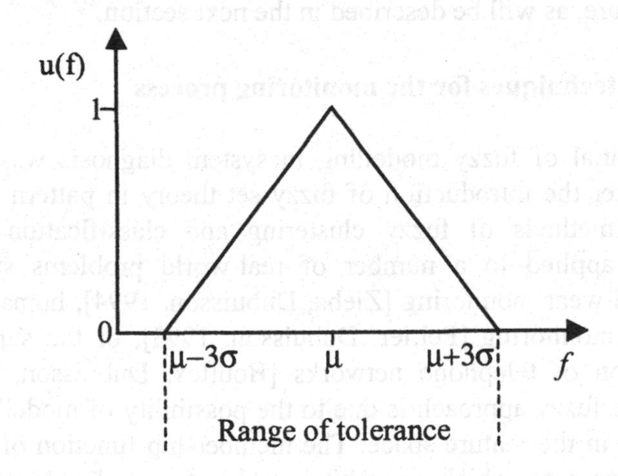

Figure 13. Fuzzy set 'good quality' defined for characteristic f

Monitoring characteristic *f* over time the fuzzy set 'good quality' is defined as either constant or time variable for characteristic *f*. As a result, a fuzzy quality control chart is obtained as a combination of crisp characteristic values of *f* with fuzzy control limits (*Figure 14* [adapted from Schleicher, 1994, p. 18]).

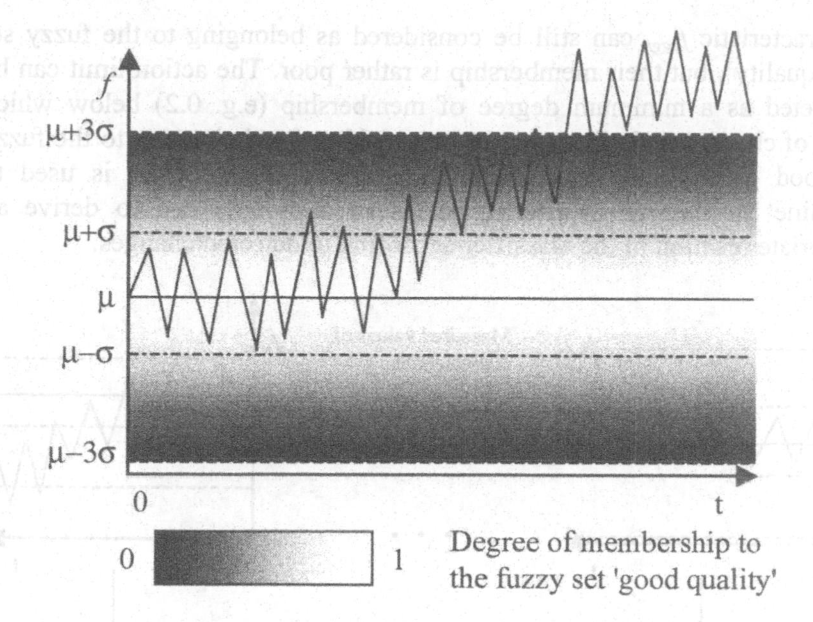

Figure 14. Fuzzy quality control chart

The advantage of using fuzzy quality control charts is that many intermediate states between good and bad quality can be distinguished, allowing a flexible reaction of the classifier. Moreover, the fuzzy representation of the range of tolerance provides a possibility to handle qualitative characteristics (e.g. good, bad) in the same way as quantitative ones (represented as real numbers).

The information processing in the monitoring procedure based on fuzzy control charts is illustrated in *Figure 15* [adapted from Schleicher, 1994, p. 19]. Suppose that several characteristics $f_1, ..., f_n$ are to be monitored. In the first step degrees of membership of values of a monitored characteristic f_i, $i=1, ..., n$, to the fuzzy set 'good quality' are calculated for each moment of time. The result of this transformation are functions $u(f_1), ..., u(f_n)$ representing a pointwise membership of monitored characteristics to the fuzzy set 'good quality' over time. In the second step these functions have to be aggregated to a single function expressing an overall degree of membership using one of the aggregation operators (e.g. in the case of discrete functions the γ-operator, the arithmetic mean or the fuzzy integral can be used, while in the case of continuous functions maximum or minimum operators can be applied). The result of the third step is an aggregated fuzzy control chart where the warning and action limits are defined on the interval [0, 1] independent from characteristics $f_1, ..., f_n$. The warning limit determines a limiting degree of membership to which values of characteristic f_{aggr} can still be considered as belonging to the fuzzy set 'good quality', but their membership is rather poor. The action limit can be interpreted as a minimum degree of membership (e.g. 0.2) below which values of characteristic f_{aggr} can not be considered as belonging to the fuzzy set 'good quality' any more. This final fuzzy control chart is used to determine the current monitoring status of a process and to derive an appropriate reaction of the classifier according to detected changes.

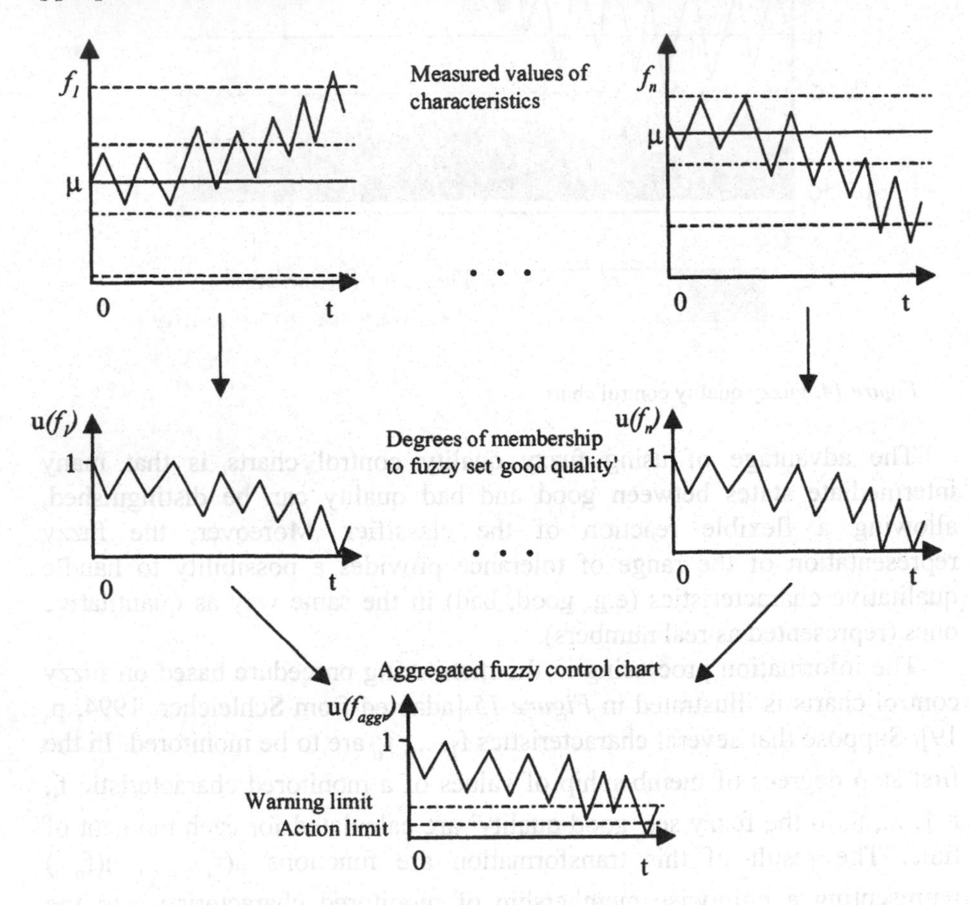

Figure 15. Processing of fuzzy quality control charts

Fuzzy control charts represent an improvement towards statistical quality control charts due to their ability to recognise gradual changes in a system's state. Both types of quality control charts are based on monitoring a general characteristic that describes the performance of the classifier (e.g. classification error rate). Another possibility to conduct the monitoring process is to consider the cluster structure itself and to evaluate the results of the classification of new objects that must enable the unambiguous assignment of objects to clusters. Ambiguous information about cluster membership indicates a poor performance of a classifier and must be rejected in order to avoid misclassification errors. Furthermore, this result can indicate the need to improve a current classifier.

3.1.2.2 Reject options in fuzzy pattern recognition

Although fuzzy methods provide a powerful framework for pattern recognition due to their ability to generate gradual memberships of objects to clusters, a number of rules have been proposed to defuzzify the classification results in order to be able to make a final (crisp) decision about a system state. This can be relevant for the design of automatic pattern recognition systems where a part of the fuzzy information that is not useful should be ignored. By managing reject options in fuzzy classification, errors of crisp assignment of objects to clusters in unclear situations can be avoided.

Reject options were introduced into the framework of statistical pattern recognition in order to decrease the misclassification risk [Chow, 1970]. Two types of rejection can be considered [Dubuisson, Masson, 1993]:

- Distance or membership reject;
- Ambiguity reject.

The idea of the first type of rejection is to avoid the assignment of an object with a very low degree of membership to one of the existing clusters. For instance, in *Figure 16* objects in the upper right-hand corner should be rejected for the cluster assignment since their distances to both clusters are equally large and their memberships to both clusters are equally low. Therefore, the objects belong to none of the clusters. It should be noted that this reject option could be handled well by possibilistic clustering algorithms such as possibilistic c-means. These algorithms do not contain a probabilistic constraint of normalisation of memberships of an object across all clusters. As a result, possibilistic degrees of membership can be interpreted as typicality of objects to clusters, or absolute degrees of membership, and in the case shown in *Figure 16* they would be very low for objects in the upper right-hand corner. On the contrary, probabilistic fuzzy algorithms (for instance, fuzzy c-means) would provide equal degrees of membership of about 0.5 for these objects.

The second type of rejection deals with the case when the clusters can not be clearly distinguished. The aim of this reject option is to avoid the assignment of an object to one of the clusters when degrees of membership contain ambiguous information. For instance, a group of objects between the two clusters in *Figure 16* belongs to both clusters to the same degree of approximately 0.5. It is not possible to separate clearly these objects into two clusters, therefore they should be ambiguity rejected. This case can be handled by possibilistic, as well as probabilistic, clustering algorithms.

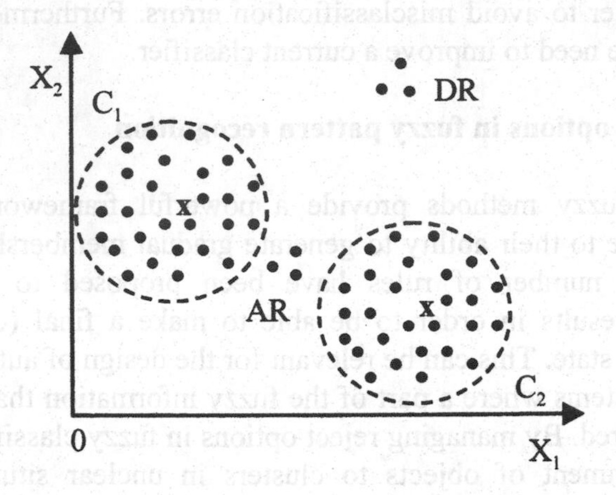

Figure 16. Ambiguity reject (AR) and distance reject (DR) options in pattern recognition

The straightforward approach to incorporate reject options in pattern recognition is to threshold degrees of membership of an object to be classified. In the following, several decision rules concerning the definition of reject options are presented and discussed.

Let us interpret an exclusive assignment to cluster C_i as an action a_i in decision theory ([Denoeux et al., 1997], [Zimmermann, 1992, p. 12]). Suppose that a set of actions corresponding to possible assignments to c clusters is given by $A=\{a_1, a_2, ..., a_c\}$ and denote the action related to an object $\mathbf{x}$ by $a(\mathbf{x})$. The most usual rule for a hard assignment of an object to one of the clusters consists in choosing the action corresponding to the highest degree of membership [Pal, 1977]:

$$a(\mathbf{x}) = a_i \quad \text{if} \quad u_i(\mathbf{x}) = \max_{j=1,c} u_j(\mathbf{x}), \tag{3.1}$$

where $u_j(\mathbf{x})$ denotes a degree of membership of object $\mathbf{x}$ to cluster C_j, j = 1, ..., c. In order to include reject options in the decision process, this rule must additionally incorporate thresholds of membership which are either arbitrarily fixed or calculated from the learning set of objects as follows:

$$u_i^o = \min_{k \in C_i} u_i(\mathbf{x}_k), \quad i = 1, ..., c, \tag{3.2}$$

where $\mathbf{x}_k$, $k \in C_i$, is a learning set of objects belonging to class C_i.

Suppose that an extended set $A = \{a_0, a_d, a_1, ..., a_c\}$ of possible actions is considered, where a_0 and a_d represent ambiguity and distance rejection, respectively. Let $J(\mathbf{x})$ be a set of candidate clusters for object $\mathbf{x}$ that is defined according to the above rule as:

$$J(\mathbf{x}) = \left\{ i \in \{1,...,c\} \middle| u_i(\mathbf{x}) > u_i^o \right\} \tag{3.3}$$

A decision rule concerning the assignment of object $\mathbf{x}$ to one of these clusters can be formulated as follows [Denoeux et al., 1997]:

$$\begin{aligned} a(\mathbf{x}) &= a_i \quad \text{if} \quad J(\mathbf{x}) = \{i\}, \\ a(\mathbf{x}) &= a_d \quad \text{if} \quad J(\mathbf{x}) = \varnothing, \\ a(\mathbf{x}) &= a_0 \quad \text{if} \quad |J(\mathbf{x})| > 1. \end{aligned} \tag{3.4}$$

The interpretation of this rule is simple. If the set of candidate clusters includes only one element i, then the assignment of object $\mathbf{x}$ to cluster C_i is unambiguous. If set $J(\mathbf{x})$ is an empty set, then the assignment of an object to one of the clusters is rejected because its degree of membership to all clusters is too low. If the set $J(\mathbf{x})$ contains more than one element this indicates that several clusters appear to be equally likely and the assignment of an object is rejected due to the ambiguity of the situation.

A drawback of this decision rule is that the same membership threshold is used for ambiguity and distance reject, which can lead to undesirable results. To avoid this problem the 'membership ratio' rule was introduced in ([Frelicot, 1992], [Frelicot et al., 1995]) additional to rule (3.4), to deal with ambiguity rejection. This ratio is defined by:

$$R(\mathbf{x}) = \frac{u_{m_2}(\mathbf{x})}{u_{m_1}(\mathbf{x})}, \tag{3.5}$$

where $u_{m_1}(\mathbf{x}) = \max_{i \in J(x)} u_i(\mathbf{x})$ and $u_{m_2}(\mathbf{x}) = \max_{i \in J(x) \setminus \{m_1\}} u_i(\mathbf{x})$.

If $R(\mathbf{x})$ is close to zero, then the degree $u_{m_1}(\mathbf{x})$ is much higher than all other degrees of membership and $\mathbf{x}$ has not to be ambiguity rejected. Action a_{m_1} can then be selected with high confidence. If $R(\mathbf{x})$ is close to one, then the assignment of an object to at least two clusters is equally likely and $\mathbf{x}$ is ambiguity rejected. To be able to make a decision it is convenient to compare R(**x**) to a predefined ambiguity threshold R^o. If $R(\mathbf{x}) \geq R^o$, then an object is rejected, otherwise action a_{m_1} is selected. This approach makes it possible to adjust distance and ambiguity reject rates independently and provides reliable decisions.

This method for determining membership and ambiguity rejection can be used in dynamic pattern recognition to detect the inadequacy of the current cluster structure. If too many objects are membership rejected this may indicate that the system is in a new state and the classifier must be re-learned with an increased number of clusters. If there are too many objects that are ambiguity rejected, provided that threshold u_i^o in equation (3.2) is fixed relatively high, this may indicate that at least two clusters can not be considered as distinct any more but as similar and should probably be merged to form a single cluster.

3.1.2.3 Parametric concept of a membership function for a dynamic classifier

Using fuzzy clustering methods, temporal changes in the cluster structure can be recognised by monitoring membership functions representing fuzzy clusters and evaluating the degrees of membership of new objects to clusters (see section 3.1, point 6). The monitoring procedure and the corresponding adaptation law of a classifier depend crucially on the type and structure of the chosen classifier.

In [Bocklisch, 1981] a fuzzy clustering algorithm based on a special parametric concept of a membership function suitable for dynamic classification was proposed. The idea of this method is to describe objects by elementary membership functions and to obtain cluster membership functions by aggregating fuzzy objects using the union operation. The aggregation procedure is controlled by a threshold value, which is used to determine similar objects and correspondingly objects belonging to the same cluster. This hierarchical procedure of classifier design is repeated for several threshold values and the best cluster configuration is chosen using the evaluation of some validity measures and expert knowledge. Due to these repeated evaluations, the method can also be characterised as iterative. After the best cluster configuration is found, a fuzzy description of clusters is derived with the help of the parametric concept of the membership function.

In contrast to the non-parametric concept where a degree of membership is assigned to each element x of the universe of discourse X independently from the neighbouring element (type A membership function), the parametric concept provides an analytical model describing a dependency between degrees of membership and elements x (type B membership function) [Zimmermann, Zysno, 1985, pp. 150-153]. The judgement of membership in the latter case is based on the comparison of element x with an ideal that results in a distance between an element and an ideal. Thus, membership is defined as a function of the distance specified by the number of parameters. In this way, the parametric concept makes it possible to change from the set-theoretic consideration of membership functions to the consideration of crisp parameters of membership functions. The parameters of a membership function are usually defined dependent on the application. The parametric concept proposed in [Bocklisch, 1981] aims to design a membership function that represents a given data structure and whose parameters are easy to interpret in the context of the classification problem.

The proposed membership function is defined by the following properties:

1. The same type of a membership function is used to model elementary events (objects) as well as global events (clusters). These membership functions are denoted by u_e and u_g, respectively.
2. The membership function $u(\mathbf{x}; \mathbf{p})$ is described by a functional relationship $u(\cdot)$ and by a parameter vector $\mathbf{p}$ which consists of two sub-vectors $\mathbf{p}_1$ and $\mathbf{p}_2$ characterising the location and the fuzziness of the membership function, respectively. The membership function is asymmetric and has a single maximum.
3. The parameters of the membership function have a clear physical meaning and can be explicitly interpreted. They include two location parameters $\mathbf{x}_o$ and a, and three parameters of fuzziness b, c, and d:

- Parameter $\mathbf{x}_o$ – the location of the membership function. This value corresponds to an element with the maximum degree of membership (e.g. best representative of a fuzzy cluster). It can be determined as an arithmetic mean or a centre of gravity of elements belonging to this fuzzy set, or defined subjectively.
- Parameter a - the maximum value of the membership function. This value must not be equal to one as it is usual for normalised membership functions. For cluster membership functions the value of parameter a can be proportional to the current importance of a cluster determined by the number of objects currently belonging to this cluster. Besides this, the value of a is influenced by a forgetting function. Therefore, the value of parameter a varies in the course of time: it increases as a cluster grows and decreases as a cluster becomes old and is no longer supported by new objects. Since the membership function is not normalised, the choice of aggregation operators that can be used is restricted to the class of operators which do not have requirements on function normalisation.

- Parameter c – the support of the membership function. This parameter determines a set of elements for which a degree of membership is higher than a predefined marginal degree b.
- Parameter b – marginal degree of membership. This value is determined on the boundary of the support of the membership function defined by parameter c.
- Parameter d – slope of the membership function. This parameter determines the form of the membership function. In case $d \to \infty$ one obtains the conventional rectangular characteristic function that can only take values from set {0, 1}.

When deriving a membership function it is assumed that a degree of membership is equivalent to a similarity measure. In pattern recognition it is usual to choose the distance between vectors as a dissimilarity measure between them. Thus, considering a cluster with a centre of gravity in the origin of the co-ordinate system, it is assumed that the larger the distance of a vector $\mathbf{x}$ from the origin, the smaller its membership to a cluster. For the proposed type of membership function an Euclidean distance measure is chosen. The parameters described above are used as weights (specific for each dimension of a vector $\mathbf{x}$) to control the form of the membership function. An M-dimensional membership function with its maximum in the origin is defined according to [Bocklisch, 1981] as follows:

$$u(\mathbf{x}) = a \cdot \left\{ 1 + \sum_{i=1}^{M} \left(\frac{1}{b_i} - 1\right) \cdot \left| \frac{x_i}{c_i} \right|^{d_i} \right\}^{-1} \tag{3.6}$$

In order to increase the flexibility of the function, parameters b, c, and d are defined differently for the left and the right side of each of the M components of the function. The resulting membership function is represented graphically in *Figure 17*, where the one-dimensional case is shown for the sake of simplicity.

This general form for a membership function is used to represent fuzzy clusters. Fuzzy objects are described by an elementary membership function obtained from the general model with the following parameter settings: $a = 1$, $b_l = b_r = 0.5$, $c_l = c_r = c_e$, $d_l = d_r = 2$. This corresponds to a normalised symmetric membership function. The parameter of fuzziness c_e can be defined by an expert context dependent.

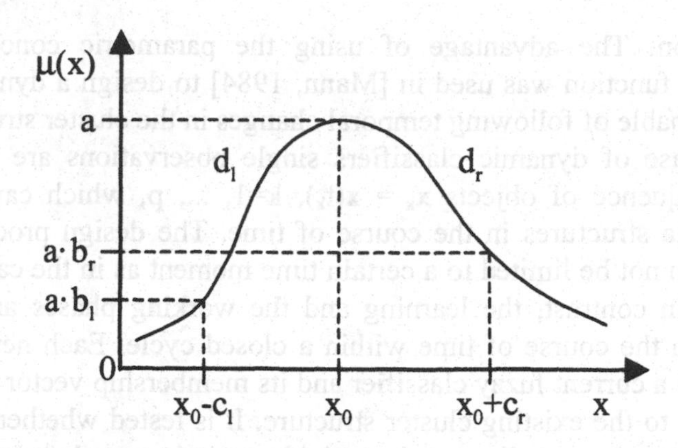

Figure 17. A parametric membership function

An application of union or intersection operations to the considered parametric elementary membership functions can lead to new types of functions that do not fit into the original concept (they can be characterised by multiple maxima and an extended parameter vector). In order to preserve the proposed parametric representation a two-step aggregation procedure is proposed in [Bocklisch, 1981, pp. 35-36]: At first, an aggregation operation is applied, then the aggregated function is transformed into the original concept by computing a new parameter vector from the intermediate results of the aggregation step.

In order to reduce the consideration of membership functions from an M-dimensional feature space to a one-dimensional case, a new co-ordinate system is built for each single cluster by shifting or by shifting and rotating the old one. The purpose of this transformation is to obtain a cluster-specific co-ordinate system so that the cluster membership function has its maximum in the origin. The M-dimensional membership function is projected on the new axes and M one-dimensional functions are considered using the parametric concept described above. In this case, the parameters of the transformation (length of shift and angles of rotation) are included in the parameter vector of a cluster membership function.

Hence, the fuzzy classifier is designed by aggregating elementary membership functions of objects and by representing clusters by global parametric membership functions. The classification of a new object is carried out by computing a vector of degrees of membership of an object to all clusters, which Bocklisch calls a 'sympathy vector'. The drawback of this classifier is, however, the problem of interpreting the resulting cluster structure since clusters are defined in different co-ordinate systems, hence making the analysis rather complicated.

The proposed clustering algorithm provides a flexible cluster model and seems to have good adaptation capabilities due to its parametric representation. The advantage of using the parametric concept of the membership function was used in [Mann, 1984] to design a dynamic fuzzy classifier capable of following temporal changes in the cluster structure.

In the case of dynamic classifiers single observations are given as a temporal sequence of objects $\mathbf{x}_k = \mathbf{x}(t_k)$, k=1, ..., p, which can represent different data structures in the course of time. The design procedure of a classifier can not be limited to a certain time moment as in the case of static classifiers. In contrast, the learning and the working phases are repeated iteratively in the course of time within a closed cycle. Each new object is classified by a current fuzzy classifier and its membership vector is analysed with respect to the existing cluster structure. It is tested whether it fits into the cluster structure well enough, and if not it is tested for the kind of changes that have occurred. These tests represent a monitoring procedure in the considered clustering algorithm. Its result is one of the following decisions: formation of a cluster; merging of clusters; splitting of clusters; or modification of clusters. The update of the classifier is carried out by changing recursively the parameter vectors of cluster membership functions and by applying set-theoretic operations to cluster membership functions, if so required by the result of the monitoring procedure. Because of the recursive calculation rule of cluster parameters Mann calls this learning method recursive classification. Finally, the classifier is devaluated by a forgetting function, which is equivalent to an ageing of the classifier. Since, according to [Peschel, Mende, 1983], the temporal development of the classifier can be considered as an evolutionary process a forgetting function can be chosen from evolutionary models of the power-product-type. The forgetting process is an important component of the classification method that provides the learning ability to the classifier and a possibility to completely re-learn the classifier as time passes.

The following algorithm describes the learning procedure combining the design of a dynamic classifier with the adaptation procedure in one learning-and-working cycle.

Algorithm 1: Learning algorithm for dynamic classifier design [Mann, 1984, p. 35].

1. *Fuzzy classification of a new object* $\mathbf{x}_k = \mathbf{x}(t_k), k = 1,...,p$, *i.e. determination of a sympathy vector* $\mathbf{u}_k = (u_{1k},...,u_{ck})$ *by calculating degrees of membership of object* $\mathbf{x}_k$ *to all c clusters.*
2. *Comparison of all components of the sympathy vector with the merging threshold* u_s. *Determination of a set of clusters for which the corresponding components of the sympathy vector exceed the given threshold:*

$$Z = \{i \mid u_{ik} \geq u_s\}$$

If Z is an empty set, go to Step 3. Otherwise go to Step 4.

3. ***Formation of a new cluster, since all components of the sympathy vector are smaller than the threshold. The membership function of a new cluster is defined by an elementary membership function of a new object:***

   $c=c+1$ and $u_c(\mathbf{x}, \mathbf{p}_c) = u_e(\mathbf{x}_k, \mathbf{p}_e)$.

   Go to Step 7.

4. ***Enlargement of a cluster i' ∈ Z with an object $\mathbf{x}_k$ by applying a union operation to combine a cluster membership function with an elementary membership function of an object. If the cardinality of set Z is equal to 1 (there is only one cluster for which the corresponding component of the sympathy vector exceeds the merging threshold), go to step 6. Otherwise, go to Step 5.***

5. ***Merging of all clusters i''∈ Z with cluster i', i'' ≠ i', by applying a union operation to cluster membership functions.***

6. ***Test whether the membership function of cluster i' satisfies a parametric concept of a membership function (this can be recognised by considering the parameter vector $p_{i'}$). If this is not the case, split off subclasses.***

7. ***Application of a forgetting function to the classifier. The learning and working cycle is completed and the next object can be considered: k = k+1. Go to Step 1.***

In this algorithm the decision concerning changes in the cluster structure is taken depending on the result of the comparison between the degree of membership of a new object and a merging threshold u_s. If this degree is smaller than a threshold for all clusters, then the object can not be assigned to any of the existing clusters and a new cluster is formed. If a degree of membership exceeds the threshold for more than one cluster, then this indicates that the corresponding clusters are overlapped and can be merged. The merging of clusters is carried out by applying the union operation to cluster membership functions. This aggregation procedure requires at first the definition of a joint aggregation space for a new cluster since all clusters are defined in different cluster-specific co-ordinate systems. In the next step the clusters to be merged are transformed into the aggregation space where membership function projections on new axes are used to calculate new parameter vectors. An aggregation of the transformed membership functions by the union operator is then performed with respect to each dimension of the aggregation space as well as for each side (left and right) of the membership function. With the help of the merging threshold u_s it is possible to control the degree of fuzziness and the fineness of the cluster structure. On the one hand, the merging threshold represents a degree of membership that a single object must possess to be assigned to a cluster. Smaller values of u_s lead to fuzzier clusters. On the other hand, the merging threshold influences the number of clusters. Decreasing the value of u_s more clusters will be merged and the number of clusters will be reduced leading to a rough cluster structure. Thus, the value of the merging threshold represents an important control parameter of the learning algorithm.

The splitting of clusters is performed if a cluster membership function does not fit into the underlying parametric concept. The reason for splitting a cluster is the dissolution of an old centre of concentration of objects and formation of new groups of objects within a cluster, which are drifting away from the former centre of gravity of the cluster. In order to detect such changes in the distribution of objects within a cluster, parameter d characterising projections of the membership functions on the axes is considered. If new objects are concentrated more and more on the boundary of a cluster the value of d grows and the form of the membership function changes from an unimodal shape to that of equal distribution. A cluster is split if the value of d exceeds some predefined threshold.

The drift of clusters is not explicitly detected by the considered algorithm. The classifier takes into account these changes and follows them due to aggregation of new objects with cluster membership functions. In this way cluster membership functions move in the direction of arrival of new objects.

Although the proposed parametric concept of the membership function seems to be suitable for the design of adaptive classifiers, it is rather simple and correspondingly too restrictive to process objects and clusters in multidimensional feature spaces. The proposed clustering algorithm is rather complicated due to the need to process cluster-specific co-ordinate systems, a lot of transformations during aggregation procedures, and the consideration of single projections of membership functions. The goal of the clustering algorithm is similar to the one of point-prototype based clustering algorithms, however only membership functions are used to represent clusters, which are moved into the origin of the co-ordinate system. Information concerning cluster centres is represented implicitly by the location parameter x_0 of the membership function. However, ideas of the monitoring procedure concerning formation, merging and splitting of clusters in the discussed clustering method of Bocklisch and Mann have a general character and can be used for the design of other types of dynamic classifiers.

3.2 The Adaptation Process

In dynamic pattern recognition systems the classifier design cannot be separated temporally from the phase of its application to the classification of new objects. The dynamic classifier must constantly be updated based on new objects in order to preserve its performance in the course of time. The update of the classifier means that its parameters are adjusted over time in order to represent the current cluster structure in the best possible way. In other words, a dynamic classifier must possess adaptive capabilities in order to follow temporal changes. The choice of parameters to be adapted depends on the type of the classifier (e.g. point-prototype based classifier, neural networks, or decision trees). In this section different strategies for the update of a dynamic classifier are discussed. The most simple strategy to update the classifier without any changes in the conventional design procedure is to re-learn[1] the classifier periodically from scratch as time passes (Section 3.2.1). Although this procedure is not always economical with respect to computing time, it provides an opportunity to apply a static classifier in a dynamic environment. In order to supply the classifier with adaptation ability, the classifier can be incrementally updated with new objects with the passing of time (Section 3.2.2). This strategy requires the recursive representation of the learning rule so that classifier parameters can be recursively updated using new objects. This adaptation law allows a classifier to follow gradual temporal changes but does not have a mechanism for adjusting a classifier to abrupt changes in the cluster structure such as changes in the number of clusters. The most flexible but also the most complicated approach is the development of an adaptation law depending on the temporal changes detected in the cluster structure by the monitoring process (Section 3.2.3). This approach provides a flexible combination of re-learning and incremental updating procedures enhanced by special elements for the intelligent and efficient design of a dynamic classifier.

In the following sections the main principles of these three strategies will be considered and their advantages and disadvantages will be discussed.

[1] The notion 'learning the classifier' is usually used in the framework of unsupervised pattern recognition whereas 'training the classifier' is applied in the framework of supervised pattern recognition. In this thesis both notions will be generalised by the term 'learning' regarded to the process of the classifier design.

3.2.1 Re-learning of the classifier

As mentioned in Section 2.3.3, one strategy to update the originally static classifier to temporal changes in the cluster structure is to re-learn the classifier from scratch after each time window in order to preserve its performance over time. The classifier can be re-learned based either on the complete history of objects (complete learning approach) or on the partial history (re-learning approach). In the first case the classifier is re-learned using a training set which includes all available objects from the past. This approach can be very time consuming because of the constantly growing number of objects. Besides, it may be insufficient always to discard the information learned in the previous phase and to learn it again, especially if adaptation is not required. In the second case the training set for re-learning consists solely of the most recent objects obtained during the last time window. This approach avoids the problem of a constantly increasing training set by disregarding old previous objects up to the current time window. However, old objects can contain some relevant information that must be used to obtain a representative classifier. This can lead to a drop in classifier performance, particularly if new objects are not representative enough and contain a lot of noise. As explained in Section 2.3.2, the choice of the length of the time window has a big influence on the adaptation ability of the classifier. As can be seen, both approaches have certain advantages and shortcomings and cannot guarantee an optimal learning strategy.

The main drawback of the updating strategy based on re-learning the classifier is that structural changes cannot be recognised explicitly, i.e. exactly which clusters have been merged or split remains unknown. The classifier learns a new cluster structure blindly based on the available information, and changes can be detected by comparing the current cluster structure with the preceding one. However, for unsupervised re-learning of the classifier certain information about temporal changes is required, that is, the correct number of clusters at the current moment. The most common technique to determine the optimal number of clusters relies on validity measures (e.g. partition entropy [Bezdek, 1981, p. 111], proportion exponent [Windham, 1981], degree of compactness and separation [Xie, Beni, 1991] etc.). Validity measures are used in cluster analysis to quantify the separation and compactness of the clusters. If the number of clusters is chosen correctly the clustering algorithm can identify well separated and compact groups within the data and the validity measure takes its maximum or minimum value (depending on the chosen measure). However, the definition of cluster separation and compactness is not unique and depends on a specific problem. Therefore, the mathematical formulation of the validity measure is extremely difficult according to [Bezdek, 1981, p. 98]. The main idea of the technique based on the validity measure is to repeat the clustering procedure several times with different numbers of clusters (between predefined minimum and maximum values) and to choose the number at which the validity measure has a local minimum or a local maximum (depending on the measure) to be 'optimal'. The interval for the search for an optimal number of clusters is determined depending on the application, where a priori information concerning the maximum possible number of clusters is given, or an expert can define a required upper limit taking into account the need for clear interpretability of results. If the classifier is re-learned after each time window the number of iterations with a different number of clusters can be reduced. The number of clusters in the current time window can be incrementally decreased or increased with respect to the number of clusters in the preceding time window until the extreme value of the validity measure is achieved. Therefore, the number of iterations of the clustering procedure can be variable for different time windows and determined depending on the behaviour of the validity measure. The reason for this strategy is the assumption that the number of clusters changes at a rather gradual rate over time.

One of the best representatives for this technique is the unsupervised optimal fuzzy clustering (UOFC) algorithm [Gath, Geva, 1989], which determines an optimal number of clusters automatically by maximising the average partition density criterion. The algorithm starts with a single cluster prototype and iterates for an increasing number of clusters, calculating a new partition of the data set and evaluating the validity measure in each iteration until the maximum predefined number of clusters is achieved. The best partitioning is obtained for the number of clusters that maximises the average density criterion plus one or sometimes two. The results reported in [Geva, Kerem, 1998] show that the UOFC algorithm is suitable for an accurate and reliable identification of bioelectric brain states based on the analysis of the EEG time series.

This approach to determine the optimal number of clusters depends, however, to a high degree on the quality and reliability of the validity measures used. As already mentioned, it does not guarantee that the optimal number of clusters will always be found since it is difficult to define a unique measure that takes into account the variability in cluster shape, density, and size. So far, none of the most frequently used validity measures provides a clear answer about the optimal number of clusters in all situations. Moreover, re-learning the classifier within each time window is computationally expensive because of the need to repeat clustering runs.

An alternative approach for unsupervised detection of the number of clusters during re-learning the classifier is presented by clustering algorithms based on the principle of merging similar clusters. In the literature a number of algorithms was proposed, which start with an over-specified number of clusters and due to merging of similar clusters terminate with an optimal number of clusters. An example of this approach is the competitive agglomeration (CA) algorithm introduced in [Krishnapuram, 1997], which produces a sequence of partitions with a decreasing number of clusters. The update equation creates an environment in which clusters compete for objects and only clusters with large cardinalities survive. The final partition has the 'optimal' number of clusters from the point of view of the objective function.

In ([Setnes, Kaymak, 1998], [Stutz, 1998]) extended versions of the fuzzy c-means (FCM) algorithm [Bezdek, 1981] with cluster merging were proposed. The main idea of these algorithms is the iterative clustering of objects and the evaluation of the similarity of all pairs of clusters at each iteration. If the similarity between two clusters exceeds a given threshold these are merged and the number of clusters is decreased. The algorithms stop when there are no more clusters that can be merged. These two algorithms differ in the definition of the similarity measure for clusters. Their performance depends on the correct choice of the threshold for merging.

3.2.2 Incremental updating of the classifier

Another alternative for updating the classifier over time in order to follow temporal changes of the cluster structure is to improve the current classifier using incremental updating based on new objects. This idea was put into practice in [Marsili-Libelli, 1998] where a classical fuzzy c-means (FCM) [Bezdek, 1981] was enhanced with an updating feature enabling it to detect departures of a system's state from normal conditions.

In the classical version of FCM the knowledge contained in the training set of objects is condensed into the cluster prototypes. They are defined as the fuzzy weighted centres of gravity of objects according to the following equation:

$$v_i = \frac{\sum_{j=1}^{N} (u_{ij})^q \mathbf{x}_j}{\sum_{j=1}^{N} (u_{ij})^q}, \quad i = 1, \ldots, c \tag{3.7}$$

where $\mathbf{x}_j$ is an M-dimensional feature vector representing the j-th object, u_{ij} is the degree of membership of the object j to cluster i, $q \in (1, \infty)$ is the fuzzy weighting exponent, N is the number of training objects, c is the predefined number of clusters. Cluster centres describe typical values of the corresponding clusters and usually represent the 'normal' state of the system and an appropriate number of faulty states. Degrees of membership u_{ij} which are components of the membership matrix **U** denote the extent to which object $\mathbf{x}_j$ is similar the corresponding cluster prototype and are calculated as follows:

$$u_{ij} = \frac{1}{\sum_{r=1}^{c}\left(\frac{d_{ij}}{d_{rj}}\right)^{\frac{2}{q-1}}} \tag{3.8}$$

During the training phase, the fuzzy c-means algorithm operates in an iterative mode computing sequentially membership matrix **U** according to (3.8), the cluster centres **V** according to (3.7) and the distances

$$(d_{ij}), i = 1,..,c,\ j = 1,...,N,$$

until the membership matrix stabilises.

In order to classify a new object $\mathbf{x}^o$ into existing clusters, memberships of an object to all clusters are calculated applying (3.8) only once. In classical fuzzy c-means cluster prototypes remain unchanged during the classification of new objects. In order to provide the classifier with some trend-following capabilities, it is proposed in [Marsili-Libelli, 1998] to update the location of cluster prototypes using degrees of membership of new objects to clusters. It is assumed that all N previous objects have already been classified and a new (N+1)-th object is considered. The cluster prototypes are computed according to the following recursive equation:

$$v_{i|N+1} = \frac{\sum_{j=1}^{N+1}(u_{ij})^q \mathbf{x}_j}{\sum_{j=1}^{N+1}(u_{ij})^q} = \frac{\sum_{j=1}^{N}(u_{ij})^q \mathbf{x}_j + (u_{i,N+1})^q \mathbf{x}_{N+1}}{\sum_{j=1}^{N}(u_{ij})^q + (u_{i,N+1})^q} = \tag{3.9}$$

$$\frac{vn(N) + (u_{i,N+1})^q \mathbf{x}_{N+1}}{vd(N) + (u_{i,N+1})^q}$$

where $vn(N) = \sum_{j=1}^{N} (u_{ij})^q \mathbf{x}_j$ and $vd(N) = \sum_{j=1}^{N} (u_{ij})^q$.

The quantities vn(N) and vd(N) are calculated recursively using degrees of membership of N existing objects and the current values of vn(N) and vd(N) are saved after each recursion. The adaptation of the cluster centres is provided by additional terms corresponding to a new (N+1)-th object. It is obvious that the computational expense of the updating procedure is moderate due to the recursive mode of calculation of the cluster centres. Using the complete knowledge base including the training data set as well as new data up to the current sample, there is no need to store the growing membership matrix **U** to calculate cluster centres since all the necessary information is contained in the scalar quantities vn(N) and vd(N). The drawback of using the complete data set for the updating procedure is, at first, that old objects may become irrelevant as time passes and this may have negative effects on the updating procedure, and secondly, that each new object can influence the location of cluster centres regardless of its relative importance.

The first drawback can be avoided by using a moving window and by replacing old data of the initial training set with new on-line data. This strategy implies the storage of the membership matrix **U** for updating cluster centres. If the degrees of membership of each object are stored in columns, then the membership matrix corresponding to the current window is obtained by deleting the column corresponding to the oldest object and by adding a new column with degrees of membership of a new object. In [Marsili-Libelli, 1998] the size of the moving window was chosen to be equal to the number N of training data records. Thus, the column dimension of the membership matrix remains fixed at N. The following algorithm was proposed for the updating procedure using a moving window:

Algorithm 2: Classifier update with a moving window [Marsili-Libelli, 1998].

1. *Classify a new object x_j, $j>N$, by calculating the degrees of membership u_{ij} of an object to all clusters $i=1, ..., c$.*
2. *Evaluate the new membership matrix at step j by dropping the leftmost column corresponding to the oldest object and by adding a rightmost column containing the degrees of membership of the new object:*

$$U_j = [U_{j-1}(2,\ldots,N) \quad u_j] \tag{3.10}$$

3. *Update the matrix of cluster centres V using a new matrix U_j according to equation (3.7).*

The updating procedure based on the moving window provides a possibility to follow the temporal development of objects by the incremental displacement of cluster centres from their initial locations corresponding to the training data set. On the other side, it allows each new object to influence the location of cluster centres, although not all objects can be considered representative enough to contribute to the knowledge base. Hence, the problem is to discriminate between 'good' and 'bad' objects depending on their significance and to decide whether the information provided by a new object is sufficient enough to be included in the knowledge base. In other words, whether a new object can improve the existing partitioning by making it crisper should be evaluated.

Different criteria were introduced into the literature to determine the degree of uncertainty of a fuzzy partition [Pal, Bezdek, 1994]. In [Marsili-Libelli, 1998] the partition entropy was selected as an evaluation criterion of the quality of the updated partition. The partition entropy is similar to the average information content of a source proposed by Shannon [Shannon, 1948] and is defined as follows [Bezdek, 1981, p. 111]:

$$H_N = -\frac{1}{N}\sum_{j=1}^{N}\sum_{i=1}^{c} u_{ij} \cdot \ln(u_{ij}) \tag{3.11}$$

If the membership matrix **U** provides a clear partition, then one has the complete information at one's disposal and consequently the entropy is zero. If the degrees of membership of an object to all clusters are equal, the entropy takes its maximum value of one. Thus, the smaller the entropy, the crisper the fuzzy partition.

An evaluation of changes of the normalised partition entropy can be used in the algorithm for updating cluster centres in order to decide whether a new object improves the fuzzy partition. The algorithm consists of the following steps:

Algorithm 3: Conditional classifier update [Marsili-Libelli, 1998].

1. *Classify a new object x_j, $j>N$, by calculating the degrees of membership u_{ij} of an object to all clusters $i=1, \ldots, c$.*

2. *Evaluate the new membership matrix at step j as in Algorithm 2.*
3. *Compute the partition* $H_N(U_j)$ *and* $H_N(U_{j-1})$ *of the partition at the current step j and at the previous step j-1, respectively.*
4. *Calculate a variation of the partition entropy:*

$$\Delta H_N(j) = \frac{H_N(U_j) - H_N(U_{j-1})}{H_N(U_{j-1})} \qquad (3.12)$$

5. *If variation of the partition entropy is positive but does not exceed the predefined limit, update the matrix V of cluster centres according to equation (3.9), otherwise discard the new matrix* U_j *and keep the previous matrix* U_{j-1} *and the corresponding cluster centres V unchanged until the next step. Allowing a small positive increase of the partition entropy has the objective of following the temporal development of objects, even if this does not always improve the partition. Negative variations always lead to an update of the cluster centre, since they indicate a crisper partition.*

The updating procedure with the evaluation of the partition entropy at each step provides more control over the displacement of cluster centres. The 'bad' objects and outliers do not affect clustering results and only 'good' objects can influence the location of cluster centres. The classifier can follow slow gradual changes of the cluster structure over time but it lacks a mechanism to be adjusted to abrupt changes characterised by a change of the number of clusters.

3.2.3 Adaptation of the classifier

Compared to the two approaches presented above a more flexible approach to designing a classifier so that it can automatically recognise temporal changes in the cluster structure is its adaptation according to the detected changes. For this purpose, the adaptation process must be coupled with the result (status) of the monitoring process in order to obtain a flexible dynamic classifier.

In [Nakhaeizadeh et al., 1997] and [Lanquillon, 1997] a number of general models for adapting dynamic classifiers in supervised learning was proposed. Although they were implemented using statistical or neural network classifiers, most of the proposed ideas have a general character and could be applied in supervised as well as in unsupervised dynamic pattern recognition systems with some modifications. According to [Nakhaeizadeh et al., 1997], two approaches to the adaptation of a classifier can be distinguished:

1. adaptation by explicitly changing the classifier's structure or
2. adaptation by changing the training data set that is used to design the classifier.

In the first approach the classifier is adapted based on the most recent objects according to the result of the monitoring process. Depending on the changes detected during the monitoring process, it could be necessary either to incrementally update the classifier or to re-learn the classifier using a new training data set. The main idea of this flexible approach is that the decision about the appropriate update of the classifier is controlled by the monitoring status and the parameters of the classifier depend on the current values of monitored characteristics. Developing an incremental learning algorithm for the classifier, the goal is to moderately adapt the classifier to gradual changes based on the most recently observed objects in such a way that the previously learned information (old classifier) is reused. Also, if the classifier must be re-learned in the case of serious changes, it seems reasonable to reuse the old classifier in some way, for instance as the initialisation for the learning algorithm. Recall that the alternative simple approach for adapting the classifier to structural changes is to re-learn the classifier completely after each batch of new data. This approach will be referred to as 'conventional updating approach' in the following. An advantage of the flexible incremental approach towards the conventional can be seen by comparing the performance of both approaches applied to different classifiers as it will be shown in the end of this section.

Another strategy for the adaptation of a classifier is concerned with the update of the training data set, which is sometimes referred to as a template set. This approach aims at obtaining a currently relevant training set that can be used to design a classifier. The requirements for the choice of the training data set are usually formulated as follows:

- it must be as small as possible to allow easy and fast classifier design,
- it must be representative, i.e. training data must contain good prototypes of each cluster so that the classifier designed using this training set has a good discriminating ability.

In static pattern recognition the training data set is chosen only once before the classifier design and remains unchanged during classification (the training set is not used any more). In dynamic pattern recognition it may be necessary to re-learn the classifier as time passes if its performance decreases due to structural changes. In this case the initial training set may not be representative any more since it was designed before structural changes occurred. Thus, it is important to update the training set over time by including new objects. A simple approach would be to include all new objects arriving over time into the training set. However, the training set would rapidly grow in this case making the classifier design very time-consuming and it would possibly contain many insignificant objects that cannot improve the classifier performance. Therefore, a better approach is to select carefully only representative objects for the training set.

The problems that arise by choosing the training set in dynamic pattern recognition can be formulated as follows:
– to reduce the constantly growing set of objects,
– to distinguish between relevant and irrelevant objects at the current moment.

In order to control the size of the training set of objects the concept of the moving time window or the more general concept of the template set can be used. The most representative and currently relevant objects for the training set can be chosen by applying the concept of usefulness. These concepts represent the main elements for updating the training set.

In the following different adaptation procedures for updating the classifier or the training set will be presented and discussed based on the results of their application to different classifiers.

3.2.3.1 Learning from statistics approach

In Section 3.1.1 the monitoring process based on the modification of Shewhart quality control charts was described. The outcome of the monitoring procedure is one of three states: 'okay', 'warning' or 'action'. Suppose that new objects are provided and examined in groups, which are called batches. The size of a batch is defined by the number of objects in the batch and supposed to be constant. The following adaptation procedure for updating the classifier depending on the current monitoring status was proposed in ([Kunisch, 1996], [Nakhaeizadeh et al., 1997]) for supervised classification algorithms and called 'learning from statistics':

Status *'okay'*: The state of the process or system is stable. Therefore, there is no need to change the classifier.

Status *'warning'*: Gradual changes of the system's state are assumed. In this situation different updating strategies can be applied to slightly adapt the classifier to suspected changes. In particular, procedures based on incremental learning are recommended where the classifier is updated based on the most recent objects. The specific procedure for incremental learning depends on the applied pattern recognition method. Some of them were discussed in Section 3.2.2.

In [Nakhaeizadeh et al., 1997] it was proposed to establish an adapted classifier as a weighted linear combination of the parameters of the old classifier and a new one, which was learned based only on new objects of the current batch. This adaptation procedure was implemented within statistical classification algorithms where the error rate was monitored. In this case weights are chosen depending on the current error rate, the mean error rate and its standard deviation or on some function thereof.

The incremental updating of classifiers based on neural networks is carried out by computing parameters of the network depending on the current monitoring status and the current values of the monitoring characteristic. For the multi-layer perceptron (MLP) implemented in [Lanquillon, 1997], two parameters must be updated for learning: the current learning rate and the current number of cycles for the backpropagation algorithm [Nauck et al., 1996]. They are calculated proportional to predefined maximum values using a piecewise linear weighting function of the current error rate, the mean error rate and its standard deviation. Moreover, to obtain a really incremental learning algorithm the connection weights of the current MLP are used as a starting solution for the backpropagation algorithm instead of random initialisation of the MLP.

Status *'action'*: A serious change in the system's state is detected. The best strategy is to re-learn the classifier from scratch based on the most recent objects, since the current classifier was designed based on objects which are not representative any more. However, according to [Nakhaeizadeh et al., 1997] the new classifier may not be very successful because a set of new objects might be not very representative, especially if the size of this set is small and could contain many outliers. It seems reasonable to re-learn the classifier based on the updated training set containing some old and new objects. Possible approaches for updating the training set are described below.

The dynamic pattern recognition process using the 'learning from statistics' adaptation procedure is illustrated on *Figure 18* [adapted from Lanquillon, 1997, p. 54].

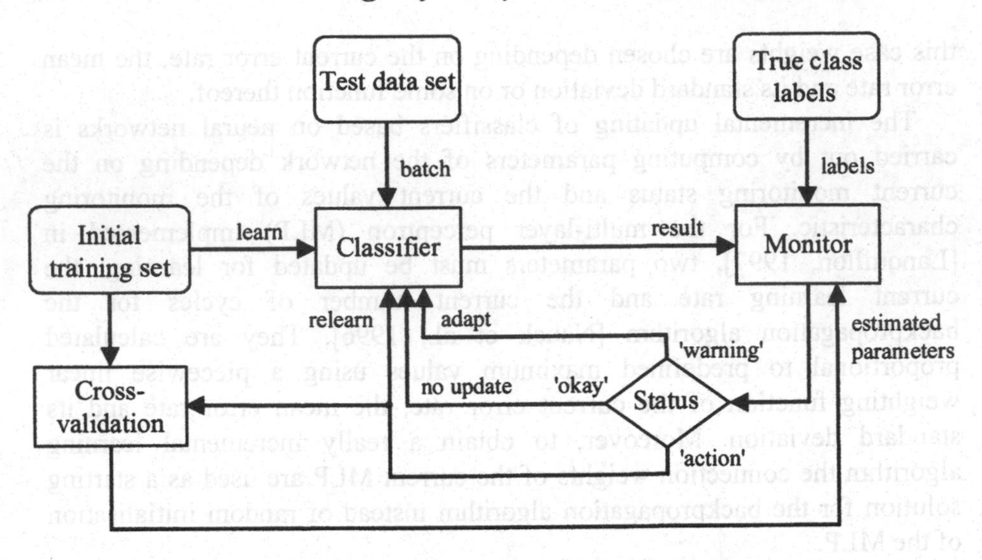

Figure 18. Adaptation of the classifier based on 'learning from statistics' approach

The classifier is designed based on the initial training data set. The monitoring procedure is initialised with estimates for the expected value μ of the characteristic *f* and its standard deviation σ. These estimates are obtained using k-fold cross validation as described in section 3.1.1, if the characteristic *f* to be monitored is based on the classification results (e.g. the error rates). If the characteristic *f* is based only on objects, it is sufficient to split the training set into k subsets and to evaluate the expected value μ and its standard deviation σ using the estimates of characteristic *f* on these subsets without learning the classifier. Since monitoring of the error rates was chosen in [Nakhaeizadeh et al., 1997] and [Lanquillon, 1997], the process of estimating the expected value μ and its standard deviation σ is referred to as cross validation. After the classifier is designed based on the initial training set, it is applied to the classification of new objects which are presented in batches. The results of the classification of the current batch are analysed within the monitoring procedure, which is performed by the monitor. Using the true class labels of the current batch, the monitor determines the current value of characteristic *f* and provides the monitoring status. The updating of the classifier is carried out depending on this status. The cycle is completed by updating the expected value μ and its standard deviation σ depending on the current value of f. If there was an 'action' status, μ and σ are re-estimated by cross validation as for the first batch. The cycle, including classification of a current batch, monitoring and updating of the classifier, is called a working-and-learning cycle.

This flexible adaptation procedure for updating the classifier, combined with the monitoring of the error rates, can be considered as a general alternative to a conventional approach of re-learning the classifier after each batch.

In the following, three adaptation procedures based on different concepts of updating the training data set are described in the framework of supervised learning. These general concepts can however be adjusted for unsupervised algorithms as well, since the classifier design (at the beginning of the pattern recognition process or due to re-learning during dynamic pattern recognition) always requires the training data set.

3.2.3.2 Learning with a moving time window

In the 'learning from statistics' approach only the current batch of objects is used to update or re-learn the classifier. Alternatively one can apply a moving time window of constant or variable length, which is used for updating the classifier. As stated in Section 2.3.2, the performance of the classifier depends crucially on the length of the time window. If the window is too small, then relevant objects are discarded too early and the classifier may not be very reliable. If the window is too large, then the update of the classifier according to structural changes may be too slow. The 'optimal' length of the window can be chosen depending on the classification problem under consideration and the underlying system dynamics.

A more flexible approach is to apply a variable window length. This idea was introduced in [Widmer, Kubat, 1993, 1996] where a window adjustment heuristic was used to determine the optimal window size. Following this idea, [Lanquillon, 1997] proposed to control the window size by a heuristic depending on the current monitoring status in the pattern recognition system. Suppose that the length (or size) of the moving time window is defined by the number n of the most recent batches. Denote the minimum and the maximum window size by n_{min} and n_{max}, respectively. The procedure for adjustment of the current window size n_{curr} is as follows:

Status *'okay'*: The performance of the classifier is sufficient and no changes are suspected. If the current window size is smaller than the maximum size $n_{curr} < n_{max}$, then n_{curr} is increased by 1 in order to obtain a more representative training set. Otherwise, the current window size remains unchanged.

Status *'warning'*: Gradual changes of a system state are suspected. If the current window size is larger than the minimum size $n_{curr} > n_{min}$, then n_{curr} is decreased by 1 in order to allow the faster adaptation of the classifier to suspected changes.

Status *'action'*: A serious change of the system's state is detected. The classifier has to be re-learned based on the most recent objects whereas old not representative objects have to be discarded. Therefore, the training set should be reduced by setting the current window size to a minimum $n_{curr} = n_{min}$.

Summarising, the current window size is determined by the function:

$$n_{curr} = N(n_{curr}, n_{min}, n_{max}, \tau) = \begin{cases} \min(n_{curr}+1, n_{max}) & \text{if } \tau = \text{okay} \\ \max(n_{curr}-1, n_{min}) & \text{if } \tau = \text{warning} \\ n_{min} & \text{if } \tau = \text{action} \end{cases} \quad (3.13)$$

In case of $n_{curr} = n_{min}$ a moving window of a fixed size is obtained.

The adaptation procedure based on the 'learning with a moving window' approach can be improved by keeping all representative objects in the template set even if they become outdated compared to the most recent ones.

3.2.3.3 Learning with a template set

The concept of a template set [Gibb, Auslander, Griffin, 1994] is a generalisation of the moving time window approach. According to the latter, the training data set is composed of the n most recent batches, although not all of objects may be considered as representative, e.g. some objects may be noisy and, thus, they must not be accepted to the template set. A general template set is designed by the careful selection of representative examples and may contain any observed object. If the template set becomes too large older, or contradictory, examples have to be discarded.

A template set is characterised by two parameters: its size and a criterion for including examples in the template set. For choosing the size of the template set the same considerations as for the choice of the time window size can be applied (Section 2.3). Moreover, if the template set is too large and the criterion for including new examples is too strict, the adaptation of the classifier to structural changes will be very slow or even impossible. On the other hand, if the template set is too small and the criterion is too soft, then new objects have a strong effect on the adaptation procedure while old relevant objects are discarded too early and the classifier may not be very representative. The formulation of a suitable criterion for including new objects in the template set (or generally to adjust the template set) is very difficult since it is almost impossible to decide whether new objects are representative or not if structural changes have occurred. A decision can be taken based on the accuracy of classification compared to a predefined threshold. For instance, in [Kunisch, 1996] the probability of membership to the true class determined by a statistical classifier is compared to a certain threshold value. This criterion allows only correctly classified objects to be included in the template set. The drawback of this criterion is however that objects that are representative for the template set but are classified wrongly by the current classifier have no chance to be selected. For instance, in the case of gradual changes in a system state misclassified objects are not added to the template set, although they can represent a tendency towards a structural change [Nakhaeizadeh et al., 1996]. Therefore, the adaptation to such changes will not be possible. To avoid this drawback, the adjustment of the template set can be coupled with the monitoring status ([Kunisch, 1996], [Lanquillon, 1997]) as follows:

Status *'okay'*: The performance of the classifier is sufficient and no changes are suspected. New objects can be included in the template set.

Status '*warning*': Gradual changes in the system's state are suspected. This case is treated depending on the previous monitoring status (after classification of the previous batch of data).

- If the previous status was *'okay'* a new status can be caused either by gradual changes in the system's state or by noise. Thus, it is preferable to adjust the template set as in the case of status *'okay'*. If gradual changes have indeed occurred, then it will be confirmed after classification of the next batch of data and then appropriate actions will be taken.
- Status '*warning*' was detected twice confirming the assumption of gradual changes. The template set is adjusted according to status *'action'* (explained below).
- If the previous status was *'action'* the corresponding actions were applied and the template set consists of new objects from the previous batch. The current status *'warning'* indicates decreasing oscillations after detected serious changes and reactions to them. There is no need to substitute the template set and therefore this case is treated as status '*okay*'.

Status '*action*': A serious change in the system's state is detected. The classifier has to be re-learned based on the most recent objects, since the current template set is not representative any more. Consequently, the complete template set is discarded and designed anew by including objects of the most recent batches.

The concept of the template set is more complicated compared to that of a moving time window but it can provide a better quality of the training data set for the adaptation of the classifier.

The adaptation of the classifier according to a 'learning with a template set' approach is illustrated in *Figure 19* [adapted from Lanquillon, 1997, p. 59], where the initial training phase and estimation of parameters for the monitoring process by cross validation are not depicted for the sake of simplicity.

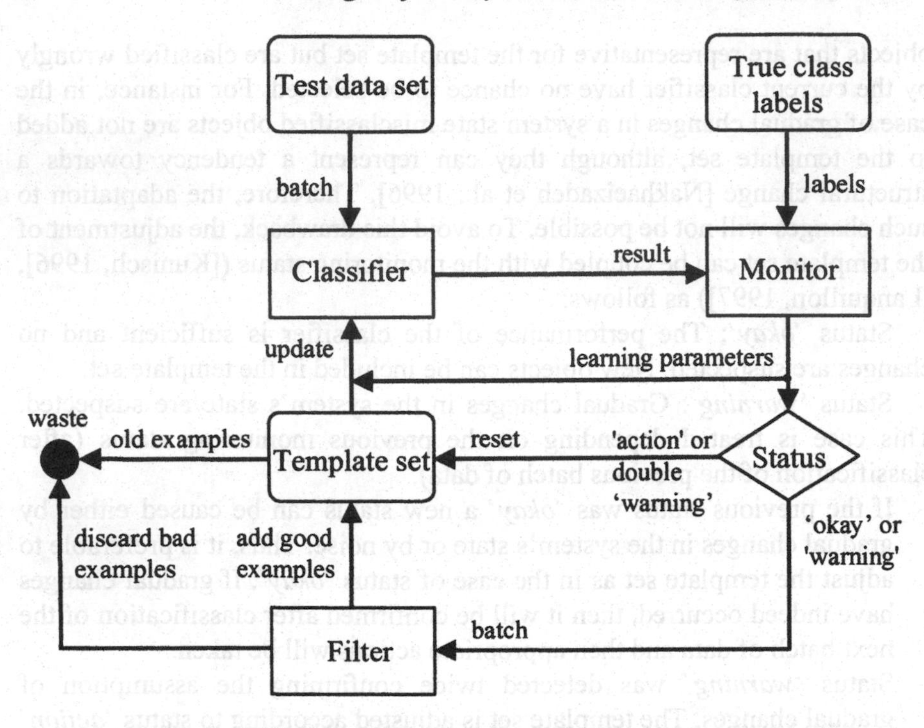

Figure 19. Adaptation of the classifier based on 'learning with a template set' approach

After classification of a new batch of data, the results are presented to the monitor where the current value of the monitored characteristic is evaluated and the current monitoring status is determined. If status *'action'* is detected or status *'warning'* has already appeared twice the current template set is discarded and substituted by objects of the last n_{min} batches (this operation is referred to as 'reset' in the *Figure 19*). Otherwise, the new batch of objects is filtered to separate good and representative examples from bad and irrelevant ones. For each object the filter checks the criterion of inclusion into the template set and objects fulfilling this criterion are added to the template set. All other objects in the batch are discarded. If the size of the template exceeds the maximum size defined by n_{max} times the batch size, old examples of the template set are also discarded. The current classifier is then updated using the current template set as training data. The updating procedure (adaptation or re-learning) depends on the monitoring status and is performed in the same way as by using the 'learning from statistics' approach (Section 3.2.3.1). However, the training set is carefully selected and contains only representative and currently relevant objects.

In order to improve the approach based on the template set, the criterion for including new objects in the template set can be extended by the concept of usefulness described in the next section.

3.2.3.4 Learning with a record of usefulness

The idea of the concept of usefulness is to give each object a weight representing its usefulness. As time passes the weight is updated according to the usefulness of an objects for classification. If the example is useful its weight is increased, otherwise it is decreased. If the weight falls below a given threshold the example is discarded. In dynamic pattern recognition it is supposed that older examples become less useful as time passes. This process is referred to as 'ageing of objects'. Different approaches have been proposed to define the usefulness of an example.

In [Nakhaeizadeh et al., 1996] a dynamic 1-nearest neighbour classifier is developed based on the record of usefulness characterising the training data set. In this approach, the nominal weight of an example as it is first observed is set to 1. As time passes, objects age at a rate of λ. Weights of objects are adapted depending on newly classified objects according to the following procedure:

1. If a new object is correctly classified by its nearest neighbour but misclassified by its second nearest neighbour, then the weight of the nearest neighbour is increased by δ.
2. If a new object is misclassified by its nearest neighbour but correctly classified by its second nearest neighbour, then the weight of the nearest neighbour is decreased by δ.
3. Otherwise, the weight is left unchanged.

The drift of the weight of example $\mathbf{x}$ due to new objects is defined by a function:

$$\text{Drift}(w(\mathbf{x}))=-\lambda+\delta p(\mathbf{x})-\varepsilon q(\mathbf{x}), \tag{3.14}$$

where $p(\mathbf{x})$ is the probability that $\mathbf{x}$ will be the nearest neighbour in the first step of the procedure, and $q(\mathbf{x})$ is the probability that $\mathbf{x}$ will be the nearest neighbour in the second step of the procedure. It is stated in [Nakhaeizadeh et al., 1996] that the weight of each example behaves like a Markov chain with an absorbing barrier at 0, whose properties can be studied in order to choose suitable values for λ, δ, and ε. Assuming the existence of an equilibrium state, the expected time until an example $\mathbf{x}$ is discarded and the expected number of examples in the template set at any point of time can be evaluated.

This approach to determining the usefulness of an example is rather simple since there is no learning process in the classification algorithm and a new example is classified by comparing it to all examples of the template set. If procedures for updating the training set and the classifier can be distinguished the concept of usefulness can be applied in different ways to both cases.

Since the definition of usefulness depends on the chosen classifier, it should be considered separately for each type of classifier. In the case of prototype-based classifiers the purpose of applying the concept of usefulness for updating the classifier is to assess the selected prototypes, in this way possibly detecting changes in the cluster structure. If the average usefulness of the set of prototypes is defined in some way, then a significant decrease of this value can indicate structural changes.

For updating the training data set the problem is to evaluate the usefulness of each example. In statistical classification algorithms, the usefulness of an observation is related to the frequency of its occurrence. The following definition of usefulness is proposed in [Nakhaeizadeh et al., 1998] for the case of a binary classification problem and on the assumption of a finite set of possible feature values. An observation is considered to be very useful for establishing a classifier if it frequently appears in one class because it is very likely to occur frequently in the future. However, if an observation has occurrences in both classes it is considered to be less useful, even though it may appear frequently. In extreme cases, if the observation appears equally frequent in both classes its true class membership is very uncertain, and this observation is rather useless. Thus, the more frequently an observation appears in only one class the more useful it is. The usefulness $u(\mathbf{x})$ of an observation $\mathbf{x}$ is defined formally as a difference between the frequencies of appearance of an observation $\mathbf{x}$ in both classes:

$$u(x) = |n_1(x) - n_2(x)| \tag{3.15}$$

where $n_i(\mathbf{x})$ denotes the number of occurrences of observation $\mathbf{x}$ in class i, i=1, 2, within a certain set of examples. This set could include either all available examples or it could be designed due to an application of the time window approach. Each observation $\mathbf{x}$, which is considered as an example, is characterised by the label of the class in which it is having its maximum number of occurrences and by its usefulness $u(\mathbf{x})$. The list of examples sorted by $u(\mathbf{x})$ is called the record of usefulness. It can be employed to select the most representative examples for the template set by choosing either the first N examples from the record or examples whose usefulness $u(\mathbf{x})$ exceeds a predefined threshold.

To express the fact that examples in dynamic classification system are getting old and less useless in the course of time the ageing function (e.g. an exponential decay), or a forgetting factor, can be applied to the examples. If the usefulness of an example due to ageing falls below a predefined threshold this example is discarded. The forgetting factor λ can be either constant or variable depending on the performance of the classification system. In [Nakhaeizadeh et al., 1997] the forgetting factor, as well as the record of usefulness, depends on the monitoring status as follows:

Status *'okay'*: The monitored process is under control. The ageing of examples seems unnecessary, therefore the forgetting factor is set to a minimum value λ_{min}. The current record of usefulness is aged by the factor $1-\lambda_{min}$ $(0 \bullet \lambda_{min} \bullet 1)$. Since the performance of the classifier is sufficient, the value of λ_{min} should be rather small keeping the record of usefulness almost unchanged.

Status *'warning'*: A gradual change of the system's state is suspected. Moderate ageing could be appropriate and the forgetting factor is set to a value between the minimum λ_{min} and maximum λ_{max} values such that $0 \bullet \lambda_{min} \bullet \lambda \bullet \lambda_{max} \bullet 1$. The current forgetting factor is defined as a linear function of monitoring parameters (current value of monitored characteristic, the expected value and its standard deviation):

$$\lambda_{curr} = \lambda_{min} + g(f_{curr}, \mu, \sigma, \tau)(\lambda_{max} - \lambda_{min}). \tag{3.16}$$

The record of usefulness is aged by the factor $1 - \lambda_{curr}$.

Status *'action'*: Serious changes in the process under consideration are assumed. The forgetting factor is set to the maximum value λ_{max} or alternatively the current training set is completely discarded as being no longer representative. In this case a new record of usefulness is determined from the observations of the most recent batch.

In order to maintain a constant size of the training set, the number of old observations that is discarded must be replaced by an equal number of new observations. This equilibrium can be achieved by choosing the minimum forgetting factor depending on the maximum size of the training set:

$$\lambda_{min} = \frac{1}{n_{max} \cdot batch\_size}, \tag{3.17}$$

where n_{max} is the number of most recent batches used to update the training set and batch_size denotes the number of objects in one batch.

The procedure of adaptation of the classifier based on updating the training set with the record of usefulness is illustrated in *Figure 20* [adapted from Lanquillon, 1997, p. 61], where the initial training phase and estimation of parameters for the monitoring process by cross validation are not depicted for the sake of simplicity. Also the initialisation of the current record of usefulness is not presented in the figure. The record of usefulness is established from the data set based on the frequencies of occurrences of single observations as described above. Each example in the record of usefulness is characterised by a weight (value of usefulness) and a class label.

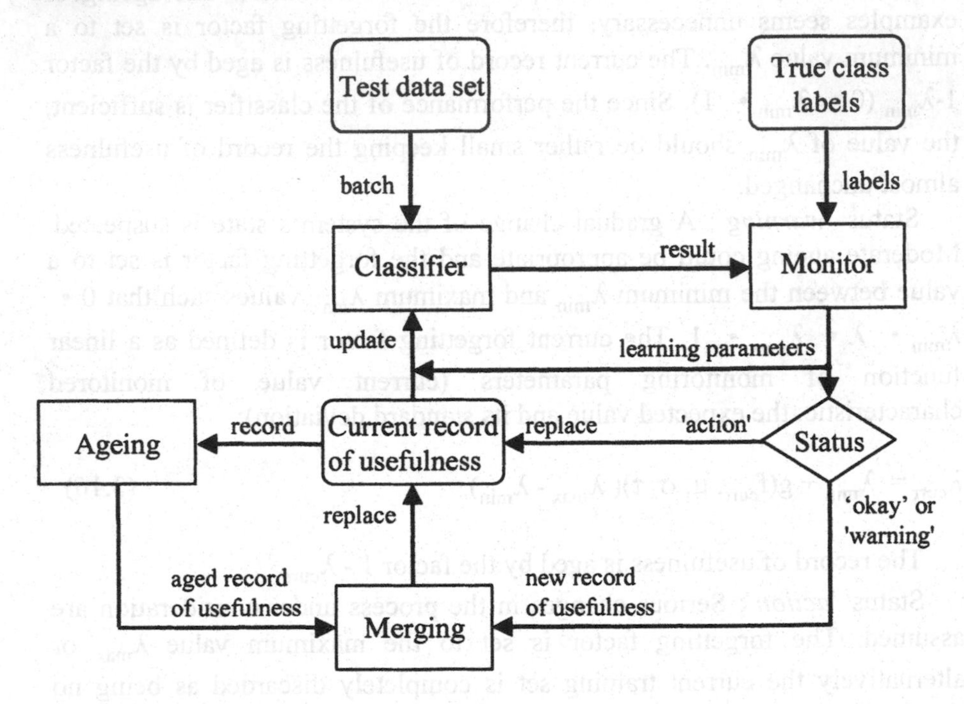

Figure 20. Adaptation of the classifier based on the record of usefulness

During the working-and-learning cycle, the results of classification of a new data batch are presented to the monitor, where the current value of the monitored characteristic is evaluated and the current monitoring status is determined. Additionally a new record of usefulness is evaluated for the new batch. If status *'action'* is detected the current record of usefulness is discarded and replaced by the new record. Otherwise, it is replaced by the aged current record which is merged with the new record obtained from the current batch. Two records of usefulness are merged in a similar way as a new record of usefulness is obtained from the data set. But instead of frequencies of occurrence in equation (3.15) the usefulness of an observation is used. That is, if an observation is present in the aged current and in the new record and has different class labels, its usefulness is determined as the absolute value of the difference of values of usefulness in both records. If an observation appears in both records with the same class label its usefulness is calculated as a sum of values of usefulness in both records. In case the obtained usefulness of an observation does not exceed the predefined threshold, this observation is discarded. Observations that appear only in one of two records are taken into the resulting record without any change.

The forgetting factor for ageing the record of usefulness depends on the monitoring status as described above, but this dependence is not shown in the *Figure 20* for the sake of simplicity. Before the merging process, the record of usefulness is aged by the factor $1 - \lambda_{curr}$, where the forgetting factor $\lambda_{curr} \in [0,1]$. A forgetting factor equal to zero causes no ageing of the record of usefulness, whereas the record will be completely forgotten and discarded if the forgetting factor is set to one.

The classifier is updated according to one of the procedures described above - flexible adaptation procedure based on 'learning from statistics' approach or conventional procedure based on re-learning the classifier after each batch. As already mentioned, the updating procedure depends crucially on the chosen classification algorithm.

3.2.3.5 Evaluation of approaches for the adaptation of a classifier

The four approaches for adaptation presented above - updating the classifier itself or updating the training data set - were implemented in dynamic classification systems based on multi-layer perceptron and on radial basis function (RBF) networks in [Lanquillon, 1997, pp. 54-62]. These approaches were investigated either separately or in combinations. In the first case, the adaptation with the 'learning from statistics' approach was based only on the most recent batch and the procedures for updating the training data set were applied together with the conventional approach for updating the classifier. Alternatively, the updating procedures for training set were combined with flexible updating of the classifier based on the 'learning from statistics' approach. The application of these different configurations of a dynamic classification system to real-world data from the credit industry has shown that these approaches applied separately improve the performance of both neural network classifiers compared to no-learn (static) or conventional re-learning approaches. However, combining the approach for updating the training set with the flexible updating procedure for the classifier does not increase the classifier performance. Particularly, the methods for updating the training set perform better when they are combined with conventional approach of re-learning the classifier after each batch [Lanquillon, 1997, pp. 68-79]. According to the author, the explanation for such a behaviour can be seen in the fact that the older relevant information is processed twice. On the one hand, the updated training set contains enough relevant objects (also older ones) to successfully learn the classifier. On the other hand, the older information is preserved by the classifier and only new objects are required for the flexible updating of the classifier. In this case, it is unnecessary to have a representative training set. Therefore, using the combination of approaches for updating the classifier and the training set leads to the increased importance of older examples and the performance of the classifier may be decreased.

Two of the presented approaches for adapting the classifier based on 'learning from statistics' and on 'learning with a template set' were implemented in [Kunisch, 1996, pp. 21 - 27, 69-73] for the following statistical classifiers: normal-based linear discriminant rule (NLDR), normal-based quadratic discriminant rule (NQDR) and logistic discriminant rule. The performance of neural and statistical classifiers for the same parameter settings was compared in [Lanquillon, 1997, pp. 82-83] showing that neural classifiers outperform statistical ones in some cases. Particularly, the RBF network which is based on the concept of usefulness provides the best results (in terms of error rate) among all tested neural and statistical approaches.

In the following chapter a new algorithm for dynamic fuzzy classifier design will be proposed, which is partly based on the ideas presented in this chapter but also uses a number of novel criteria to establish the monitoring procedure. The proposed algorithm is introduced into the framework of unsupervised learning and allows the design of an adaptive classifier capable of recognising automatically gradual and abrupt changes in the cluster structure as time passes and adjusting its structure to detected changes. The adaptation law for updating the template set is extended by additional features that should guarantee a more reliable and efficient classifier.

4 DYNAMIC FUZZY CLASSIFIER DESIGN WITH POINT-PROTOTYPE BASED CLUSTERING ALGORITHMS

In this section a dynamic fuzzy clustering algorithm is developed, which provides a possibility to design a dynamic classifier with an adaptive structure. The main property of a dynamic classifier is its ability to recognise temporal changes in the cluster structure caused by new objects and to adapt its structure over time according to the detected changes. The design of a dynamic fuzzy classifier consists of three main components: monitoring procedure, adaptation procedure for the classifier and adaptation procedure for the training data set. The monitoring procedure consists of a number of heuristic algorithms, which allow the recognition of abrupt changes in the cluster structure such as the formation of a new cluster, the merging of several clusters or the splitting of a cluster into several new clusters. The outcome of the monitoring procedure is a new number of clusters and an estimation of the new cluster centres. The adaptation law of the classifier depends on the result of the monitoring process. If abrupt changes were detected, the classifier is re-learned with its initialisation parameters obtained from the monitoring procedure. If only gradual changes were observed by the monitoring procedure, the classifier is incrementally updated with the new objects. The adaptation procedure is controlled by a validity measure, which is used as an indicator of the quality of fuzzy partitioning. This means that an adaptation of the classifier is carried out if this leads to an improvement of the current partitioning. The improvement can be determined by comparing the value of a validity measure for the current partitioning (after re-learning) with its previous value (before re-learning). If the validity measure indicates an improvement of the partitioning a new classifier is accepted, otherwise the previous classifier is preserved.

This chapter is organised as follows: first, the formulation of the problem of dynamic clustering is derived and the model of temporal development of the cluster structure is introduced. Thereafter, possible abrupt changes in the cluster structure are discussed and different criteria for their recognition during the monitoring procedure are formulated and then summarised into corresponding heuristic algorithms (Sections 4.3-4.5). The adaptation laws for updating the classifier and the template set in response to detected changes are proposed in Sections 4.7 and 4.8. This will be followed by the examination of different validity measures which can be relevant to control the adaptation of the classifier (Section 4.8.2). Finally, the entire algorithm for the dynamic fuzzy classifier design within a learning-and-working cycle is presented.

4.1 Formulation of the Problem of Dynamic Clustering

This section begins with the formulation of the classical (static) clustering problem, which will be used afterwards as a background to derive a formulation of a dynamic clustering problem.

Consider a set of objects $X=\{\mathbf{x}_1, ..., \mathbf{x}_N\}$, where each object is represented as an M-dimensional feature vector $\mathbf{x}_j=[x_{j1}, ..., x_{jM}]$, $j=1, ..., N$. The task of fuzzy clustering is to partition such a set of objects into c clusters (c is an integer, $2 \le c < N$) and to estimate a set of cluster prototypes $V=\{\mathbf{v}_1, ..., \mathbf{v}_c\}$. A fuzzy c-partition is given by a matrix $U=[u_{ij}]$ which satisfies the following conditions [Bezdek, 1981, p. 26]:

1. $u_{ij} \in [0,1]\,,\ 1 \le i \le c,\ 1 \le j \le N$

2. $\sum_{i=1}^{c} u_{ij} = 1,\ 1 \le j \le N$ (4.1)

3. $0 < \sum_{j=1}^{N} u_{ij} < N, 1 \le i \le c$

Elements of matrix u_{ij} denote a degree of membership of object j to cluster i. Due to Condition 1 each object can belong to several clusters with different degrees of membership. Condition 2 requires that the total membership of an object over all clusters be normalised to 1. Condition 3 forbids any cluster to be empty or to contain all objects.

Cluster prototypes, or cluster centres, represent the location of the clusters. The number of clusters is not usually known in advance and is normally determined using cluster validity measures [Zimmermann, 1996, p. 260]. Thus the problem is to determine the number of prototypes and their location so that the obtained partition fits the underlying data structure in the most precise way possible.

One of the most frequently used criteria in fuzzy clustering is the variance criterion, which for each c measures the dissimilarity between the objects in the cluster and its cluster centre by the distance measure. The variance criterion corresponds to minimising the overall within-group sum of squared errors (variances between objects and cluster prototypes) weighted by degrees of membership of objects to clusters. This clustering criterion is called a minimum variance objective [Bezdek, 1981, p. 47] and yields the following basic formulation of the fuzzy partitioning problem:

$$\min J_m(U,v) = \sum_{j=1}^{N}\sum_{i=1}^{c}(u_{ij})^q \left\|x_j - v_i\right\|_A^2, \text{ such that } \mathbf{U} \in L_{fc}, \mathbf{v} \in \Re^{cp} \quad (4.2)$$

where

$$d(\mathbf{x}_j, \mathbf{v}_i)^2 = \left\|\mathbf{x}_j - \mathbf{v}_i\right\|_A^2 = (\mathbf{x}_j - \mathbf{v}_i)^T \mathbf{A}(\mathbf{x}_j - \mathbf{v}_i) \quad (4.3)$$

is the distance between each object $\mathbf{x}_j$ and a fuzzy prototype $\mathbf{v}_i$, **A** is a (M×M) symmetric positive-definite matrix, $\|\cdot\|_A$ is an inner product norm induced by **A** on $\Re^M$, L_{fc} is the set of all matrices **U** satisfying conditions (4.1) and $q \in [1, \infty)$ is the weighting exponent.

The objective functional (4.2) may be globally minimised only if degrees of membership and cluster prototypes are given by:

$$v_i = \frac{\sum_{j=1}^{N}(u_{ij})^q \mathbf{x}_j}{\sum_{j=1}^{N}(u_{ij})^q}, \quad i = 1,\ldots,c \quad (4.4)$$

$$u_{ij} = \frac{1}{\sum_{r=1}^{c}\left(\frac{d_{ij}}{d_{rj}}\right)^{\frac{2}{q-1}}}, \quad i = 1,\ldots,c,\ j = 1,\ldots,N. \quad (4.5)$$

Objective function algorithms of form (4.2), where **A** is the identity matrix and d is the Euclidean distance, are considered as the most suitable for problems with hyperspherical clusters or clusters of roughly equal proportions. The variation of matrix **A** makes it possible to obtain clusters of different shapes. Through a modification of the objective functional and equation (4.5) a possibilistic partitioning problem can be obtained.

Below, the static clustering problem will be used as a basis for the formulation of the dynamic clustering problem.

Consider a set of dynamic objects $\mathbf{X}(t) = \{\mathbf{x}_1(t), ..., \mathbf{x}_N(t)\}$, t = 1, ..., t_p, given as a temporal sequence of observations. The time interval of observations can in general be unlimited $t \in [1, \infty)$. Each object is represented by an M-dimensional trajectory in the feature space, which contains a history of temporal development of each feature. Although process or system variables observed can be continuous in nature, measurements of these variables are usually carried out discretely with a certain sampling rate. Therefore supposing that dynamic objects are observed at discrete time instants, a trajectory can be given by a discrete vector-valued function of the form:

$$\mathbf{x}_j(t) = [\mathbf{x}_j(t_1), \mathbf{x}_j(t_2), ..., \mathbf{x}_j(t_p)]^T, \; j = 1, ..., N, \tag{4.6}$$

where p is the number of observations in a trajectory and $\mathbf{x}_j(t_k)$, k = 1, ..., p, is an observation of a feature vector at time instant t_k. Substituting an M-dimensional feature vector into the components of this function, a matrix representation of a trajectory of a dynamic object is obtained:

$$\mathbf{x}_j(t) = \begin{bmatrix} x_{j1}(t_1)\, x_{j2}(t_1) \; ... \, x_{jM}(t_1) \\ x_{j1}(t_2)\, x_{j2}(t_2) ... \, x_{jM}(t_2) \\ .... \\ x_{j1}(t_p)\, x_{j2}(t_p) ... \, x_{jM}(t_p) \end{bmatrix}, \tag{4.7}$$

where columns correspond to trajectories of single features and rows correspond to feature vectors at a certain time instant. A trajectory which explicitly contains time as an additional feature so that single features are time-dependent can be called time series. Therefore, the problem of dynamic pattern recognition is sometimes referred to in the literature as time-series classification ([Petridis, Kehagias, 1997], [Schreiber, Schmitz, 1997], [Das et al., 1997], [Struzik, Siebes, 1998]). In general, trajectories can describe a dependence of features from another variable implicitly related to time.

Dynamic data set $\mathbf{X}(t)$ can be viewed as a three dimensional matrix whose dimensions are objects, features and time (*Figure 21* [adapted from Sato et al., 1997, p. 4]).

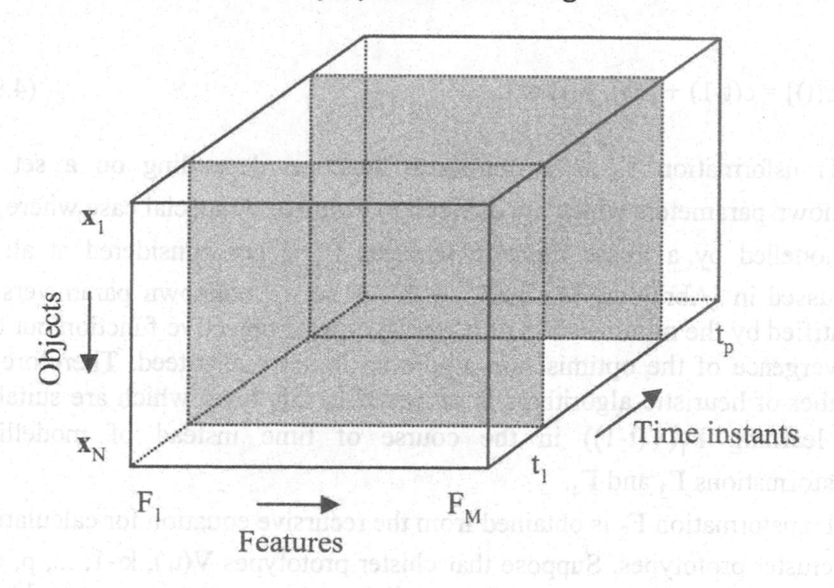

Figure 21. **3-dimensional matrix representation of a dynamic data set**

The task of dynamic fuzzy clustering is to determine the number of clusters c(t), to partition the data set into c(t) clusters and to estimate a set of cluster prototypes $V(t) = \{v_1(t),...,v_{c(t)}(t)\}$ in order to approximate the data structure at time instant t taking into account the history of temporal development of feature vectors. It should be stressed that fuzzy partition matrix $\mathbf{U}(t) = [u_{ij}(t)]$, as well as cluster prototypes V(t), evolve temporally as new observations become available.

The temporal development of cluster prototypes can be represented by the following model:

$$V(t) = \alpha(t)\,\Gamma_1[V(t-1)] + (1-\alpha(t))\,\Gamma_2[(V(t-1)]\,,\ \alpha(t)\in[0,1] \tag{4.8}$$

where Γ_1 is a transformation due to abrupt changes in the cluster structure (formation, merging, splitting or destruction of clusters) and Γ_2 is a transformation due to gradual temporal changes in the cluster structure (Section 2.3.2).

Transformation Γ_1 consists of two further transformations: a transformation Γ_3 regarding a change in the number of clusters and a transformation Γ_4 regarding a change in the locations of the cluster prototypes. According to the four types of abrupt changes considered here, a change in the number of clusters can be modelled by a linear function:

$$\Gamma_3\,[c(t)] = c(t\text{-}1) + \beta(t),\ \beta(t) \in I. \tag{4.9}$$

Transformation Γ_4 is a non-linear function depending on a set of unknown parameters which are difficult to estimate. A special case where Γ_4 is modelled by a linear function whereas Γ_3 is not considered at all is discussed in [Abrabtes, Marques, 1998]. A set of unknown parameters is identified by the minimisation of a special type of objective function but the convergence of the optimisation algorithm is not guaranteed. Therefore, a number of heuristic algorithms is proposed in this book which are suitable for learning $\Gamma_1(V(t\text{-}1))$ in the course of time instead of modelling transformations Γ_3 and Γ_4.

Transformation Γ_2 is obtained from the recursive equation for calculating the cluster prototypes. Suppose that cluster prototypes $V(t_k)$, $k=1, \ldots, p$, are determined at each time instant t_k according to the equation represented by a ratio:

$$V(t_k) = \frac{VN(t_k)}{VD(t_k)}, \tag{4.10}$$

where $VN(t_k)$ and $VD(t_k)$ are components calculated based on objects $X(t)$, $t=1, \ldots, t_k$ obtained until the time instant t_k. These components can be calculated recursively for each time instant t_k using the values of the components at the previous time instant:

$$VN(t_k) = VN(t_k - 1) + \ddot{a}(t_k), \tag{4.11}$$

$$VD(t_k) = VD(t_k - 1) + \tilde{a}(t_k), \tag{4.12}$$

where $VN(t_k\text{-}1)$ and $VD(t_k\text{-}1)$ are components calculated based on objects $X(t)$, $t=1, \ldots, t_k\text{-}1$ obtained until the time instant $t_k\text{-}1$, and $\delta(t_k)$ and $\gamma(t_k)$ are parameter vectors whose calculation is based on $X(t_k) = [x_1(t_k), \ldots, x_N(t_k)]$ new observations of objects at time instant t_k and the degrees of membership $u_{ij}(t_k)$ of objects $X(t_k)$ to the clusters, whose components are given by:

$$\delta_i(t_k) = \sum_{j=1}^{N} u_{ij}(t_k)^m x_j(t_k) \tag{4.13}$$
,

$$\gamma_i(t_k) = \sum_{j=1}^{N} u_{ij}(t_k)^m, \ i = 1, \ldots, c \tag{4.14}$$
.

Thus transformation Γ_2 for calculating the cluster prototypes can be written as:

$$\Gamma_2(V(t-1)) = \frac{VN(t-1) + \ddot{a}(t)}{VD(t-1) + \tilde{a}(t)}, \ t = 1, \ldots, t_p. \tag{4.15}$$

In the following sections an algorithm for dynamic fuzzy clustering based on the adaptation process for estimating cluster prototypes V(t) over time is proposed. The adaptation law for cluster prototypes consists of a combination of transformation Γ_2 for the case of gradual changes in the cluster structure with a number of heuristic techniques in the case of abrupt changes. The heuristic algorithms proposed look for indicators of structural re-organisation of the data allowing the monitoring of temporal changes in the data structure. Since these algorithms do not provide an optimal solution to the dynamic fuzzy clustering problem, it is necessary to solve a static clustering problem at the current time instant after abrupt changes have appeared. In order to take into account the history of the temporal development of objects, a clustering algorithm is applied either to a set of most representative objects selected from the complete history or to trajectories of objects. In the latter case a clustering algorithm must be based on a dissimilarity measure for trajectories. At this point the problem of choosing an appropriate clustering algorithm, which can be used during the adaptation process for solving the dynamic clustering problem, arises and will be addressed in the next section.

Based on the above considerations the following requirements concerning the performance of a dynamic fuzzy classifier can be formulated:

1. A dynamic classifier must be able to recognise gradual changes in the cluster structure and adapt its structure to these changes,
2. A dynamic classifier must be able to recognise abrupt changes in the cluster structure and adapt its structure to these changes,
3. Re-learning of a classifier and classification must take into account the history of temporal development of objects,

4. The training data set must be preserved over time and completed with the most representative new objects,
5. The adaptation of a classifier must lead to an improvement of the fuzzy partition.

When developing a dynamic fuzzy clustering algorithm the following problems must be solved:

1. Choice of criteria for recognising abrupt changes,
2. Choice of cluster validity criteria to control the adaptation process,
3. Adaptation of the training data set based on the usefulness of objects,
4. Choice of the length of the time window or the size of the template set,
5. Similarity measure for trajectories to consider the history of temporal development of objects.

These problems will be discussed in detail in the following sections. First, the problem of choosing an appropriate clustering algorithm, which can be used for dynamic clustering will be addressed in Section 4.2 and the necessary requirements on the properties of a clustering algorithm will be presented. This will be followed by the consideration of different criteria for recognising changes in the cluster structure and the formulation of corresponding algorithms based on some of these criteria. Then, an approach for adapting a dynamic classifier and the training data set will be introduced, followed by a general representation of an algorithm for dynamic classifier design. The problem of defining a similarity measure for trajectories used as a clustering criterion within dynamic classifiers will be addressed in Chapter 5.

4.2 Requirements for a Clustering Algorithm Used for Dynamic Clustering and Classification

The method for dynamic classifier design developed in the following sections provides a general technique that can be incorporated into dynamic pattern recognition. Nevertheless, it has to be adapted to a certain structure of the classifier used. Due to the huge number of existing clustering algorithms, it is not possible to discuss all their modifications with the purpose of their dynamisation. Therefore, the consideration is restricted to prototype-based fuzzy clustering algorithms, which seem to be particularly suited to dynamic clustering. Depending on the structure and working principles of a dynamic fuzzy classifier, the following requirements for the choice of a clustering algorithm can be formulated:

1. It should be possible to apply an algorithm for possibilistic classification of objects,
2. The clustering algorithm must be able to recognise different shapes of clusters (spherical as well as elliptical clusters),

3. The clustering algorithm must be able to deal with different sizes and densities of clusters.

Although a method for dynamic classifier design can generally be combined with an arbitrary point-prototype based clustering algorithm, some of these algorithms have a number of limitations regarding their use for dynamic clustering. For instance, the fuzzy c-means (FCM) algorithm [Bezdek, 1981] can only deal with spherical clusters and will fail to recognise ellipsoidal clusters that are very likely to appear in the course of time after merging a pair of clusters. Another shortcoming of this algorithm applied to dynamic clustering is that it cannot detect whether objects with an ambiguous assignment to clusters are located in the neighbourhood of existing clusters or far away from them so that they represent a separate group. As a result the FCM algorithm has problems detecting new clusters due to its probabilistic nature. This shortcoming can be avoided by the use of the possibilistic c-means (PCM) algorithm [Krishnapuram, Keller, 1993], which generates possibilistic membership degrees of objects to clusters representing typicality of objects in the clusters. Consequently objects located far away from clusters get low degrees of membership in all clusters. With the help of a membership reject option new clusters can be detected. However, since the PCM algorithm is a modification of the FCM algorithm it also provides a partition into spherical clusters. Recognition of ellipsoidal clusters of different sizes can be achieved by changing the distance norm in the clustering algorithm. This idea is realised in the algorithm of Gustafson and Kessel [Gustafson, Kessel, 1979], which determines a particular distance function for each cluster and is able to deal with different ellipsoidal forms of clusters. In order to treat a different size of clusters, prior knowledge about the clusters is required to set values of a corresponding parameter. Although this algorithm represents an improvement compared to the FCM and PCM algorithms with respect to a larger flexibility of cluster forms, it still fails to recognise clusters with different density. This is a common drawback of all three algorithms discussed here, which is expressed by an undesirable shift of cluster centres in the direction of clusters with higher densities.

The unsupervised optimal fuzzy clustering (UOFC) algorithm of Gath and Geva introduced in [Gath, Geva, 1989] is an extension of the algorithm of Gustafson and Kessel and is suitable for the recognition of spherical and ellipsoidal clusters of different sizes and densities. The ability of this algorithm to recognise clusters of different densities is its main advantage compared to the algorithms described above (the FCM, PCM, and Gustafson-Kessel algorithms) and is very important for clustering dynamic objects. Since changes in the dynamic cluster structure have to be recognised as early as possible, it is necessary for a classifier to be able to detect small clusters with high density at the moment of their appearance and that big clusters after merging or before splitting do not influence the remaining clusters by shifting their cluster centres.

The unsupervised optimal fuzzy clustering algorithm partitions objects by a combination of the fuzzy c-means and a modified maximum-likelihood estimation (MLE) algorithm. The purpose of this combination is to take advantage of the high speed of convergence of the FCM algorithm by computing the cluster centres and the ability of the MLE algorithm to deal with unequal clusters. As a result, the UOFC algorithm can recognise spherical as well as ellipsoidal clusters with a large variability of cluster sizes and densities.

The calculation scheme of the UOFC algorithm is presented in Appendix.

Due to the high flexibility of the UOFC algorithm, it seems reasonable to apply it to dynamic pattern recognition. In the following sections the UOFC algorithm will be used in a reduced version as a basic clustering algorithm in the process of dynamic classifier design. In the UOFC algorithm the optimal number of clusters is determined by evaluating cluster validity measures for different partitions and choosing the partition with the best value of the validity measures. If this algorithm is used for dynamic clustering, Steps 3 to 6 of the algorithm seem to be unnecessary since during the dynamic classifier design an optimal partition is obtained by merging or splitting clusters or by the formation of new clusters. However, cluster validity measures are still used as an indicator for adapting a classifier to guarantee an improvement or at least the same quality of the partition over time.

4.3 Detection of New Clusters

The idea of the proposed procedure for detecting new clusters is to use a distance rejection option described in Section 3.1.2.2.

When a new object is presented to the pattern recognition system it is classified by the current fuzzy classifier designed in the previous time window. An object obtains degrees of membership to all existing clusters. For the analysis of the cluster structure, however, it may be useful to define crisp borders of clusters using α-cuts of membership functions describing fuzzy clusters. Then, one can speak about absorption of objects into clusters. A chosen value for an α-cut is used as a threshold of absorption. If a maximum degree of membership of a new object over all clusters exceeds a given threshold the object is absorbed by a cluster. This absorption criterion can be formulated as follows:

$$\text{if} \quad u_{m_1}(x) \geq u^o \quad \text{then} \quad x \in C_{m_1} \subset \{C_1,...,C_c\}$$
$$\text{otherwise} \quad x \notin \{C_1,...,C_c\} \tag{4.16}$$

where u_{ml} is the maximum degree of membership of an object **x** over all clusters and u^o is the threshold of absorption.

If the maximum degree of membership is not unique (there is more than one degree with equal maximum values), then an object is assigned arbitrarily to one of these clusters. The ambiguity of such an assignment is irrelevant by considering the membership (distance) threshold criterion. If degree u_{ml} of an object is lower than absorption threshold u^o an object is membership rejected for the absorption. Objects that cannot be absorbed by any cluster due to their low degrees of membership are called free objects.

In order to be able to detect free objects using the absorption criterion (4.16) objects must be classified by a possibilistic classification algorithm. In this case degrees of membership of an object to all clusters do not have to sum up to one, therefore they can take low values expressing that an object is atypical for all existing clusters. In the case of probabilistic clustering the degrees of membership of an object to all clusters can be at minimum 1/c, if an object is equally atypical for all clusters.

It should be noted that the absorption criterion is used as an intermediate decision rule for the detection of new clusters, but not as a final criterion for the assignment of objects to clusters for the purpose of semantic labelling of the latter. The absorption of objects is one of several procedures performed during the monitoring process in order to recognise changes in the current cluster structure. The final decision about the assignment of an object to a certain cluster can be made after the monitoring and adaptation processes are finished.

In general an object rejected for absorption in the current time window can be absorbed later on when clusters have grown due to the absorption of other objects. The absorption procedure is repeated for one time window as long as there are new objects to be absorbed.

4.3.1 Criteria for the detection of new clusters

After classification and absorption of new objects, the problem is to decide whether free objects constitute a new cluster or stray data. In order to declare new clusters within free objects the following three criteria must be satisfied.

Criteria for the detection of new clusters:

1. *Existence of free objects:* there is a number of objects which can not be assigned to any of the existing clusters due to their low degrees of membership,
2. *Sufficient number of free objects:* the number of free objects must be large enough compared to existing clusters,
3. *Compactness of free objects:* free objects must constitute compact groups.

The first criterion is examined using the absorption procedure described above, which results in a number of free objects.

In order to decide whether the number of free objects is enough to form a new cluster, the sizes of existing clusters must be considered. The size of a cluster is defined as a number of objects absorbed in this cluster. Defining a criterion for a minimum number of free objects, which is sufficient to form a new cluster, two situations can be distinguished:

1. If all existing clusters are approximately equally large, i.e. the pairwise deviations of the cluster sizes lie within a predefined limit, the number of free objects is compared to an average cluster size. A threshold for a minimum number of free objects necessary to declare a new cluster is defined as a share of average cluster size $\alpha^{cs} n^{av}$, where coefficient α^{cs} is chosen from the interval [0, 1] and the average cluster size is given by:

$$n^{av} = \frac{1}{c}\sum_{i=1}^{c} N_i^{u^o} \tag{4.17}$$

where $N_i^{u^o} = \left|\{\mathbf{x}_j \mid u_{ij} \geq u^o\}\right|$ is the number of objects $\mathbf{x}_j$ whose degrees of membership to class i exceed the absorption threshold u^o. The criterion of the desired number of free objects can be formulated as follows: if the condition

$$n^{free} \geq \alpha^{cs} n^{av} \tag{4.18}$$

is satisfied, the number of objects is sufficient to assume new clusters. Otherwise free objects are considered as stray data.

2. If cluster sizes are very different the number of free objects is compared with the minimum size of a non-empty cluster. A threshold for a minimum number of free objects is defined as a share of minimum cluster size $\alpha^{cs} n^{min}$, $\alpha^{cs} \in [0, 1]$, where the minimum cluster size is given by:

$$n^{min} = \min(N_1^{u_o},...,N_c^{u_o}) \text{ so that } N_i^{u_o} \neq \emptyset, i=1,...,c \tag{4.19}$$

The criterion of the desired number of free objects is formulated as follows: if the condition

$$n^{free} \geq \alpha^{cs} n^{min} \tag{4.20}$$

is satisfied, the number of objects is enough to assume new clusters. Otherwise free objects are considered as stray data.

As a result of examining the second criterion, the number c_t^{new} of new clusters that may possibly appear is calculated as follows:

$$c_t^{new} = \left[\frac{n^{free}}{\alpha^{cs} \cdot n^{av}}\right]_{rounded} \text{ in case of appr. equally sized clusters} \tag{4.21}$$

$$c_t^{new} = \left[\frac{n^{free}}{\alpha^{cs} \cdot n^{min}}\right]_{rounded} \text{ in case of unequally sized clusters} \tag{4.22}$$

If this number is larger than zero, than a maximum of c_t^{new} new clusters can be assumed. Stray data are considered for classification in the next time window together with new objects.

In order to examine the third criterion for forming new clusters, that is to verify whether free objects represent compact groups, the partitioning of free objects into c_t^{new} groups or clusters is proposed in order to evaluate some measure of compactness for each new cluster and to compare its values with those of existing clusters.

Free objects can be partitioned into c_t^{new} clusters using, for example, the fuzzy c-means algorithm [Bezdek, 1981, p. 65]. Since most of the measures of compactness take into account degrees of membership of objects to clusters, after clustering it seems reasonable to calculate possibilistic degrees of membership of free objects to the new cluster centres using the possibilistic c-means algorithm [Krishnapuram, Keller, 1993] in order to obtain degrees of typicality of free objects to these new clusters. One of the most obvious measures of compactness is the partition density of clusters introduced in [Gath, Geva, 1989] which will be used here with some modifications. For each cluster, the partition density is calculated as the ratio between the number of 'good' objects belonging to a cluster and its hypervolume:

$$pd(i) = \frac{n_i^{good}}{h_i}, \quad i = 1, \ldots, c_t^{new} \tag{4.23}$$

where the hypervolume of cluster i is defined as $h_i = \sqrt{\det(\mathbf{F}_i)}$ and $\mathbf{F}_i$ is the fuzzy covariance matrix of cluster i given by:

$$\mathbf{F}_i = \frac{\sum_{j=1}^{N} u_{ij} (\mathbf{x}_j - \mathbf{v}_i)(\mathbf{x}_j - \mathbf{v}_i)^T}{\sum_{j=1}^{N} u_{ij}} \tag{4.24}$$

In [Gath, Geva, 1989] 'good' objects are defined as those whose distance to the cluster centre does not exceed the standard deviation of features for this cluster and the number of objects is determined by a sum of their degrees of membership to cluster i:

$$n_i^{good} = \sum_{j=1}^{N} u_{ij} \quad \forall \mathbf{x}_j \in \left\{ \mathbf{x}_j \,\middle|\, (\mathbf{x}_j - \mathbf{v}_i)^T \mathbf{F}_i^{-1} (\mathbf{x}_j - \mathbf{v}_i) < 1 \right\} \tag{4.25}$$

In order to avoid time consuming calculations due to inversion of fuzzy covariance matrices and to provide a degree of freedom to the definition of 'good' objects, they are chosen based on their degree of membership to cluster i:

$$n_i^{good} = \sum_{j=1}^{N} u_{ij} \quad \forall \mathbf{x}_j \in \left\{ \mathbf{x}_j \,\middle|\, u_{ij} \geq \alpha^{good} \right\} \tag{4.26}$$

This means that an object is considered to be 'good' for cluster i if its degree of membership to this cluster exceeds a predefined threshold α^{good}. This threshold can either be equal to the absorption threshold or determined independently. Obviously, the higher threshold α^{good} is, the smaller the number of 'good' objects will be.

It is important to note that 'good' objects represent kernels of new clusters, which must also be the most dense regions in clusters. However, due to a number of stray objects that may be present the cluster centres may be moved away from the centres of dense regions. In order to localise dense groups within free objects and to avoid the influence of stray objects, the re-clustering of 'good' free objects with the number c_t^{new} of clusters using the fuzzy c-means algorithm is proposed. Thus, the procedure for localising dense groups includes two steps: the choice of 'good' free objects and the re-clustering of 'good' free objects. These two steps are repeated iteratively until the deviation of cluster centres calculated in two successive iterations falls below a given threshold (i.e. the cluster centres stabilise). The localisation procedure leads to a movement of new cluster centres towards the centres of dense regions. After finishing this procedure, the partition density of new clusters of 'good' free objects can be evaluated. It is to be expected that the density of new clusters detected using the localisation procedure will be higher than the density of new clusters detected by clustering free objects just once. The advantage of the localisation procedure for the detection of dense groups within free objects is illustrated in two examples below.

In order to check the criterion of compactness, or sufficient density, of new clusters, the partition density is calculated for all existing clusters and for each new cluster using equations (4.23), (4.24) and (4.26). If the partition density of a new cluster $pd^{new}(i)$, i=1, ..., c_t^{new} exceeds a predefined threshold defined as a share of the average partition density of the existing clusters, then a new cluster can be declared with a high degree of confidence. Otherwise the assumed number of new clusters decreases by one. This criterion can be formulated as follows:

$$\begin{aligned}&\text{if}\quad pd^{new}(i) \geq \alpha^{dens} \cdot pd^{av},\ i=1,\ldots,c_t^{new},\quad \text{then } c_t^{new} \text{ is unchanged}\\&\text{otherwise}\quad c_t^{new} = c_t^{new} - 1\end{aligned} \tag{4.27}$$

where coefficient α^{dens} is chosen from the interval [0, 1] and the average partition density is determined by:

$$pd^{av} = \frac{1}{c}\sum_{i=1}^{c} pd(i)\,. \tag{4.28}$$

4.3.2 Algorithm for the detection of new clusters

The three criteria proposed for the detection of new clusters within free objects represent the main steps of the monitoring procedure used to recognise new clusters in the course of time. This monitoring procedure detects groups of objects that cannot be assigned to any of existing clusters and whose size and density are comparable with those of existing clusters. Due to the number of thresholds that need to be defined by an expert, it is possible to adapt this procedure to the requirements of a concrete application. Low values of thresholds of cluster size α^{cs} and cluster density α^{dens} make the monitoring procedure very sensitive to outliers and stray objects and lead to the early recognition of new clusters. Using high values for these thresholds in the monitoring procedure allows only new clusters similar to existing ones to be detected. The described steps regarding the verification of the proposed criteria are summarised into an algorithm for detecting new clusters which is used within the monitoring procedure during dynamic classifier design.

The outcome of this algorithm is the number of new clusters and estimates of new cluster centres. The flow chart of this algorithm is presented on *Figure 22*.

The following examples illustrate the efficiency of the proposed algorithm for detecting new clusters and show its capacity for distinguishing between new clusters and stray data.

Example 1. Detection of a new cluster due to the localisation of dense groups within free objects.

Suppose that during each time window 100 new objects described by two features X_1 and X_2 are observed. For the sake of simplicity and better visualisation suppose that the length of a temporal history of objects is equal to 1 time instant, i.e. objects are represented by two-dimensional feature vectors at the current moment. After four time windows t=4 the classifier with c=2 clusters is designed and the cluster centres are determined at $v_1 = (1.95; 1.96)$ and $v_2 = (-0.06; -0.05)$. The number of absorbed objects so far and correspondingly the sizes of two clusters are equal to $N_1^{0.3} = 162$ and $N_2^{0.3} = 160$, whereas the number of stray objects is equal to 78. In time window t=5, 100 new objects are observed, which are rejected for absorption to any of the two clusters due to their low degrees of membership. Thus, the total number of free objects (including stray objects from the previous time window) is equal to $n^{free} = 178$. *Figure 23* shows the current cluster structure for t=5 in the two-dimensional feature space, where objects absorbed in one of the two clusters are represented by circles and free objects by crosses.

For the monitoring procedure for the detection of new clusters the following parameter settings are chosen:

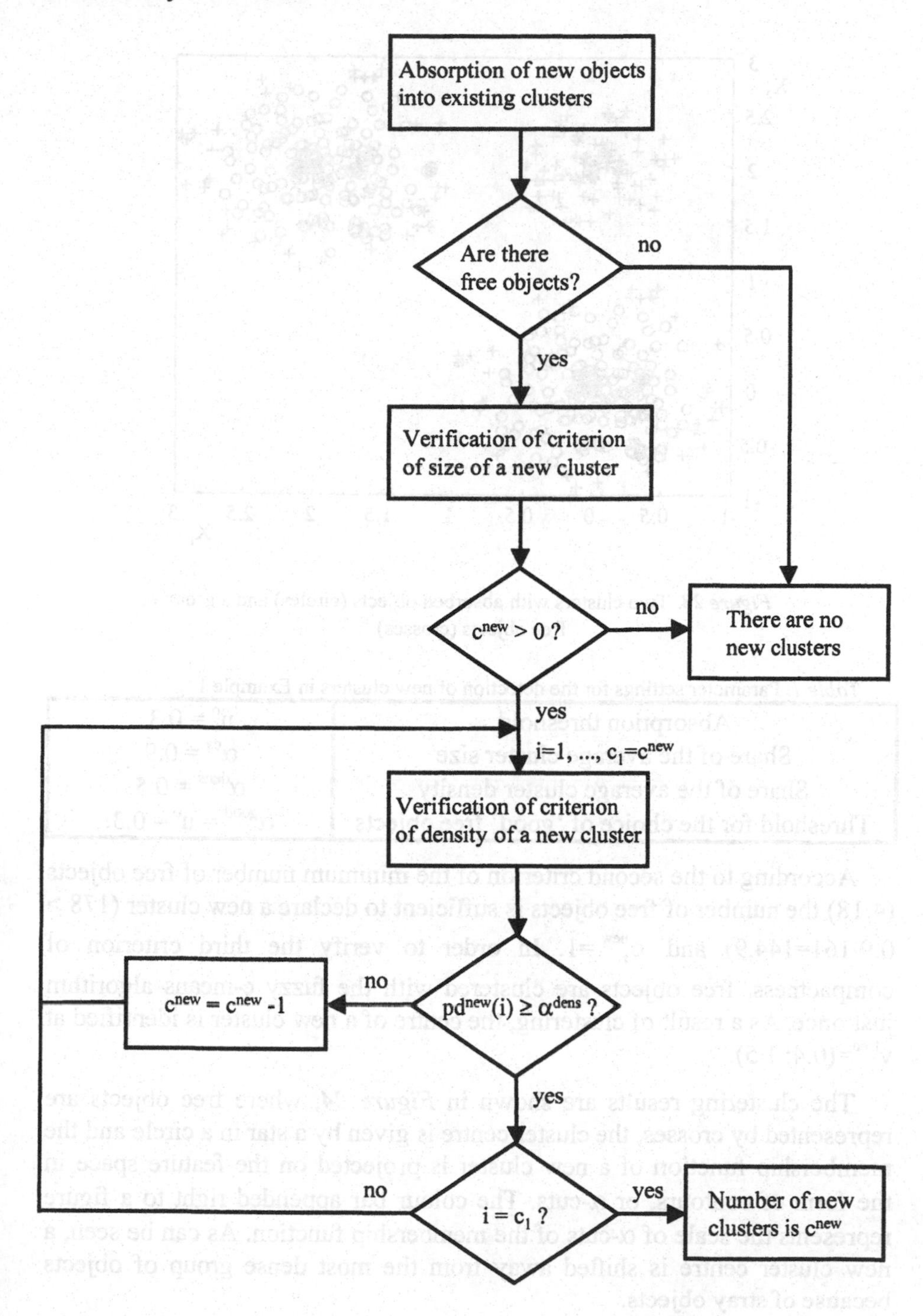

Figure 22. Flow chart of an algorithm for the detection of new clusters

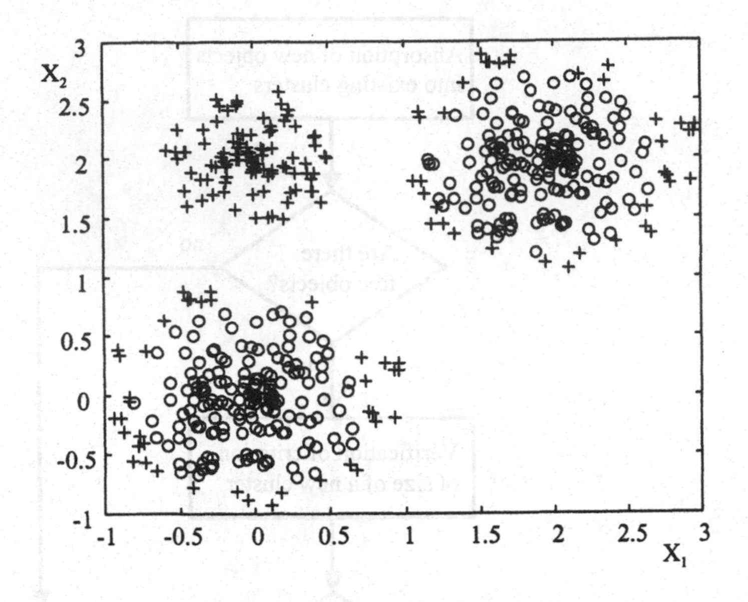

Figure 23. Two clusters with absorbed objects (circles) and a group of free objects (crosses)

Table 1. Parameter settings for the detection of new clusters in Example 1

Absorption threshold	$u^{o} = 0.3$
Share of the average cluster size	$\alpha^{cs} = 0.9$
Share of the average cluster density	$\alpha^{dens} = 0.5$
Threshold for the choice of 'good' free objects	$\alpha^{good} = u^{o} = 0.3.$

According to the second criterion of the minimum number of free objects (4.18) the number of free objects is sufficient to declare a new cluster ($178 > 0.9 \cdot 161 = 144.9$) and $c_t^{new} = 1$. In order to verify the third criterion of compactness, free objects are clustered with the fuzzy c-means algorithm just once. As a result of clustering, the centre of a new cluster is identified at $v^{free} = (0.4; 1.5)$.

The clustering results are shown in *Figure 24*, where free objects are represented by crosses, the cluster centre is given by a star in a circle and the membership function of a new cluster is projected on the feature space in the form of contours, or α-cuts. The colour bar appended right to a figure represents the scale of α-cuts of the membership function. As can be seen, a new cluster centre is shifted away from the most dense group of objects because of stray objects.

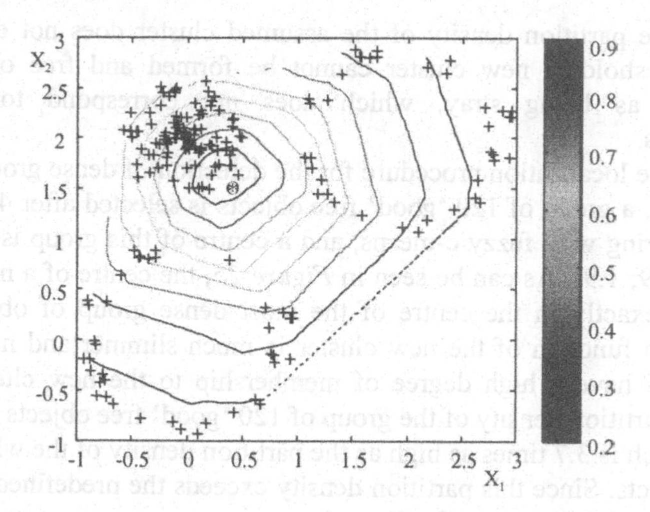

Figure 24. **Projections of the membership functions obtained for a group of free objects on the feature space after single clustering of free objects**

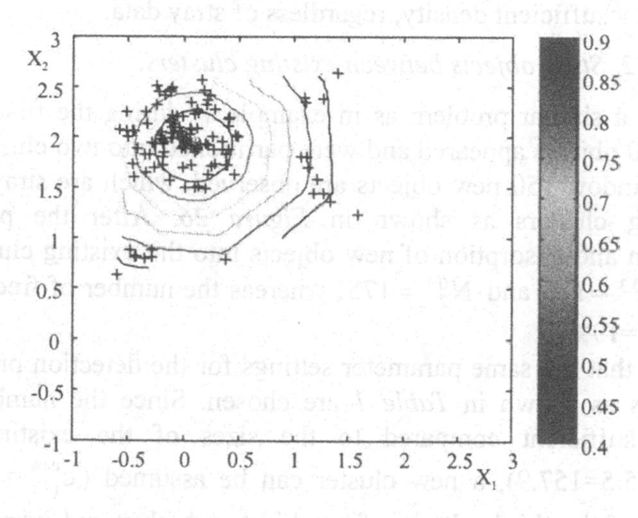

Figure 25. **Projections of the membership functions obtained for a group of free objects on the feature space after localisation of a dense group of free objects**

The evaluation of the third criterion provides the following results:

Table 2. Partition densities of new and existing clusters in Example 1

Partition densities of two existing clusters	pd(1) = 785.43; pd(2) = 796.02
Average partition density	$pd^{av} = 790.72$
Density Threshold	$\alpha^{dens} \cdot pd^{av} = 395.36$
Partition density of the group of 178 free objects	$pd^{new} = 144.67$

Since the partition density of the assumed cluster does not exceed the density threshold, a new cluster cannot be formed and free objects are considered as being stray, which does not correspond to intuitive expectations.

Using the localisation procedure for the detection of dense groups within free objects, a group of 120 'good' free objects is selected after 4 iterations of re-clustering with fuzzy c-means, and a centre of this group is identified at $v^{free}=(0.09; 1.9)$. As can be seen in *Figure 25*, the centre of a new cluster is located exactly in the centre of the most dense group of objects. The membership function of the new cluster is much slimmer and most of the free objects have a high degree of membership to the new cluster. As a result the partition density of the group of 120 'good' free objects is equal to 537.18 which is 3.7 times as high as the partition density of the whole group of free objects. Since this partition density exceeds the predefined threshold of 395.36, a new cluster can be formed.

This example shows the importance of localising dense groups of free objects in the third step of the monitoring procedure in order to detect new clusters with a sufficient density, regardless of stray data.

Example 2. Stray objects between existing clusters.

Consider a similar problem as in example 1: during the first four time windows 400 objects appeared and were partitioned into two clusters. In the fifth time window 150 new objects are observed, which are strayed among two existing clusters as shown in *Figure 26*. After the possibilistic classification and absorption of new objects into the existing clusters, their sizes are $N_1^{0.3}=176$ and $N_2^{0.3}=175$, whereas the number of free objects is equal to $n^{free}=199$.

Suppose that the same parameter settings for the detection procedure of new clusters as shown in *Table 1* are chosen. Since the number of free objects is sufficient compared to the sizes of the existing clusters $(199>0.9\cdot 175.5=157.9)$, a new cluster can be assumed ($c_t^{new}=1$). For the verification of the third criterion free objects are clustered iteratively with the fuzzy c-means algorithm and applying the localisation procedure. The localisation and clustering of a group of 110 'good' free objects converges to a new cluster centre at $v^{free}=(1.16; 1.14)$ and results in a membership function shown on *Figure 27*.

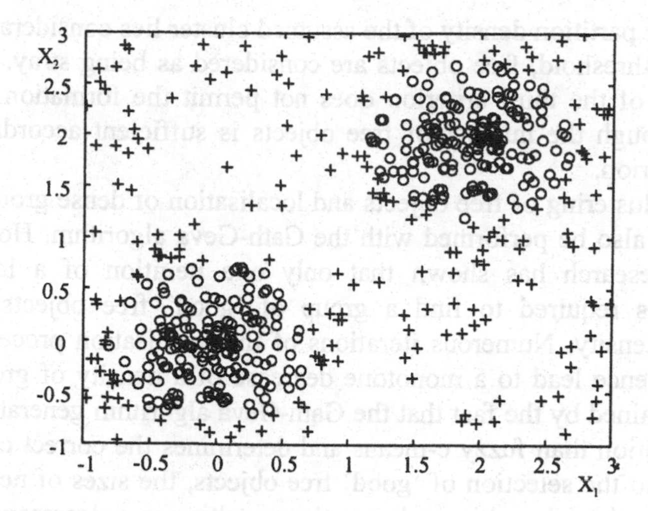

Figure 26. Free objects (crosses) are strayed between two existing clusters

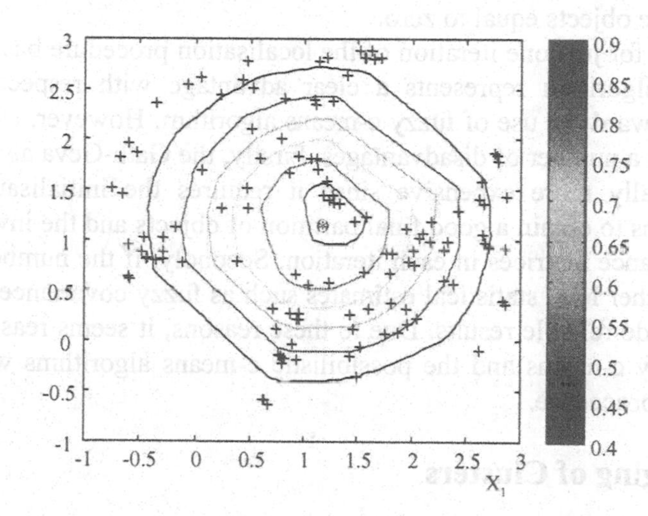

Figure 27. Detection of stray objects based on projections of 'good' free objects and the membership function obtained after clustering of free objects using the localisation procedure

The evaluation of the partition densities of the new and existing clusters provides the following results:

Table 3. Partition densities of new and existing clusters in Example 2

Partition densities of two existing clusters	pd(1) = 780.31; pd(2) = 842.14
Average partition density	$pd^{av} = 811.23$
Density Threshold	$\alpha^{dens} \cdot pd^{av} = 405.61$
Partition density of the group of 110 'good' free objects	$pd^{new} = 58.2$

Since the partition density of the assumed cluster lies considerably below the density threshold, free objects are considered as being stray. Thus, the verification of the third criterion does not permit the formation of a new cluster although the number of free objects is sufficient according to the second criterion.

The re-clustering of free objects and localisation of dense groups of free objects can also be performed with the Gath-Geva algorithm. However, an empirical research has shown that only one iteration of a localisation procedure is required to find a group of 'good' free objects with the maximum density. Numerous iterations of the localisation procedure until the convergence lead to a monotone decrease of a density of groups. This can be explained by the fact that the Gath-Geva algorithm generates a much crisper partition than fuzzy c-means and determines the correct cluster size itself. Due to the selection of 'good' free objects, the sizes of new clusters are significantly reduced in each iteration resulting in a decreasing density. In an extreme case the localisation procedure may be stopped with a number of 'good' free objects equal to zero.

The need for just one iteration of the localisation procedure based on the Gath-Geva algorithm represents a clear advantage with respect to time efficiency toward the use of fuzzy c-means algorithm. However, this can be coupled with a number of disadvantages. Firstly, the Gath-Geva algorithm is computationally more expensive since it requires the initialisation with fuzzy c-means to obtain a good final partition of objects and the inversion of fuzzy covariance matrices in each iteration. Secondly, if the number of free objects is rather low, statistical estimates such as fuzzy covariance matrices cannot provide reliable results. Due to these reasons, it seems reasonable to use the fuzzy c-means and the possibilistic c-means algorithms within the localisation procedure.

4.4 Merging of Clusters

In Section 3.1.2.2 it was stated that in order to obtain a clear partition and to avoid classification errors, objects with ambiguous degrees of membership must be rejected by a classifier for assignment to one of clusters. This rejection approach is however not suitable for dynamic pattern recognition whose purpose is to learn the correct cluster structure over time by adapting a classifier. On the contrary the ambiguity of the membership of an object to existing clusters can be used to recognise changes in the cluster structure and can indicate two different situations. If a degree of membership is equally low for two or more clusters this means that an object does not belong to any cluster. If alternatively a degree of membership of an object is equally high for two or more clusters, it means that an object belongs to several clusters. If the number of objects with high and ambiguous degrees of membership is rather large, it can be assumed that two or more clusters are overlapping being very close to each other. In this case clusters cannot be considered as heterogeneous any more and it seems reasonable to merge these clusters.

4.4.1 Criteria for merging of clusters

In order to detect close and overlapping clusters to be merged in the existing cluster structure the following criteria must be satisfied:

Criteria for merging of clusters.

1. *Existence of objects with high and ambiguous degrees of membership to clusters,*
2. *A sufficient number of objects with high and ambiguous degrees of membership to clusters OR*
3. *Closeness of cluster centres to each other.*

These criteria are examined during the monitoring procedure in each time window after classification and absorption of new objects. To improve the efficiency of the monitoring procedure, it is sufficient to consider those clusters that have absorbed new objects in the current time window as candidates for merging with other clusters. This means that overlapping clusters can appear in the current time window if at least one cluster has grown due to the absorption of new objects. In order to verify this condition, cluster sizes in the current and in the previous time windows are compared. If a cluster size has grown the cluster is included into the set of candidate clusters that will be considered for merging:

$$\text{if} \quad n_t^{cs}(i) > n_{t-1}^{cs}(i), \text{ then} \quad i \in A^{merg}, i = 1,\ldots,c \tag{4.29}$$

where $A^{merg} = \{1, \ldots, r\}$, $r \leq c$, a set of clusters to be considered as candidates for merging and $n_t^{cs}(i)$ is the number of objects in cluster i in time window t.

An exception to this rule should be made in the first time window, where there is no possibility to compare results with ones from the previous time window. The analysis of the cluster structure must involve all clusters in order to recognise the overlapping or almost empty clusters that can appear if the initial number of clusters is over-specified.

In the literature different measures for verifying criteria for merging two clusters were introduced. Most of them are based on the definition of a similarity measure for clusters, which takes into account either the degree of overlapping of clusters or the distance between cluster centres. Until now these measures were used during static classifier design to determine the optimal number of clusters as an alternative solution to using the validity measures. The general idea of the cluster merging approach is to repeat several iterations of clustering with a variable number of clusters starting with a high over-specified number (an upper bound) and reducing the number gradually until an appropriate number is found. In each iteration similar clusters are merged and the clustering of objects is performed with decreased number of clusters. This procedure is repeated until no more clusters can be merged and finished with the optimal number of clusters.

During dynamic classifier design measures for the verification of merging criteria are applied within the monitoring procedure to recognise overlapping clusters that can appear due to temporal changes in the cluster structure. If this is the case, the classifier must be adapted to a new cluster structure by merging similar clusters. Below, different formulations of criteria for merging of clusters are presented, which can be used during dynamic clustering. The time index t for cluster centres and membership functions is dropped since clusters are considered for merging at the same time instant.

Consider two fuzzy clusters C_i and C_j represented by their membership functions $u_i(\mathbf{x}_k)$ and $u_j(\mathbf{x}_k)$, $k=1, ..., N$, where N is the number of objects. In [Setnes, Kaymak, 1998] the similarity between two clusters is evaluated based on the inclusion measure, which is defined as the ratio of the cardinality of the intersection area of the two fuzzy sets to the cardinality of each fuzzy set:

$$I_{ij} = \frac{\sum_{k=1}^{N} \min(u_{ik}, u_{jk})}{\sum_{k=1}^{N} u_{ik}}. \tag{4.30}$$

where u_{ik} and u_{jk} are degrees of membership of object k to clusters i and j, respectively. This definition is illustrated in *Figure 28*, where two intersecting one-dimensional fuzzy sets with membership functions u_i and u_j are shown.

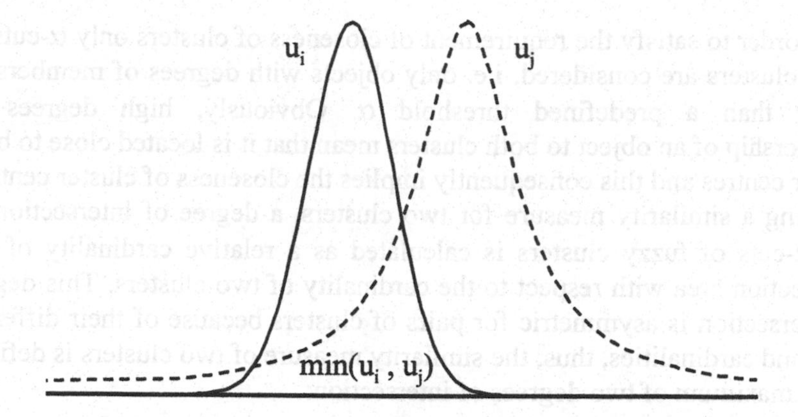

Figure 28. Merging of fuzzy clusters based on their intersection

Since fuzzy sets have generally different supports, the inclusion measure is asymmetric, and therefore the similarity measure of two clusters is defined as the maximum of two inclusion measures:

$$s_{ij} = \max\,(I_{ij}, I_{ji}). \tag{4.31}$$

According to [Setnes, Kaymak, 1998], clusters are merged if the similarity measure of two clusters s_{ij} exceeds a predefined merging threshold λ.

In general, decreasing values of λ lead to increased cluster merging and vice versa. For λ=0 all clusters will be merged, whereas for λ=1 no clusters will be merged. The choice of this threshold depends on the requirements of a concrete application and requires tuning.

The drawback of this cluster similarity measure is that it does not require that objects in the intersection area of two clusters have high degrees of membership to both clusters. That is to say, this similarity measure takes into account only Criterion 2 for cluster merging, which requires just a sufficient number of objects in the intersection area. Because of this drawback, the results of cluster merging may be sometimes unsatisfactory. For instance, it is possible that using similarity measure (4.31) two clusters are merged, although cluster centres are rather far away from each other. If clusters are large and there is a sufficient number of objects with relatively low degrees of membership to both clusters located in the intersection area of two fuzzy sets, the value of the similarity measure can become larger than a threshold leading to the merging of these clusters.

In this book the use of the similarity measure for clusters based on the degree of intersection (overlapping) of two clusters, provided that these are relatively close to each other, is proposed. This similarity measure can be considered as a modification of similarity measure (4.31).

In order to satisfy the requirement of closeness of clusters only α-cuts of fuzzy clusters are considered, i.e. only objects with degrees of membership higher than a predefined threshold α. Obviously, high degrees of membership of an object to both clusters mean that it is located close to both cluster centres and this consequently implies the closeness of cluster centres. Defining a similarity measure for two clusters, a degree of intersection of two α-cuts of fuzzy clusters is calculated as a relative cardinality of the intersection area with respect to the cardinality of two clusters. This degree of intersection is asymmetric for pairs of clusters because of their different sizes and cardinalities, thus, the similarity measure of two clusters is defined as the maximum of two degrees of intersection:

$$s_{ij} = \max\left(\frac{\text{card}(H_\alpha(u_i) \cap H_\alpha(u_j))}{\text{card}(H_\alpha(u_i))}, \frac{\text{card}(H_\alpha(u_i) \cap H_\alpha(u_j))}{\text{card}(H_\alpha(u_j))}\right), \qquad (4.32)$$

where u_i and u_j are membership functions of fuzzy clusters C_i and C_j and $H_\alpha(u) = \{\mathbf{x} \in X | u(\mathbf{x}) \geq \alpha\}$ is the α-cut of a fuzzy set with membership function $u(\mathbf{x})$, card(A) is the cardinality of a fuzzy set A defined as a sum of the degrees of membership of objects belonging to fuzzy set A.

The criterion for merging two clusters can be formulated in terms of α-cuts of fuzzy clusters as follows. Two clusters can be merged if the number of objects in the intersection area of α-cuts of these clusters is large compared to the number of objects in corresponding α-cuts of clusters (*Figure 29*). In other words, clusters can be merged if the similarity measure (4.32) of two clusters exceeds a predefined merging threshold λ: $s_{ij} \geq \lambda$.

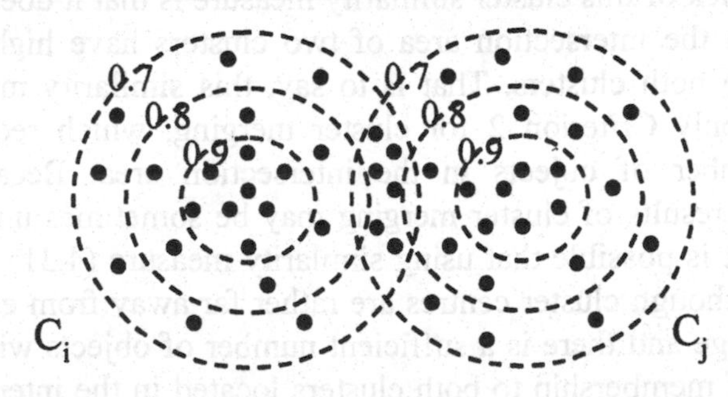

Figure 29. Merging of fuzzy clusters based on the degree of overlapping of α-cuts

Figure 29 shows the intersection of α-cuts of fuzzy clusters, where α=0.7 is chosen. Obviously, parameter α determines the strength of the merging criterion: the higher the chosen value of parameter α, the higher the degree of intersection for merging and in general the lower the number of merged clusters. The choice of parameter α depends on the goals of a concrete application, but it seems reasonable to take values $\alpha \geq 0.5$.

Another merging criterion based on the closeness of cluster centres is introduced in [Stutz, 1998]. Assuming that the clusters are spherical, an estimate for the radius of spherical clusters in the form of the standard deviation (sd) of a fuzzy cluster is defined as:

$$sd_i = \sqrt{\frac{1}{\sum_{k=1}^{N} u_{ik} - 1} \cdot \sum_{k=1}^{N} u_{ik} \left\| \mathbf{x}_k - \mathbf{v}_i \right\|^2}, \tag{4.33}$$

where $\mathbf{v}_i$ is the centre of cluster C_i and $\mathbf{x}_k$, k=1, ..., N, are considered objects.

Clusters C_i and C_j may be merged if the following criterion is fulfilled [Stutz, 1998]:

$$\left\| \mathbf{v}_i - \mathbf{v}_j \right\| < \begin{cases} k \cdot sd_i, & \text{if } sd_i \neq 0 \text{ and } sd_j = 0 \\ k \cdot sd_j, & \text{if } sd_j \neq 0 \text{ and } sd_i = 0 \\ k \cdot \overline{sd}, & \text{if } sd_i = 0 \text{ and } sd_j = 0 \\ k \cdot \min(sd_i, sd_j), & \text{otherwise} \end{cases} \tag{4.34}$$

where $k > 0$ is a coefficient for weighting a degree of closeness of two clusters and $\overline{sd}$ is the mean standard deviation of clusters.

Conditions of criterion (4.34) can be interpreted as follows:

- Cluster C_j is empty or a point located inside the cluster C_i ($k \leq 1$) or close to cluster C_i ($k > 1$); the analogues formulation is given in the second condition for cluster C_i being a point or empty, (*Figure 30*, a).
- Clusters C_i and C_j are empty or points and the distance between their centres is smaller than the weighted mean radius of clusters (distance threshold). This case can occur if the number of clusters is over-specified.
- The distance between two cluster centres is smaller than the weighted minimum of two standard deviations (*Figure 30*, b), i.e. both cluster centres are located inside the other cluster respectively ($k \leq 1$). If only one centre or even none of centres lie inside the other cluster, then two clusters can be merged, provided that the values of k chosen are considerably higher than 1.

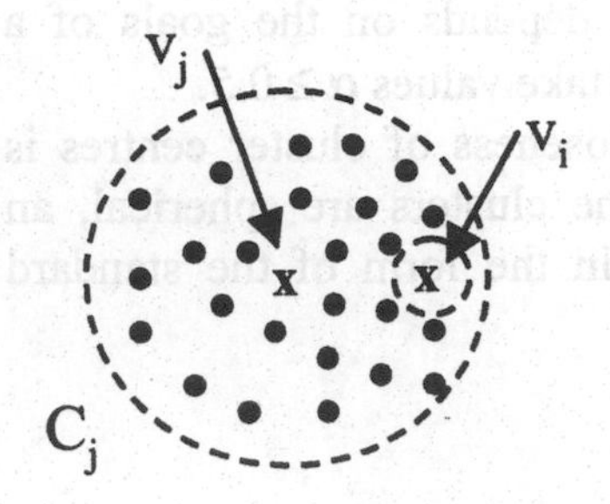

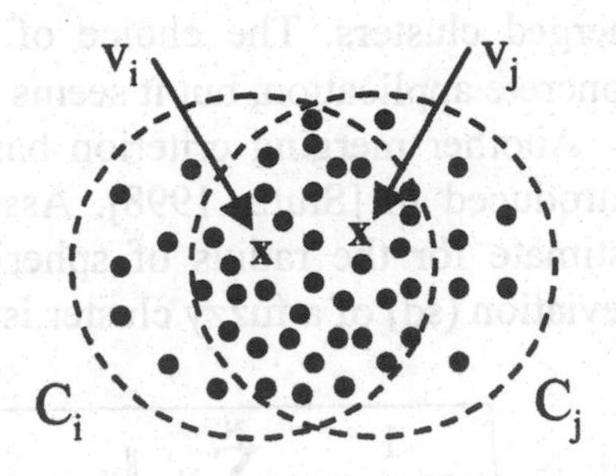

Figure 30. Merging of fuzzy clusters based on their standard deviation

a) Cluster i is empty and inside the cluster j ($k \leq 1$)

b) Both cluster centres are inside the other cluster respectively ($k \leq 1$)

In the fourth condition it may be reasonable to use the sum of two standard deviations instead of the minimum as the weaker criterion and to compare the distance between two cluster centres with $k \cdot (sd_i + sd_j)$. This condition relies on the degree of overlapping of two clusters sufficient for merging. For $k \leq 1$ clusters must overlap, for $k > 1$ they may be separate (which is not reasonable for merging). This condition is better suited as the merging criterion than the fourth condition in criterion (4.34) in particular in cases where clusters have different size.

As can be seen in *Figure 30* the criterion of closeness of cluster centres is equivalent to the degree of intersection of fuzzy clusters.

Criterion (4.34) was proposed for the use within the fuzzy c-means algorithm and therefore it is suitable only for spherical clusters of similar size. It can not be used for ellipsoidal clusters or clusters of different size leading to inconsistent or undesirable results in these situations.

4.4.2 Criteria for merging of ellipsoidal clusters

In order to define merging criteria for elliptical clusters, it is necessary to consider not only the distance between centres of clusters or their degree of intersection but also their relative orientation with respect to each other.

In [Kaymak, Babuska, 1995] the compatibility criteria for merging ellipsoidal clusters, which are used to quantify the similarity of clusters, were introduced. Consider two clusters C_i and C_j with centres at v_i and v_j. Suppose that the eigenvalues of the covariance matrices of two clusters are given by $[\lambda_{i(1)}, ..., \lambda_{i(m)}]$ and $[\lambda_{j(1)}, ..., \lambda_{j(m)}]$, whose components are arranged in descending order denoted by indexes (1), ..., (m). Let the corresponding eigenvectors be $[e_{i(1)}, ..., e_{i(m)}]$ and $[e_{j(1)}, ..., e_{j(m)}]$. The compatibility criteria for ellipsoidal clusters are defined as follows:

1. *Parallelism:* the smallest eigenvectors of two clusters must be sufficiently parallel:

$$\xi^1_{ij} = \left| e_{i(m)} \cdot e_{j(m)} \right|, \; \xi^1_{ij} \text{ close to } 1, \tag{4.35}$$

2. *Closeness:* the distance between cluster centres must be sufficiently small:

$$\xi^2_{ij} = \left\| \mathbf{v}_i - \mathbf{v}_j \right\|, \; \xi^2_{ij} \text{ close to } 0. \tag{4.36}$$

The idea of these criteria is illustrated in *Figure 31* [adapted from Kaymak, 1998, p. 244]. According to the first criterion, two clusters are compatible if they are parallel on the hyperplane. The second criterion requires that the cluster centres are sufficiently close.

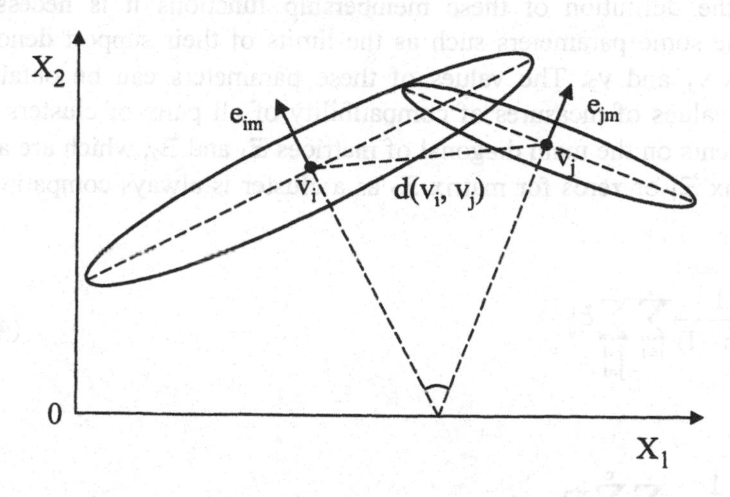

Figure 31. Illustration of criteria for merging ellipsoidal clusters

In order to obtain the overall similarity measure for clusters, measures of compatibility must be aggregated. [Kaymak, Babuska, 1995] suggest a decision-making step for aggregation, whose final goal is to determine which pairs of clusters can be merged. Possible pairs of clusters represent the decision alternatives and their number is given by c(c-1)/2 if c clusters are considered. Compatibility criteria are evaluated for each pair of clusters resulting in two matrices Ξ_1 and Ξ_2. The decision goals for each criterion are defined in the form of fuzzy sets. For criterion (4.35) of parallelism of clusters and for criterion (4.36) of closeness of clusters the fuzzy sets 'close to 1' and 'close to zero' are defined respectively. The membership functions of these fuzzy sets are determined on the intervals [0, 1] and [0, ∞) for fuzzy sets 'close to 1' and 'close to zero', respectively, as shown in *Figure 32.*

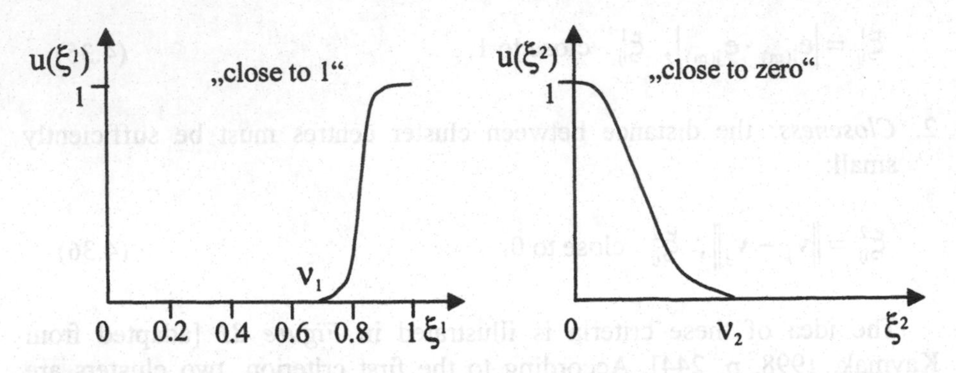

Figure 32. **Membership functions of fuzzy sets 'close to 1' and 'close to zero'**

For the definition of these membership functions it is necessary to determine some parameters such as the limits of their support denoted by variables ν_1 and ν_2. The values of these parameters can be obtained as average values of measures of compatibility of all pairs of clusters except the elements on the main diagonal of matrices Ξ_1 and Ξ_2, which are all ones for matrix Ξ_1 or zeros for matrix Ξ_2 as a cluster is always compatible with itself.

$$\nu_1 = \frac{1}{c(c-1)} \sum_{i=1}^{c} \sum_{\substack{j=1 \\ j \neq i}}^{c} \xi_{ij}^1 \qquad (4.37)$$

$$\nu_2 = \frac{1}{c(c-1)} \sum_{i=1}^{c} \sum_{\substack{j=1 \\ j \neq i}}^{c} \xi_{ij}^2 \,. \qquad (4.38)$$

The variable values of support of membership functions provide the possibility to adapt the merging criteria to specific problems.

After calculating the values of measures of compatibility ξ_{ij}^1 and ξ_{ij}^2 the degree of parallelism u_{ij}^1 and the degree of closeness u_{ij}^2 are determined for each pair C_i and C_j of clusters using the membership functions 'close to 1' and 'close to zero'. In order to obtain the overall degree of cluster compatibility, degrees u_{ij}^1 and u_{ij}^2 must be aggregated using one of the aggregation operators. It should be noted that criteria of parallelism and closeness compensate each other, i.e. clusters that are very close but not sufficiently parallel can be merged and vice versa. Therefore, [Kaymak, 1998, p. 246] proposes using the generalised averaging operator as an aggregation operator. The decision-making procedure results in the overall compatibility matrix **S**, whose elements are given by:

$$s_{ij} = \left(\frac{(u_{ij}^1)^q + (u_{ij}^2)^q}{2} \right)^{1/q}, q \in \Re \quad (4.39)$$

where s_{ij} is the compatibility, or similarity, of clusters C_i and C_j. By definition, compatibility matrix **S** is symmetric with elements equal to one on its main diagonal. The choice of parameter q in the aggregation operator influences the merging behaviour in such a way that increasing values of q lead to too much compensation and the over-merging of clusters. According to [Kaymak, 1998, p. 250], the value of q=0.5 is empirically determined as the best suited for most applications.

The criterion for merging ellipsoidal clusters based on the compatibility matrix is formulated in the same way as the aforementioned merging criteria: clusters can be merged if the degree of compatibility exceeds the pre-defined threshold λ: $s_{ij} \geq \lambda$. In general, the choice of threshold λ requires tuning depending on the requirements of a concrete application, but a value between 0.3 and 0.7 seems to be most appropriate. It should be noted that it is possible to merge more than two clusters during one merging step if there is more than one element s_{ij} exceeding threshold λ. In this case, the computational time can be considerably reduced.

Depending on the choice of the aggregation operator and the value of threshold λ the merging procedure based on the compatibility criteria may provide undesirable results. For instance, the aggregation procedure may lead to the compatibility matrix according to which two parallel clusters must be merged even though there is another cluster located between them. In order to avoid impermissible merging of clusters, an additional merging condition for compatible clusters was introduced in [Kaymak, 1998, p. 249]. The idea of this condition is to define the mutual neighbourhood as the region in the antecedent product space, which is located within a certain distance of compatible cluster centres. The merging condition must prohibit the merging of compatible clusters if there is an incompatible cluster in their mutual neighbourhood. This condition is formulated as follows:

$$\min_{v_k \notin A^{co}} \max_{v_i \in A^{co}} d_{ik} > \max_{v_i \in A^{co}} \max_{v_j \in A^{co}} d_{ij}, \quad (4.40)$$

where A^{co} is a group of compatible clusters, and

$$d_{ij} = \left|P(v_i) - P(v_j)\right| \tag{4.41}$$

with $P(\cdot)$ denoting the projection of the cluster centres into the antecedent product space.

The merging condition requires that the minimum distance between the cluster centre, which does not belong to the group of compatible clusters, and the one of the compatible clusters is larger than the maximum distance between compatible cluster centres. Compatible clusters can be merged if this condition is satisfied.

The effect of the merging condition is illustrated in *Figure 33*. Suppose that after the verification of compatibility criteria two clusters with centres v_1 and v_2 were recognised as compatible. In the case of the cluster structure shown in *Figure 33*, a, these compatible clusters cannot be merged since there is an incompatible cluster with centre v_3 between them and the merging condition (4.40) is violated: $d_{13} < d_{12}$. In contrast, compatible clusters with centres v_1 and v_2 shown on *Figure 33*, b, can be merged since the incompatible cluster lies outside the mutual neighbourhood of compatible clusters and the merging condition is satisfied: $d_{13} > d_{12}$.

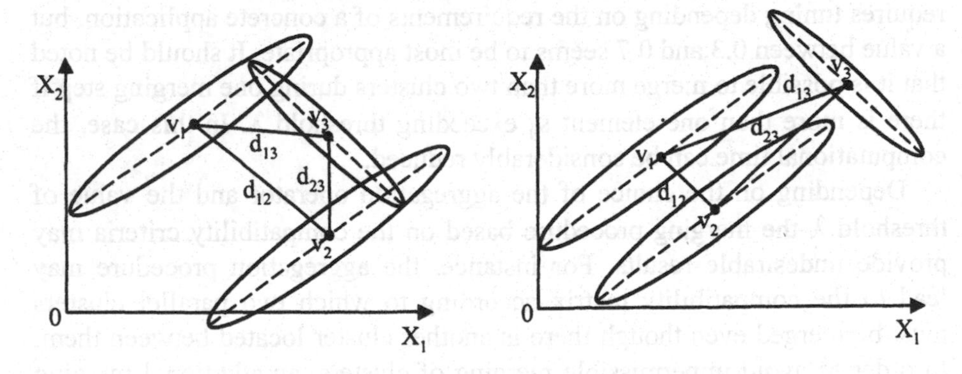

Figure 33. Application of the merging condition in order to avoid impermissible merging

a) The merging condition is violated; b) The merging condition is satisfied

The aforementioned criteria for merging of ellipsoidal clusters and the corresponding similarity measure for ellipsoidal clusters have a number of drawbacks. Value v_2 of the support of the fuzzy set 'close to zero' is defined based on relative pairwise distances between centres of clusters without taking into consideration the absolute values of distances between clusters. As a result the smaller the distance between two clusters in comparison to distances for other pairs of clusters, the higher the degree of closeness of these two clusters independently of the real value of this distance. The consequence of this definition is that for all cluster structures with a similar relation of pairwise distances between cluster centres similar degrees of closeness for pairs of clusters are obtained.

For instance, if the pairwise distances for one group of clusters are $d_{12}^1 = 2, d_{13}^1 = 10, d_{23}^1 = 12$ and the pairwise distances for another group of clusters are $d_{12}^2 = 6, d_{13}^2 = 30, d_{23}^2 = 36$, then the degrees of closeness $u(d_{12}^1)$ and $u(d_{12}^2)$ for two pairs of clusters will be approximately the same depending on the chosen membership function of the fuzzy set 'close to zero'. In the case of the triangular membership function $u(d_{12}^1) = u(d_{12}^2) = 0.5$ and in case of the exponential membership function $u(d_{12}^1) = 0.314$ and $u(d_{12}^2) = 0.316$. The degree of closeness for other pairs of clusters will be around zero since the corresponding pairwise distances are greater than support v_2 of the fuzzy set: $d_{13}^1, d_{23}^1 > v_2 = 4$ and $d_{13}^2, d_{23}^2 > v_2 = 12$.

Moreover, if only two clusters are considered for merging c=2 their degree of closeness will be always zero independently of the distance between cluster centres, since according to (4.38) the value of the support of the fuzzy set 'close to zero' will be $v_2 = 0.5 \cdot d_{12}$.

Since the compatibility criteria are considered by [Kaymak, 1998, p. 246] to be compensatory, the merging of parallel but non-intersecting clusters is possible, that is neither intuitively clear nor desirable for some applications. In general the compatibility criterion based on the closeness of clusters does not take into account the size of clusters as in criterion (4.34) neither does it verify whether the centre of a cluster is within another cluster or whether there is an intersection between clusters. It means that the criterion of closeness does not require that clusters overlap. The lack of this requirement can be considered as the reason for the possible impermissible merging of clusters. To circumvent undesirable merging results an additional merging condition is used [Kaymak, 1998, p. 249], which evaluates the mutual location of compatible and incompatible clusters.

In this section different criteria for merging spherical and elliptical clusters were considered. For the recognition of similar spherical clusters the similarity measure based on the degree of intersection of α-cuts of fuzzy clusters was proposed. For the detection of similar ellipsoidal clusters the similarity measure based on degrees of parallelism and closeness of clusters was introduced. In dynamic clustering problems it is necessary to deal with both types of clusters, which can appear in the course of time. Therefore, it seems to be advantageous to define general merging criteria independently of the cluster shape, which can be applied for recognising similar spherical as well as ellipsoidal clusters.

4.4.3 Criteria and algorithm for merging spherical and ellipsoidal clusters

It must be noted that applying the merging criterion (4.32) based on the intersection area of α-cuts to ellipsoidal clusters instead of the compatibility criterion based on the closeness of cluster centres avoids disadvantages of the latter criterion and provides an intuitively clear reason for cluster merging. The merging criterion based on the degree of intersection implies the closeness of cluster centres and automatically excludes the case of an incompatible cluster between compatible clusters. Obviously this criterion is independent of the form of clusters and relies only on degrees of membership of objects to fuzzy clusters. Therefore, the following general merging criteria for spherical and ellipsoidal clusters are proposed in this book:

Criteria for merging spherical and ellipsoidal clusters.

1. *Intersection:* there must be a sufficient number of objects in the intersection area of α-cuts of fuzzy clusters (4.32),
2. *Parallelism:* the smallest eigenvectors of two clusters must be sufficiently parallel (4.35).

In the case of spherical or almost spherical clusters the decision about merging clusters depends only on the first criterion since the second criterion is always satisfied. In the case of elliptical clusters both criteria must be satisfied in order to make a decision about merging clusters. This means that the similarity measure with respect to the intersection of clusters, as well as the degree of parallelism of clusters, must exceed predefined thresholds: $s_{ij}^{int} \geq \lambda$, $u(\xi_{ij}^{1}) \geq \eta$, where s_{ij}^{int} is the similarity measure defined by (4.32), $u(\xi_{ij}^{1})$ is the degree of parallelism obtained using the fuzzy set 'close to 1', ξ_{ij}^{1} is given by (4.35), and η is the threshold for parallelism of clusters. The value of the support of the fuzzy set 'close to zero' can be determined by setting the threshold for the maximum angle between vectors of the minimum principal components of two clusters, for which clusters cannot be considered as parallel any more, and calculating the inner product of these vectors. Values for thresholds λ and η are defined by the expert.

If both conditions are satisfied a total similarity measure s_{ij} between pairs of clusters can be calculated by aggregating the partial measures with respect to intersection and parallelism to the overall value using, for instance, (4.39) with q=0.5. Elements s_{ij} are then summarised in the similarity matrix **S**.

The problem of using the aforementioned merging criteria for different types of clusters is that the second criteria of parallelism is not always relevant. In practice there are seldom ideally spherical clusters with equal first and second principal components. It is more probable that there will be a slight difference between the principal components so that most spherical clusters can be considered as ellipses (in general, a sphere is a special case of an ellipse). Hence, when applying merging criteria it is formally necessary to consider to which degree a cluster can be assumed to have an ellipsoidal form. This problem is illustrated in *Figure 34*. If two clusters with a clear ellipsoidal form (the ratio between the first and the second principal components is much greater than 1) are overlapping but their principal components are perpendicular (*Figure 34*, a) these clusters cannot be merged due to the criterion of parallelism. In contrast, if two clusters look similar to ellipses as well as to spheres (the ratio between the first and the second principal components is close to 1) and the intersection area is sufficient with respect to cluster sizes (*Figure 34*, b), whether their principal components are perpendicular to each other or not is insignificant. These clusters can be merged due to the first criterion and the second criterion can be ignored.

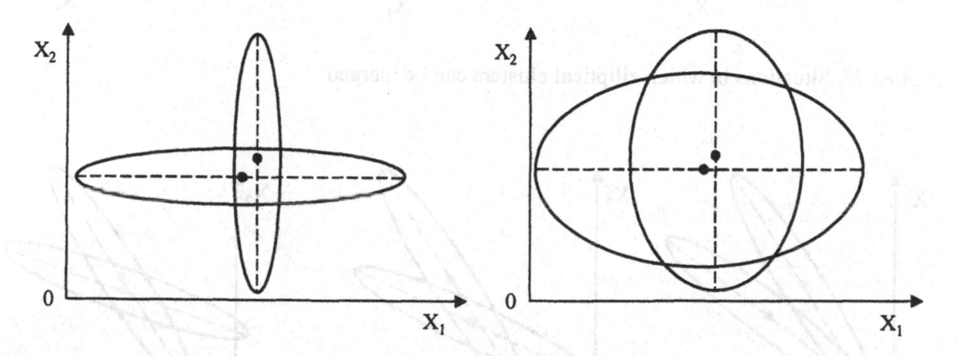

Figure 34. The relevance of criterion of parallelism for merging clusters

a) Ellipsoidal clusters with perpendicular principal components cannot be merged;

b) Ellipsoidal clusters looking similar to spheres can be merged although their principal components are perpendicular

In order to handle both situations the definition of an additional threshold λ_{max} for the similarity measure based on the intersection of clusters is proposed. If the value of this measure is sufficiently large $s_{ij}^{int} \geq \lambda_{max}$, then there is no need to verify the second criterion of parallelism. The reason for this condition is that the satisfaction of the merging criterion based on the intersection of α-cuts of fuzzy clusters implies the closeness of the cluster centres as well as the parallelism of the clusters. For long, slim elliptical clusters the degree of intersection can be large if and only if clusters are more or less parallel. If clusters are not parallel but the degree of intersection is high with respect to the clusters' sizes, it means that the clusters are not really elliptical and their parallelism does not have to be considered as a criterion for merging.

Figure 35 shows different situations in which two elliptical clusters can be merged due to a high degree of intersection caused by their closeness and parallelism. In contrast, in the examples shown in *Figure 36* the degree of intersection is not sufficient (does not exceed threshold λ_{max}) and both criteria for merging must be verified. Since these elliptical clusters are not parallel enough they are rejected for merging.

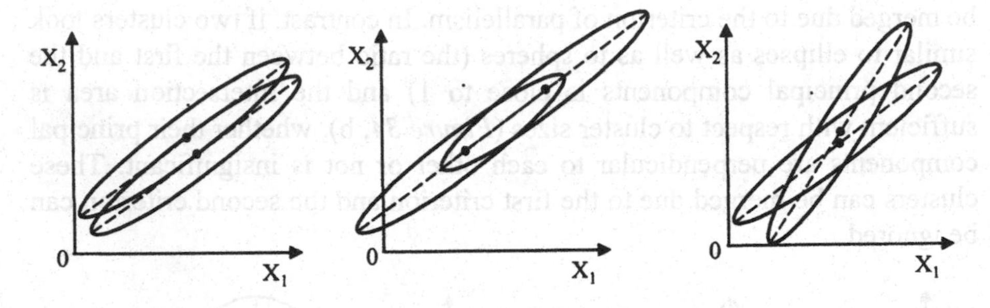

Figure 35. Situations in which elliptical clusters can be merged

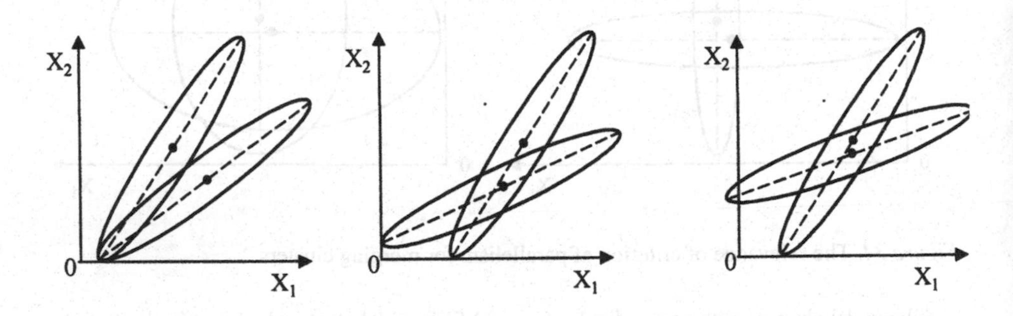

Figure 36. Situations in which elliptical clusters cannot be merged

If the decision about merging clusters is made it is necessary to estimate a new cluster centre. In [Stutz, 1998] summing up degrees of membership of objects belonging to two clusters which are merged is proposed. In order to guarantee that new degrees of membership lie in the interval [0, 1], they must be normalised using the maximum degree obtained over all objects of both clusters. Suppose that clusters C_i and C_j have to be merged to a new cluster $C_{i\cup j}$. New degrees of membership for objects belonging to cluster $C_{i\cup j}$ are calculated according to the following equation:

$$u_{i\cup j,k} = \frac{u_{ik} + u_{jk}}{\max_k (u_{i\cup j,k})} \tag{4.42}$$

where u_{ik} and u_{jk} are degrees of membership of object k to clusters C_i and C_j, respectively.

Based on new degrees of membership a new cluster centre is estimated as a weighted arithmetic mean of objects using the fuzzy c-means algorithm. The obtained cluster centre is then used as an initialisation parameter for re-learning the dynamic classifier during the adaptation procedure.

In the following an algorithm for detecting and merging similar clusters based on the proposed general merging criteria is summarised.

Algorithm 4: Detection of similar clusters to be merged.

1. *Set $c^{merg}=0$. Find clusters that have absorbed a lot of new data since the previous time window by verifying condition (4.29) for each cluster and determine set A^{merg} of candidate clusters for merging.*
2. *Calculate a similarity measure for each pair of clusters in set A^{merg} using merging criterion (4.32) based on the degree of intersection.*
3. *Choose a pair of clusters C_i and C_j with the maximum value of the similarity measure.*
4. *Check whether the maximum value of the similarity measure based on the intersection exceeds a predefined threshold λ_{max}. If this is so, proceed with Step 6.*
5. *Check whether the similarity measure with respect to the intersection of clusters given by (4.32) and the degree of parallelism of clusters given by (4.35) exceed predefined thresholds $s_{ij}^{int} \geq \lambda, u(\xi_{ij}^{|}) \geq \eta$. If at least one condition is violated, clusters cannot be merged; proceed with Step 8.*
6. *Clusters C_i and C_j can be merged. The number of cluster pairs to be merged is increased by one: $c^{merg}=c^{merg}+1$.*
7. *Estimate a new cluster centre for the initialisation of the re-learning procedure of the classifier.*
8. *If there are more pairs in set A^{merg} to be examined proceed with Step 3 of the algorithm.*
9. *Steps 3-7 are repeated until no more merging is possible.*

Using this algorithm, clusters are merged pairwise and iteratively. The outcome of the algorithm is the number of pairs to be merged and estimates of new cluster centres. The flow chart of the algorithm is presented in *Figure 37*. The presented algorithm for detecting and merging similar clusters is integrated into the monitoring procedure used for the dynamic fuzzy classifier design.

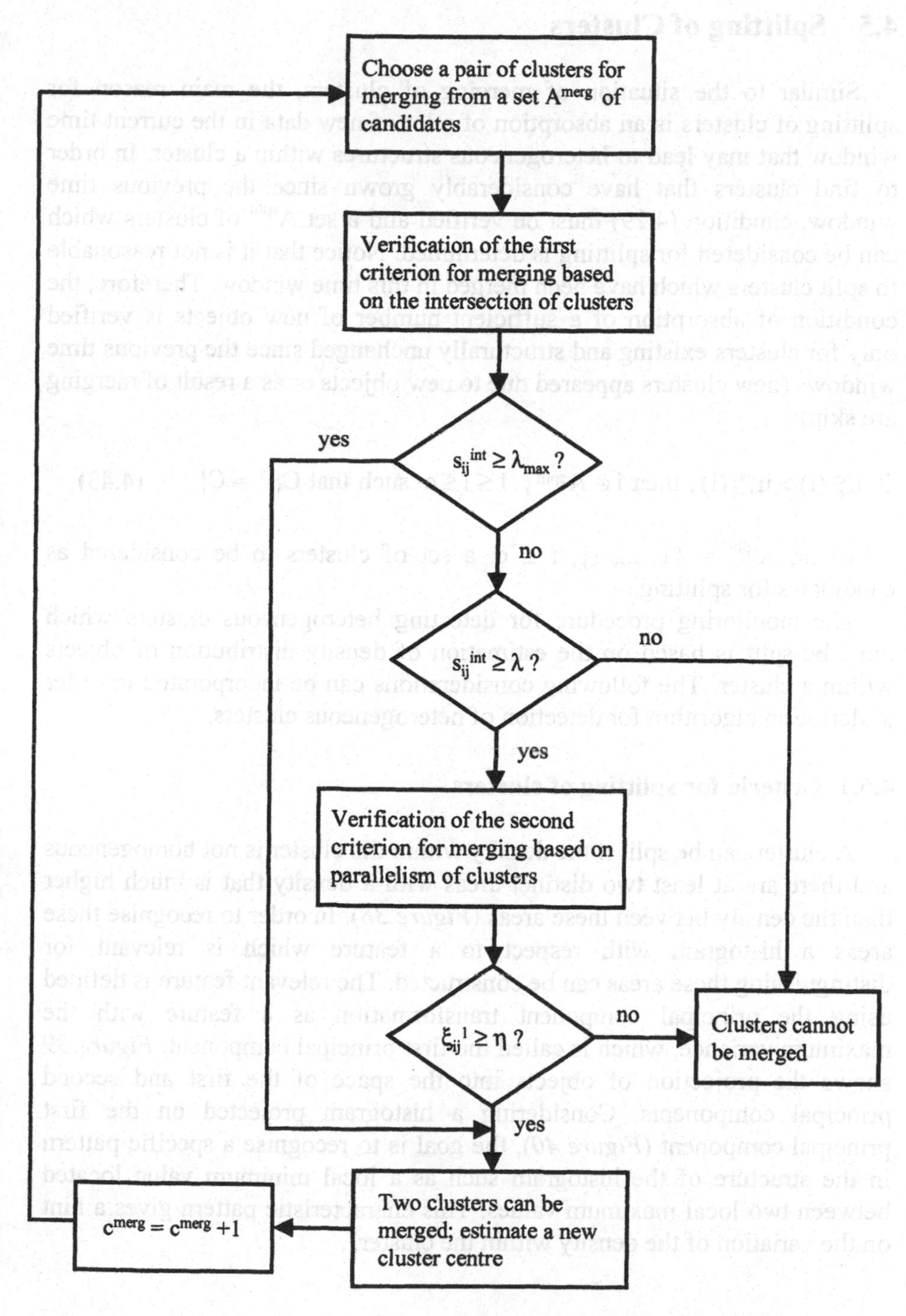

Figure 37. A flow chart of an algorithm for detecting similar clusters to be merged

4.5 Splitting of Clusters

Similar to the situation of merging of clusters, the main reason for splitting of clusters is an absorption of a lot of new data in the current time window that may lead to heterogeneous structures within a cluster. In order to find clusters that have considerably grown since the previous time window, condition (4.29) must be verified and a set A^{split} of clusters which can be considered for splitting is determined. Notice that it is not reasonable to split clusters which have been merged in this time window. Therefore, the condition of absorption of a sufficient number of new objects is verified only for clusters existing and structurally unchanged since the previous time windows (new clusters appeared due to new objects or as a result of merging are skip):

$$\text{if } n_t^{cs}(i) > n_{t-1}^{cs}(i), \text{ then } i \in A^{split}, \ 1 \le i \le c \ \text{ such that } C_i^{t-1} \approx C_i^t \qquad (4.43)$$

where $A^{split} = \{1, ..., r\}$, $r \le c$, a set of clusters to be considered as candidates for splitting.

The monitoring procedure for detecting heterogeneous clusters which must be split is based on the estimation of density distribution of objects within a cluster. The following considerations can be incorporated in order to derive an algorithm for detection of heterogeneous clusters.

4.5.1 Criteria for splitting of clusters

A cluster can be split if the density within the cluster is not homogeneous and there are at least two distinct areas with a density that is much higher than the density between these areas (*Figure 38*). In order to recognise these areas a histogram with respect to a feature which is relevant for distinguishing these areas can be constructed. The relevant feature is defined using the principal component transformation as a feature with the maximum variance, which is called the first principal component. *Figure 39* shows the projection of objects into the space of the first and second principal components. Considering a histogram projected on the first principal component (*Figure 40*), the goal is to recognise a specific pattern in the structure of the histogram such as a local minimum value located between two local maximum values. This characteristic pattern gives a hint on the variation of the density within the cluster.

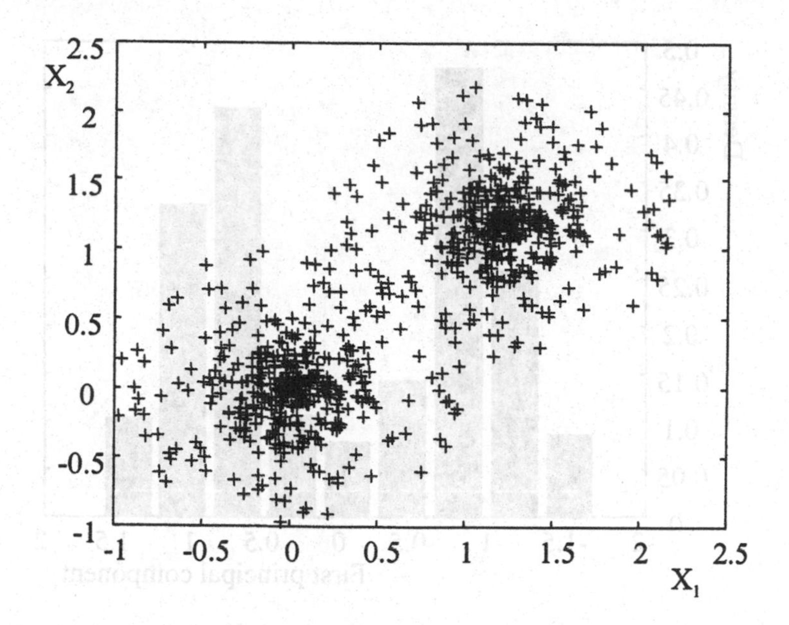

Figure 38. A heterogeneous cluster in the two-dimensional feature space

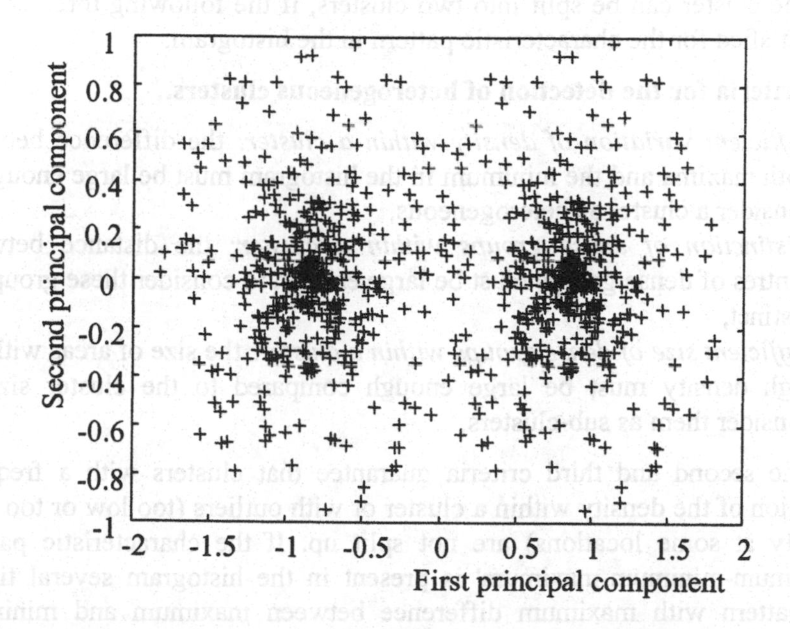

Figure 39. A heterogeneous cluster in the space of the two first principal components

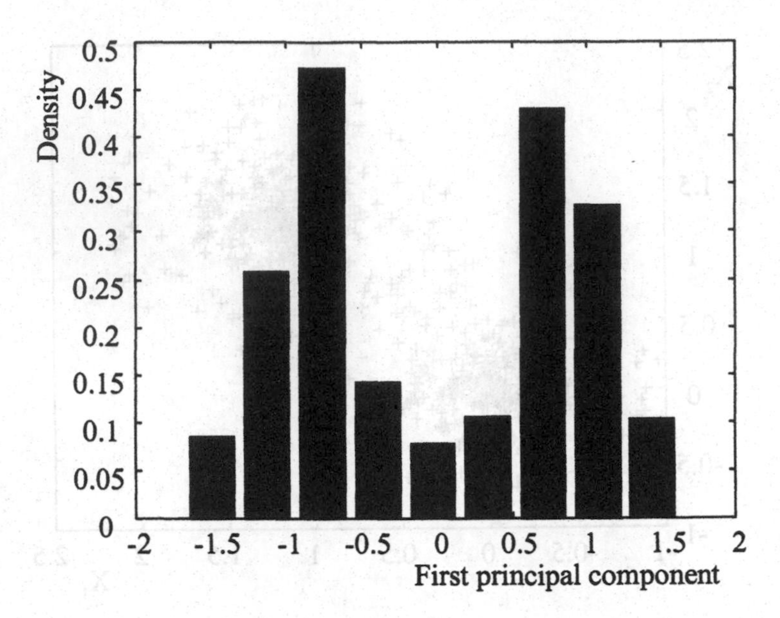

Figure 40. Histogram of objects' density with respect to the first principal component

The cluster can be split into two clusters, if the following three criteria are satisfied for the characteristic pattern in the histogram:

Criteria for the detection of heterogeneous clusters.

1. *Sufficient variation of density within a cluster:* the difference between both maxima and the minimum in the histogram must be large enough to consider a cluster as heterogeneous,
2. *Distinction of dense groups within a cluster:* the distance between centres of dense groups must be large enough to consider these groups as distinct,
3. *Sufficient size of dense groups within a cluster:* the size of areas with the high density must be large enough compared to the cluster size to consider them as sub-clusters.

The second and third criteria guarantee that clusters with a frequent variation of the density within a cluster or with outliers (too low or too high density at some locations) are not split up. If the characteristic pattern 'maximum-minimum-maximum' is present in the histogram several times, the pattern with maximum difference between maximum and minimum values of the histogram is chosen for the analysis. Splitting into more than two clusters can be performed iteratively considering new appearing clusters for the splitting test. Notice that it is convenient for the analysis to construct a histogram with a relatively small number of bars (from 8 to 12) so that small variations of the density do not have big influence on the histogram's shape.

The proposed criteria for detection of heterogeneous clusters can be applied to spherical as well as elliptical clusters, since they consider objects in the space of principal components.

In order to verify these three criteria for splitting, a number of thresholds must be set. Let r^{dens} be a coefficient expressing how many times the density of objects must differ within a cluster to be sufficient for splitting a cluster. The higher the value of this coefficient, the stronger is the criterion of density and the smaller a number of clusters which are split. The threshold r^{reld} for a relative difference between the maximum h_{max} and the minimum value h_{min} of the histogram can be derived using the relation $h^{max} = r^{dens} \cdot h^{min}$ in the following way:

$$r^{reld} = \frac{h^{max} - h^{min}}{h^{max}} = \frac{r^{dens} - 1}{r^{dens}}. \tag{4.44}$$

In order to compare the size of dense groups of objects with the cluster size, consider a diameter of dense groups or a cluster defined as a value range of the feature with the maximum variance (the first principal component). It means the diameter of a cluster corresponds to the whole value range of the histogram shown on *Figure 40* and the diameter of a dense group of objects is given by the value range of some bars of the histogram. Let r^{diam} be a size threshold expressing a minimum diameter of dense groups of objects as a share of the diameter of a cluster which is sufficient to consider these groups as separate sub-clusters within a single cluster (*Figure 41*). This threshold must be defined in the interval [0, 0.5] to be able to recognise at least two new groups within a cluster. The smaller the value of r^{diam}, the weaker the second criterion for splitting and the smaller the size of dense groups which can be detected by the splitting procedure.

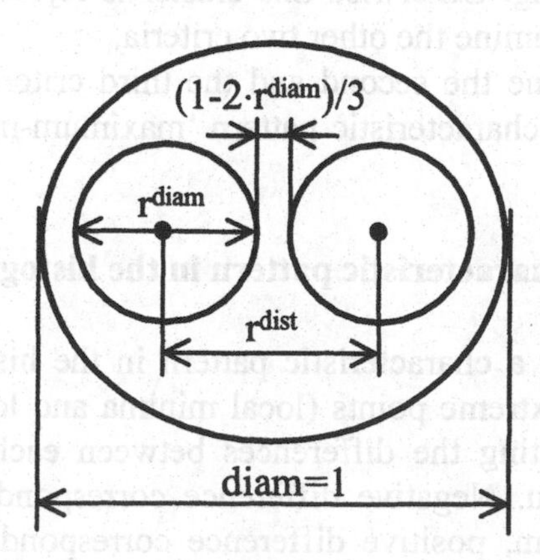

Figure 41. Illustration of thresholds of size and distance between centres of density areas

If the value of the size threshold is chosen smaller than 0.5 and the number of density groups to be detected is assumed to be equal to two, then there can be a distance equal to $(1 - 2 \cdot r^{diam}) / 3$ between dense groups and between these groups and the borders of the cluster (i.e. there can be areas of low density along the border of the cluster as it is illustrated on *Figure 41*). The third criterion requires a sufficient distance between dense groups of objects so that they can be clearly distinguished and splitting in case of random variation of density can be prevented. Denote the threshold for a sufficient distance between dense groups by r^{dist} expressing the ratio of the distance between centres of dense groups to the size of a cluster. The value of this threshold can be obtained depending on the assumed number of dense groups within a cluster and using the size threshold for these dense groups in the following way:

$$r^{dist} = r^{diam} + \frac{1 - 2 \cdot r^{diam}}{3} = \frac{r^{diam} + 1}{3}. \tag{4.45}$$

The criteria for splitting are examined for each cluster from the set A^{split} based on the histogram of the density of objects belonging to a cluster projected on the first principal component. For the verification of the first criterion the maximum and the minimum values of the histogram are evaluated and the relative difference of densities of objects within a cluster is calculated as:

$$\Delta dens = \frac{h^{max} - h^{min}}{h^{max}}. \tag{4.46}$$

If $\Delta dens \geq r^{reld}$, the first criterion is satisfied and a cluster can be considered for splitting. Otherwise this cluster is rejected for splitting and there is no need to examine the other two criteria.

In order to examine the second and the third criteria for splitting it is necessary to find the characteristic pattern 'maximum-minimum-maximum' in the histogram.

4.5.2 Search for a characteristic pattern in the histogram

When looking for a characteristic pattern in the histogram, one has to find the number of extreme points (local minima and local maxima) in the histogram by calculating the differences between each two neighbouring bars of the histogram. Negative difference corresponds to the increasing trend in the histogram, positive difference corresponds to the decreasing trend. The extreme points can be detected by considering changes of the sign of this difference. A change from minus to plus corresponds to a local maximum and a change from plus to minus corresponds to a local minimum in the histogram. The number of changes of the difference sign provides the total number of extreme points in the histogram. If the first and/or the last extreme point is the minimum, it seems reasonable to add to the list of extreme points the first and/or the last points of the histogram as additional maxima. These boarder points of the histogram can also be centres of dense regions in a cluster satisfying the criteria for splitting (*Figure 43*, b). Therefore they must be considered along with the extreme points. Depending on the number of extreme points, three cases can be distinguished:

1. If there is only one extreme point, it means that there is only one density area within the objects (if it is a maximum) or the objects are concentrated on the boarder (if it is a minimum). Thus, the cluster does not need to be split up.
2. If there is exactly three extreme points and their order corresponds to 'maximum-minimum-maximum' (*Figure 43*, a), then they are labelled as P_1, P_2 and P_3 and the criteria for splitting must be verified. If the order of extreme points corresponds to 'minimum-maximum-minimum', then there is only one density area within the objects, thus a cluster does not need to be split up.
3. If there are more than three extreme points, then the search for a characteristic pattern in the histogram is continued until the pattern 'maximum-minimum-maximum' satisfying all three criteria is recognised or there is no patterns any more to be investigated. The idea of the proposed search procedure is to narrow iteratively the search interval by the extreme points of a detected pattern. For instance, consider a histogram shown in *Figure 42*. The search interval is limited by the most last points of the histogram included to the list of extreme points as maxima. Suppose that the first pattern characterised by the maximum density difference and detected by the search algorithm is P_1-P_2-P_3. Suppose that the thresholds for splitting are chosen so that the second criterion of distinction of dense groups and the third criterion of cluster size are not satisfied for this pattern. The search is continued left from the pattern and the search interval is limited by points P_3 and P_4. The next pattern found by the search procedure is P_4-P_5-P_6, for which the criteria for splitting are not satisfied as well. The search interval is narrowed by the point P_6 and the search is continued on the interval between P_3 and P_6. The third pattern detected by the procedure is P_3-P_7-P_6, for which all three criteria for splitting are satisfied. It must be noted that the search procedure looks for the maximum density difference in the interval and therefore as soon as points P_3 and P_6 are detected as two maxima the deepest local minimum between these points is looked for.

As a result point P_7 can be recognised despite of the frequent small deviation of the density between points P_3 and P_6. Thus, in each iteration if the characteristic pattern is detected, the search interval is divided into two ones restricted by the extreme points of the detected pattern and the extreme points of patterns detected in previous iterations, and the search is carried out on each of the intervals. The search procedure is illustrated on *Figure 42*.

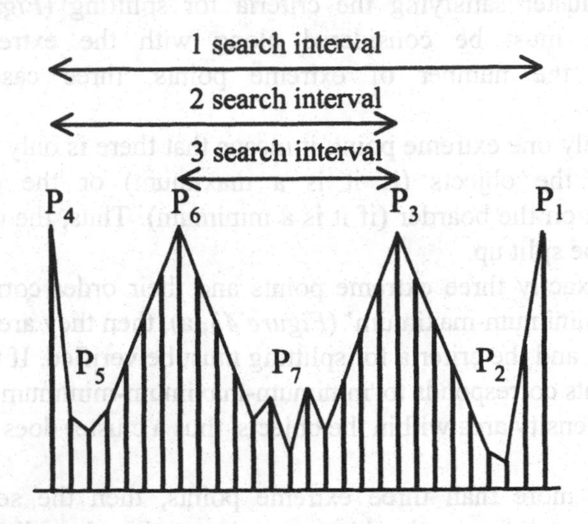

Figure 42. Illustration of the search procedure for a density histogram

Possible patterns of a histogram are shown on *Figure 43*. The ideal characteristic pattern for splitting 'maximum-minimum-maximum' for which all criteria for splitting are satisfied is illustrated in *Figure 43*, a. *Figure 43*, b shows that including the last points of the histogram to the set of extreme points a pattern for splitting can be recognised. *Figure 43*, c and d illustrate cases where the correct pattern of splitting can be detected by the search algorithm despite of the high boundary deviation (*Figure 43*, c) and small variation (*Figure 43*, d) of the density. Notice that the zigzag pattern between the two maxima cannot prohibit the procedure to detect the correct pattern for splitting as long as the difference of the density in the zigzag pattern does not exceed corresponding values at maximum points.

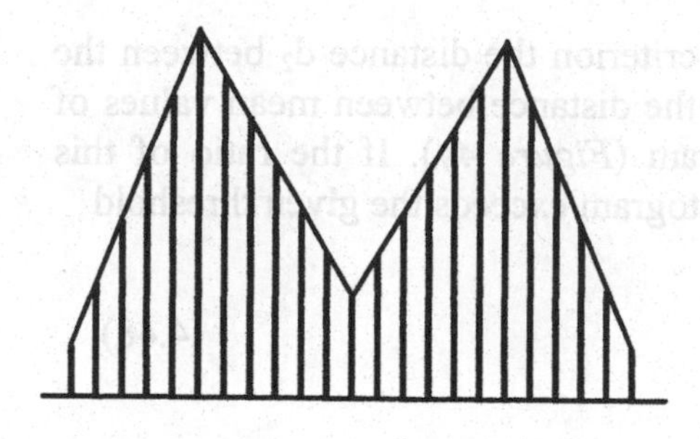

a) ideal pattern 'maximum- minimum-maximum',

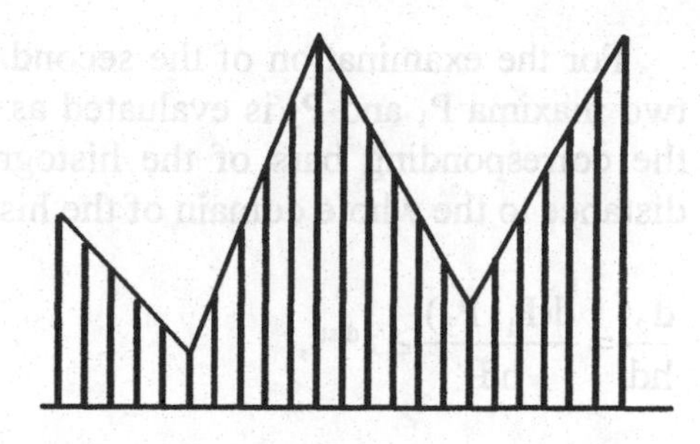

b) pattern containing the most left and right points of a histogram as additional extreme points,

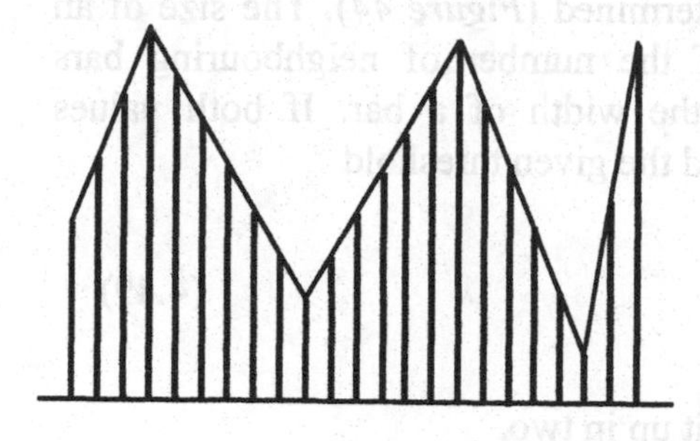

c) pattern containing a peak of density at the last point of a histogram,

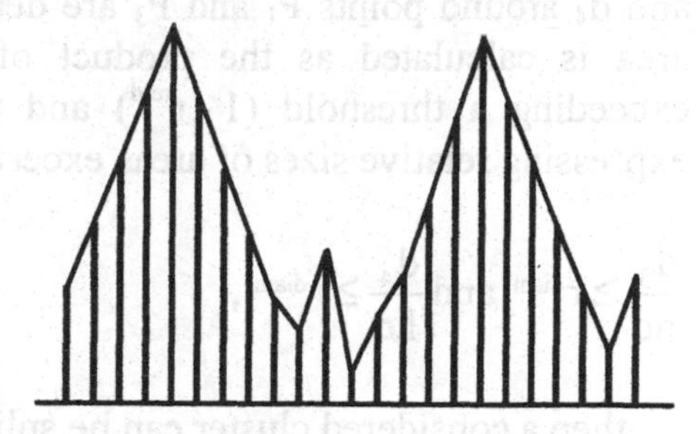

d) pattern containing the variation of the density between two maxima

Figure 43. Examples of patterns in the histogram

4.5.3 Algorithm for the detection of heterogeneous clusters to be split

If a characteristic pattern is detected in the histogram, the criteria for splitting are verified. In order to examine the first criterion of density the difference d_1 between two extreme points corresponding to the second largest value and the minimum value is calculated (*Figure 44*). The criterion can be formulated as follows: if the relative difference of the density exceeds the given threshold

$$\Delta \mathrm{dens} = \frac{d_1}{h^{\max 2}} \geq r^{\mathrm{reld}}, \tag{4.47}$$

then the criterion of density is satisfied for a cluster and the next criterion can be verified. Otherwise the cluster is rejected for splitting.

For the examination of the second criterion the distance d_2 between the two maxima P_1 and P_3 is evaluated as the distance between mean values of the corresponding bars of the histogram (*Figure 44*). If the ratio of this distance to the whole domain of the histogram exceeds the given threshold

$$\frac{d_2}{hd} = \frac{d(P_1, P_3)}{hd} \geq r^{dist}, \tag{4.48}$$

then the criterion of distinction of dense groups is satisfied for a cluster and the next criterion can be examined. Otherwise the cluster is rejected for splitting.

When verifying the last criterion, the sizes of areas with high density d_3 and d_4 around points P_1 and P_3 are determined (*Figure 44*). The size of an area is calculated as the product of the number of neighbouring bars exceeding a threshold $(1 - r^{reld})$ and the width of a bar. If both values expressing relative sizes of areas exceed the given threshold

$$\frac{d_3}{hd} \geq r^{diam} \text{ and } \frac{d_4}{hd} \geq r^{diam}, \tag{4.49}$$

then a considered cluster can be split up in two.

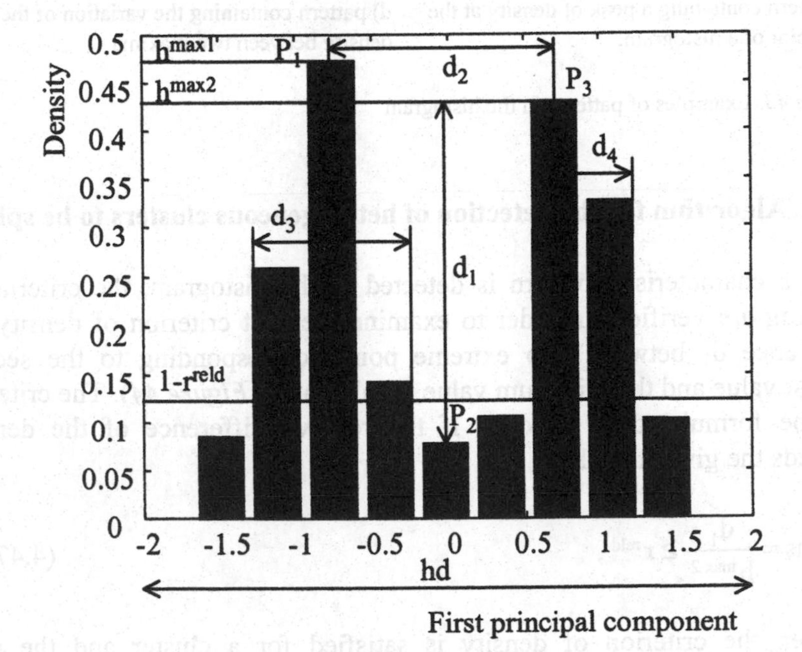

Figure 44. Verification of criteria for splitting a cluster based on the density histogram

If the number of bars in the histogram is chosen relative large compared to the number of objects, the histogram may contain a lot of patterns with small variations of the density. In order to simplify the search for a characteristic pattern and to avoid the risk to stick in the pattern which is not the best with respect to criteria for splitting, it is proposed to smooth the histogram in the same way as it is usually done for time series.

Consider a histogram with r bars $h_1, \ldots, h_p$ such that a centre of bar h_i is located at x_i, $i=1, \ldots, p$. Using the method of a moving average the bar values of a new histogram $\overline{h}$ are calculated as mean values over r neighbouring bar values in the following way:

if r is an uneven number, (r-1)/2 left and (r-1)/2 right neighbouring values are considered

$$\overline{h}_i = \frac{1}{r} \sum_{j=i-(r-1)/2}^{i+(r-1)/2} h_j \ , \qquad \text{where} \quad i = 1 + \frac{r-1}{2}, \ldots, p - \frac{r-1}{2}, \tag{4.50}$$

if r is an even number, r/2 left and r/2 right neighbouring values are considered

$$\overline{h}_i = \frac{1}{r} \sum_{j=i-r/2}^{i+r/2} h_j \ , \qquad \text{where} \quad i = 1 + \frac{r}{2}, \ldots, p - \frac{r}{2}. \tag{4.51}$$

The example in *Figure 45* illustrates the effect of smoothing of histograms. The original histogram (above) consists of 100 bars and contains a significant part of density fluctuations. The first characteristic pattern found by the search procedure P_1-P_3-P_2 and corresponding to the bars at points x=1, 59, 100 does not satisfy criteria for splitting (the third criterion of size is violated). The second characteristic pattern detected in the histogram P_4-P_3-P_5 corresponds to the bars at points x=35, 59, 85 and satisfies all criteria for splitting. Applying a moving average of length 5, a smooth histogram (below) filtered from the most part of fluctuations is obtained so that a clear characteristic pattern P_1-P_3-P_2 (x=36, 58, 84) can be found in the first run of the search procedure. In general the number of patterns which must be examined by the algorithm for the original histogram (without smoothing) can be considerably higher. Thus, applying a method for smoothing a histogram before starting the search procedure allows to find the characteristic pattern much faster. The value of r should be chosen depending on the number of bars in the histogram: the larger the number of bars, the higher the value of r. For the number of bars below 100 it is recommended to choose the value of r between 3 and 12.

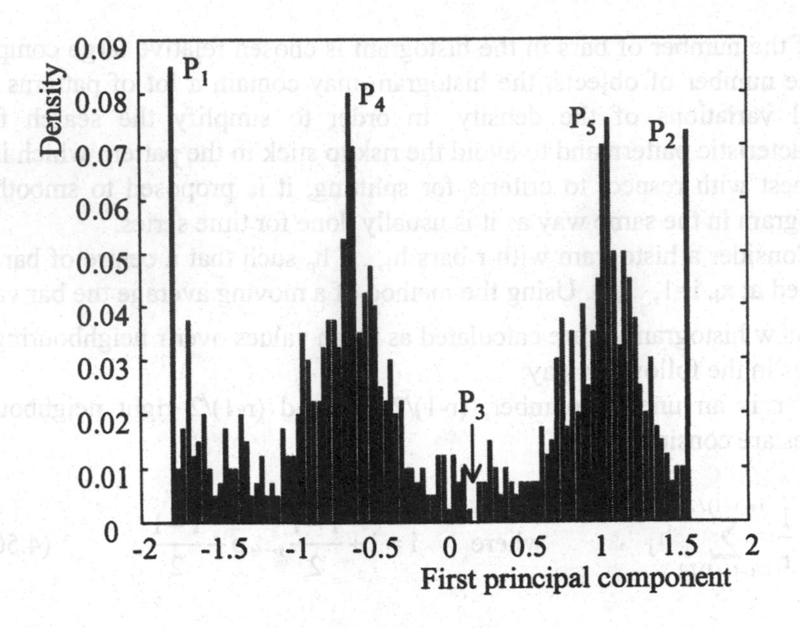

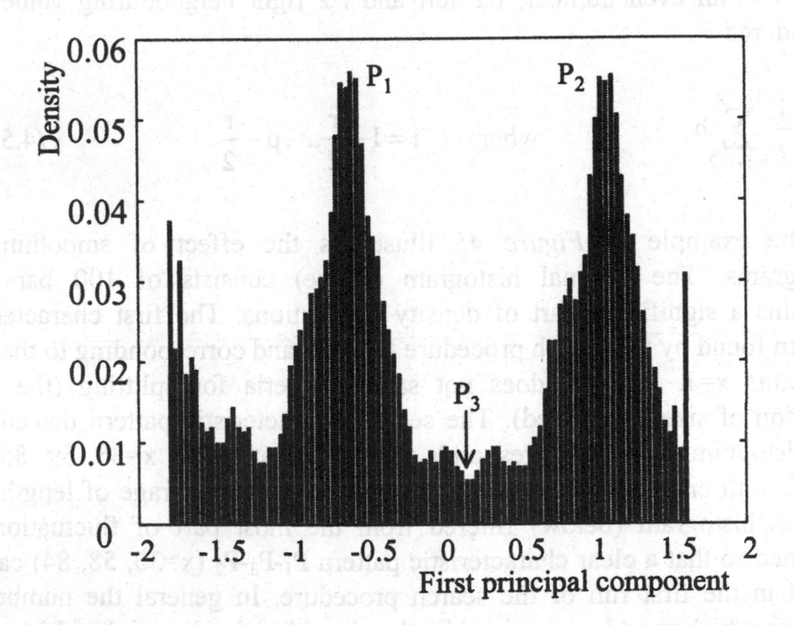

Figure 45. The effect of smoothing the density histogram

After the decision about splitting the cluster is made, it is necessary to estimate centres of new detected clusters. The easiest way is to apply the fuzzy c-means algorithm or the algorithm of Gath-Geva (if elliptical clusters must be detected) to the objects of a considered cluster in order to partition them into two clusters. The obtained cluster centres can be used as initialisation parameters for re-learning the dynamic classifier during the adaptation procedure.

The proposed splitting procedure is summarised in the following algorithm.

Algorithm 5: Detection of heterogeneous clusters to be split.

1. *Set $c^{split}=0$. Find clusters that have absorbed a lot of new objects since the previous time window by verifying condition (4.43) for each cluster and determine a set A^{split} of candidate clusters for splitting.*
2. *Choose a cluster from the set A^{split}.*
3. *Calculate principal components to find the feature with maximum variance.*
4. *Calculate a histogram with respect to the first principal component and smooth it.*
5. *Set the threshold r^{reld} for a relative difference between the maximum and the minimum value of the histogram. Set the size threshold r^{diam} for a minimum diameter of dense groups of objects within a cluster. Calculate the threshold r^{dist} for a sufficient distance between centres of dense groups of objects using equation (4.45).*
6. *Find maximum and minimum values of the histogram and calculate their relative difference according to (4.46). If the relative difference between maximum and minimum density within objects is not sufficient, i.e. $\Delta \text{dens} < r^{reld}$, then this cluster should not be split up. Go to step 13.*
7. *Calculate the number of extreme points (local minima and local maxima) in the histogram.*

7.1 *If there is only one extreme point, then the cluster does not need to be split up. Go to step 13.*

7.2 *If there is exactly three extreme points and their order corresponds to 'maximum-minimum-maximum', then label them as P1, P2 and P3 and investigate this characteristic pattern for splitting in the next step of the algorithm. If the order of extreme points corresponds to 'minimum-maximum-minimum', then the cluster does not need to be split up; go to step 13.*

7.3 *If there are more than three extreme points, then the search for a characteristic pattern in the histogram is continued until the pattern 'maximum-minimum-maximum' satisfying all three criteria is recognised or there is no patterns any more to be investigated. If a characteristic pattern is detected the following conditions are verified.*

8. *Verify the first criterion for splitting (4.47). If the criterion is satisfied, go to the next step. Otherwise this cluster is rejected for splitting; go to step 13.*

9. *Verify the second criterion for splitting (4.48). If the criterion is satisfied, go to the next step. Otherwise this cluster is rejected for splitting; go to step 13.*

10. *Verify the third criterion for splitting (4.49). If the criterion is satisfied, go to the next step. Otherwise this cluster is rejected for splitting; go to step 13.*

11. *A cluster can be split up. The number of clusters that can appear as a result of splitting is increased by one:* $c^{split}=c^{split}+1$.

12. *Estimate two new centres in a cluster using the fuzzy c-means algorithm for the initialisation of the re-learning procedure of the classifier.*

13. *Go to step 3 of the algorithm.*

This algorithm allows to detect iteratively heterogeneous clusters which can be split up into two ones. The outcome of the algorithm is the number of clusters to be split up and estimates of new cluster centres. The flow chart of this algorithm is presented on *Figure 46*.

The presented algorithm for detection of heterogeneous clusters for splitting is integrated into the monitoring procedure used for a design of a dynamic fuzzy classifier.

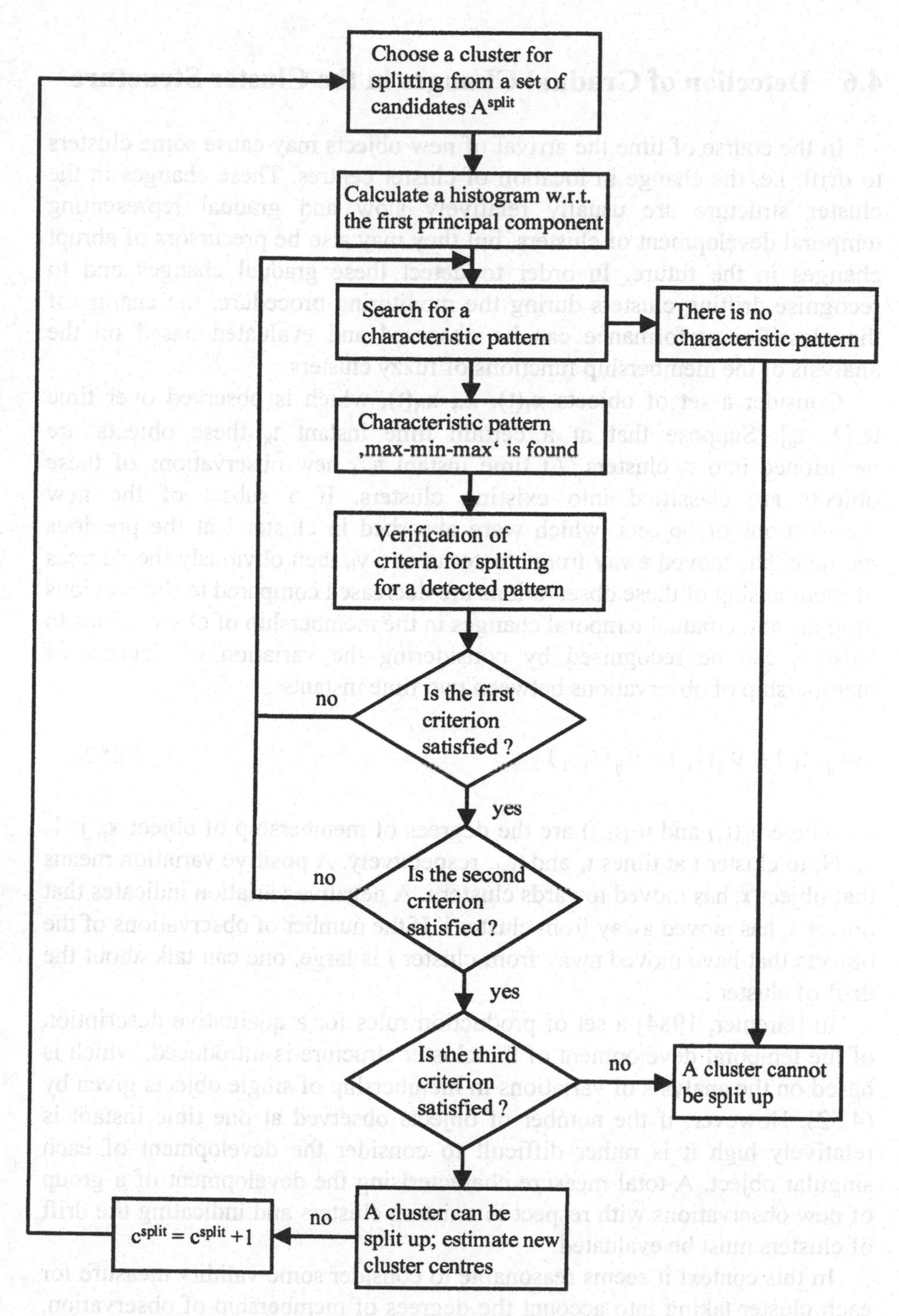

Figure 46. A flow chart of an algorithm for detecting heterogeneous clusters to be split

4.6 Detection of Gradual Changes in the Cluster Structure

In the course of time the arrival of new objects may cause some clusters to drift, i.e. the change in location of cluster centres. These changes in the cluster structure are usually relatively slow and gradual representing temporal development of clusters, but they may also be precursors of abrupt changes in the future. In order to detect these gradual changes and to recognise drifting clusters during the monitoring procedure, the change of the classifier performance can be observed and evaluated based on the analysis of the membership functions of fuzzy clusters.

Consider a set of objects $\mathbf{x}_1(t), \ldots, \mathbf{x}_N(t)$, which is observed over time $t \in [1, t_p]$. Suppose that at a certain time instant t_k these objects are partitioned into c_t clusters. At time instant t_{k+1} new observations of these objects are classified into existing clusters. If a subset of the new observations of objects, which were absorbed in cluster i at the previous moment, has moved away from cluster centre v_i, then obviously the degrees of membership of these observations are decreased compared to the previous time instant. Gradual temporal changes in the membership of observations to clusters can be recognised by considering the variation of degrees of membership of observations between two time instants:

$$\Delta u_{ij}(t_k) = u_{ij}(t_k) - u_{ij}(t_{k-1}) \tag{4.52}$$

where $u_{ij}(t_k)$ and $u_{ij}(t_{k-1})$ are the degrees of membership of object $\mathbf{x}_j$, j=1, ..., N, to cluster i at times t_k and t_{k-1}, respectively. A positive variation means that object $\mathbf{x}_j$ has moved towards cluster i. A negative variation indicates that object $\mathbf{x}_j$ has moved away from cluster i. If the number of observations of the objects that have moved away from cluster i is large, one can talk about the drift of cluster i.

In [Grenier, 1984] a set of production rules for a qualitative description of the temporal development of the cluster structure is introduced, which is based on the analysis of variations in membership of single objects given by (4.52). However, if the number of objects observed at one time instant is relatively high it is rather difficult to consider the development of each singular object. A total measure characterising the development of a group of new observations with respect to existing clusters and indicating the drift of clusters must be evaluated.

In this context it seems reasonable to consider some validity measure for each cluster taking into account the degrees of membership of observation, as well as the properties of the data structure. In [Bensaid et al., 1996] the compactness index for fuzzy clusters is proposed, which evaluates the correspondence of the data structure to the cluster structure. By contrast to the usual application of the validity measure to compare different partitions, the compactness measure is used during the monitoring procedure to compare the quality, or correctness, of the same cluster between two time instants. Therefore, for evaluating the current compactness of a cluster there is no need to consider all new observations, but only those that were absorbed in the corresponding cluster. These can be considered as good representatives of this cluster and to check whether they are well matched by the current cluster centre is intended.

The compactness of fuzzy cluster i with respect to absorbed objects is defined in the following way:

$$\pi_i^{u^o} = \frac{\sum_{j=1}^{N_i^{u^o}} (u_{ij})^2 \left\| x_j - v_i \right\|_A^2}{n_i} \qquad \forall x_j \in \left\{ x_j \middle| u_{ij} \geq u^o \right\} \tag{4.53}$$

where $N_i^{u^o}$ is the number of objects absorbed in cluster i,

$$n_i = \sum_{j=1}^{N_i^{u^o}} u_{ij}$$

is the fuzzy cardinality of cluster i taking into account only objects absorbed by this cluster, and A is an arbitrary M×M symmetric positive definite matrix.

If the compactness of a cluster is decreased compared to the previous time instant, a drift of a cluster can be assumed and the centre of this cluster must be updated based on new observations. In other words, if the variation of the compactness between two time instants t_k and t_{k-1}

$$\Delta\pi_i^{u^o}(t_k, t_{k-1}) = \pi_i^{u^o}(t_k) - \pi_i^{u^o}(t_{k-1}) \tag{4.54}$$

is negative, a cluster drift is assumed and the cluster in question is added to a set A^{gr} to be updated during the adaptation procedure.

4.7 Adaptation Procedure

In Sections 4.3, 4.4 and 4.5 three types of abrupt changes that can appear in the cluster structure in the course of time were described and the corresponding algorithms for recognising these changes by the monitoring procedure were proposed. If in time window t abrupt changes have been detected within the monitoring procedure, the problem is to adapt the classifier to the detected changes in the cluster structure. If such abrupt changes take place, the most reasonable solution is to re-learn the classifier with a new number of clusters in order to identify a new partition. Estimated centres of new clusters calculated during the monitoring procedure are used together with the existing unchanged cluster centres for the initialisation of the re-learning clustering procedure. Thus, deriving an adaptation law for a dynamic classifier the problem is to determine a new number of clusters summarising the results of the monitoring procedure.

Suppose that the number of clusters corresponding to the current classifier designed in the previous time window t-1 is given by c_{t-1}. If c_t^{new} new clusters have been detected in the time window t, the number of existing clusters must be increased by c_t^{new} before re-learning. If due to a significant growth of some clusters in the current time window t and their overlapping with other clusters c_t^{merg} pairs of similar clusters to be merged are recognised by the monitoring procedure, then the number of existing clusters must be reduced by c_t^{merg} before re-learning. If the absorption of a lot of new objects in the current time window t has lead to the formation of c_t^{split} heterogeneous clusters to be split, the number of existing clusters must be increased by c_t^{split} before re-learning the classifier. Obviously, three types of abrupt changes – formation of new clusters, similar clusters, heterogeneous clusters - can appear in seven combinations. The general adaptation law of a dynamic classifier when all abrupt changes are observed can be formulated as follows:

$$c_t = c_{t-1} + c_t^{new} - c_t^{merg} + c_t^{split} \tag{4.55}$$

where c_t is the new number of clusters that must be used to re-learn the classifier in the current time window t.

The other six formulations of the adaptation law are obtained from (4.55) if one or two of the three last components are equal to zero. The re-learning of the classifier is carried out if at least one of the three components takes a non-zero value. For the initialisation of the classifier the old centres of the unchanged clusters and the estimates of the centres of the new clusters obtained during the monitoring procedure are used.

After adapting a dynamic classifier to abrupt changes the new cluster structure is evaluated by a validity measure. If a new classifier provides a better fuzzy partitioning it is accepted, otherwise the previous classifier is preserved. In Section 4.8.2 a variety of validity measures is considered in order to choose the most suitable one for a dynamic classifier.

If no abrupt changes are detected and only gradual changes are recognised in the cluster structure during the monitoring procedure described in Section 4.6, a classifier must be incrementally updated based on new objects. The general form of an updating rule for cluster centres was formulated in Section 4.1 in equation (4.15). The specific updating rule depends on the type of clustering algorithm applied. For the classifier of Gath and Geva the updating procedure involves the cluster prototypes and the fuzzy covariance matrices of all clusters. Assume that N previous objects have already been classified and a new (N+1)-th object is considered. The location of the i-th cluster centre, i=1, ..., c, is updated in the same way as in the FCM algorithm according to the recursive equation (3.9).

The recursive equation for updating the fuzzy covariance matrix of cluster i can be formulated as follows:

$$F_i = \frac{\sum_{j=1}^{N+1} u_{ij}(x_j - v_i)(x_j - v_i)^T}{\sum_{j=1}^{N+1} u_{ij}} =$$

$$\frac{\sum_{j=1}^{N} u_{ij}(x_j - v_i)(x_j - v_i)^T + u_{N+1,j}(x_{N+1} - v_i)(x_{N+1} - v_i)^T}{\sum_{j=1}^{N} u_{ij} + u_{N+1,j}} = \tag{4.56}$$

$$\frac{FN_i(N) + u_{N+1,j}(x_{N+1} - v_i)(x_{N+1} - v_i)^T}{FD_i(N) + u_{N+1,j}}$$

where

$$FN_i(N) = \sum_{j=1}^{N} u_{ij}(x_j - v_i)(x_j - v_i)^T \quad \text{and} \quad FD_i(N) = \sum_{j=1}^{N} u_{ij}$$

Terms FN(N) and FD(N) are calculated from the previous N objects and their values are saved after each recursion step. The update of the matrix is achieved due to the additional terms corresponding to the new object N+1.

The main problem during incremental updating is to decide which new objects should be used to update a classifier. The most reasonable solution is to take into account only 'good' objects. In [Marsilli-Libelli, 1998] the use of a validity measure (in particular the entropy) to judge the quality of a new object was suggested: an object can be considered as 'good' if it improves the fuzzy partition. The use of such measure is, however, disadvantageous if it is assumed that gradual changes may lead to abrupt changes in the course of time. For instance, if two clusters are moving towards each other as time passes there are a lot of objects with high and ambiguous degrees of membership and according to a validity measure the quality of the partition deteriorates so that these objects are not selected for updating. But these objects are actually good representatives of both clusters that do indeed indicate a slow gradual deterioration of the cluster partition. If these objects are rejected for updating a classifier the recognition of overlapping clusters and merging clusters will be impossible.

Thus, it is suitable to define 'good' objects as good representatives of clusters based on their degrees of membership, i.e. as objects whose degrees of membership to clusters exceed a predefined threshold α^{good} (as was proposed in equation (4.26)):

$$X_i^{good} = \left\{ x_j \,\middle|\, u_{ij} \geq \alpha^{good} \right\}. \tag{4.57}$$

This means that one object can be considered as good for more than one cluster and consequently it can influence several clusters leading to gradual changes in the cluster structure. On the other hand, a classifier can follow gradual changes that may lead to abrupt changes due to the update of cluster prototypes.

Threshold α^{good} can be chosen equal to the absorption threshold so that all objects absorbed by a cluster can influence it. Obviously, the higher the threshold α^{good}, the smaller the number of 'good' objects. 'Bad' objects rejected for updating a classifier can be considered as candidates for a new cluster.

After adapting a classifier to gradual and abrupt changes it is necessary to update the training data set, or template set, which contains the best cluster representatives and the most recent objects observed. On the one hand, the template set extended by new objects in each time window is considered during the monitoring procedure to recognise temporal changes in the dynamic cluster structure. On the other hand, it is used to re-learn a classifier if adaptation to abrupt changes is needed. Thus, the template set must combine the most up-to-date information with the most useful old information.

4.8 Updating the Template Set of Objects

In Section 3.2.3 three different approaches for updating the training set of objects were discussed. Using a moving time window the training set consists of the objects in the current time window, which are used during the monitoring procedure to recognise temporal changes in the cluster structure and to adapt a classifier. Using a concept of the template set all representative objects can be included in the template set. As long as no changes, or only slight gradual changes, are observed, new objects of the current time window are added to the template set. As soon as abrupt changes appear, the template set is substituted by the set of the most recent objects. Using the concept of usefulness the representative objects for the template set are selected from new objects based on the value of their usefulness, which can be defined depending on the type of the classifier. If gradual changes are observed in the course of time, objects are getting old and less useful, and if the degree of usefulness falls below a certain threshold they are discarded. In the case of abrupt changes the template set is substituted by the most recent objects for which a new record of usefulness is derived.

A common feature of all these approaches is that the training set, or template set, of objects includes only recent objects that are representative of, and useful for, the current window. In the case of abrupt changes in the cluster structure the old template set is discarded and substituted by a set of new objects arriving in the current time window. In this case only the current cluster structure is considered for classifier design and the information about the old cluster structure is lost. However, it is not sufficient to discard and forget clusters that have already been learned. They may appear again in the future and it is easier and much quicker to identify an object by classifying it to existing clusters than detecting a new cluster.

Thus, by contrast to the aforementioned approaches the main idea of updating the template set in the dynamic classifier design proposed in this book is to preserve all clusters detected in the course of time by including their best representatives in the template set. The best representatives for the template set are defined by choosing 'good', or useful, objects. The degree of usefulness of an object can be defined as the highest degree of membership of an object to one of the clusters:

$$u(\mathbf{x}_j) = \max_{i=1,\ldots,c} u_{ij}, \quad j = 1,\ldots,N \tag{4.58}$$

In this way the template set must contain the best representatives of clusters ever detected as well as the most recent objects. These two types of objects can be considered as two sub-sets in the template set. If an existing cluster absorbs new objects over and over again, the sub-set of its best representatives in the template set is constantly updated to be able to follow possible gradual changes of a cluster. Otherwise, if a cluster is not up-to-date any more the sub-set of its representatives remains unchanged as time passes.

The sub-set of the most recent objects includes all new objects observed during the last ρ time windows. As stated in [Nakhaeizadeh et al., 1996] it is not sufficient to consider only 'good' and useful objects, since in this case the classifier will always ignore 'bad' objects with low degrees of membership to all clusters which may, for instance, build a new cluster in the course of time. Since the criteria for detection of new clusters require a sufficient number of free objects not already absorbed by existing clusters, it seems reasonable to keep these objects in the template set for a while so that supplemented by new objects they may lead to the formation of new clusters.

Thus, the structure of the template set can be represented as follows (*Figure 47*).

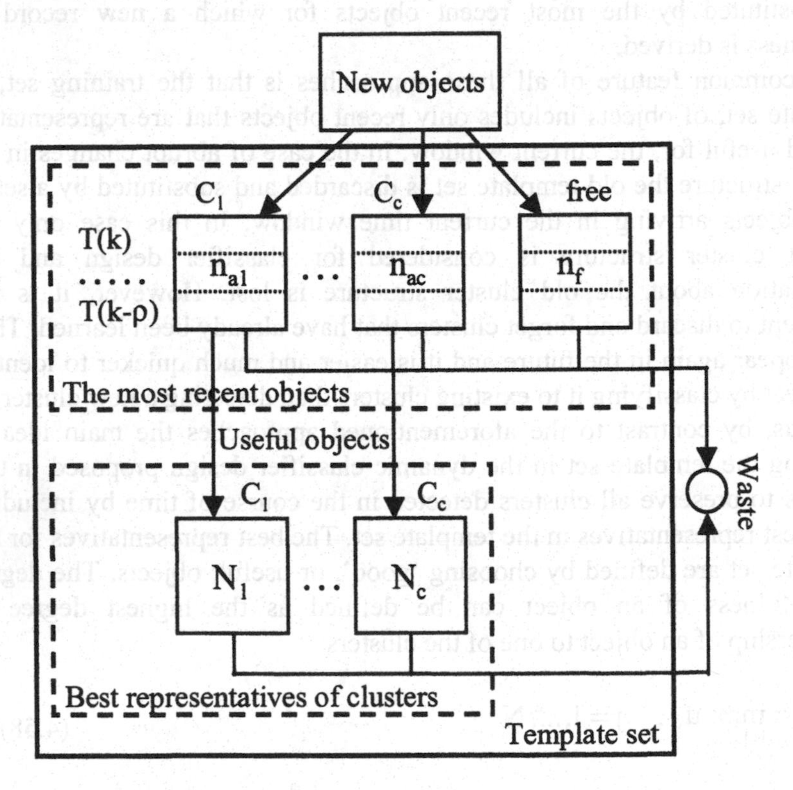

Figure 47. The structure of the template set

Suppose that the current cluster structure consists of c clusters. The template set includes a set of the most recent objects observed during the last ρ time windows and a set of the best representatives of existing clusters. The set of the most recent objects is organised in c sub-sets of objects absorbed by the corresponding clusters and a sub-set of free objects shown as boxes with ρ layers in *Figure 47*. Each layer of c boxes contains $n_{ai}(T(k))$,

$i=1, ..., c$, objects absorbed during time window T(k) by the corresponding cluster i. The set of best representatives of clusters is also separated into c sub-sets for c existing clusters each containing N_i, $i=1, ..., c$, objects selected over time according to their degrees of usefulness.

4.8.1 Updating the template set after gradual changes in the cluster structure

Based on the above considerations the following procedure for updating the template set is proposed. Suppose that the maximum size of the template set is chosen ρ times larger than the size of the time window and is restricted by the maximum number N_{max} of objects that can be contained in the set. This means that if the length of the time window is given by N_{tw} objects the current size of the template set N_{ts} can be

$$\rho \cdot N_{tw} \leq N_{ts} \leq N_{max}. \tag{4.59}$$

If $N_{T(k)}=N_{tw}$ new objects arrive in the time window T(k) and after their classification only gradual changes can be observed, new objects are saved in the template set in c+1 sets of the most recent objects according to their assignment to clusters. The oldest objects arrived in time window T(k-ρ) are discarded from these sets and the most useful of them (determined using (4.58)) are added to the sets of cluster representatives. Hence the updating of the template set follows the principle 'first in, first out' of queue theory.

Obviously, the sets of cluster representatives grow as time passes and the template set can become larger than N_{max}. In this case superfluous objects must be discarded from these sets according to the queue principle as well. The number of representatives for each cluster is reduced in such a way that the relation between the numbers of representatives for each cluster remains unchanged. This is very important in order to preserve the information about the existing cluster structure taking into consideration different sizes of clusters, since this information is used during the monitoring procedure to recognise new clusters.

In order to calculate the number of objects that must be discarded from each set of cluster representatives, or inversely the number of objects that can remain in the sets, the principle of proportional allocation used in the stratified sampling approach can be applied [Pokropp, 1998, p. 56]. Consider c sets of cluster representatives as a population N_{rep} consisting of c clusters, or stratas, $N_1, ..., N_c$. The size of the population is given by

$$N_{rep} = \sum_{i=1}^{c} N_i \,. \tag{4.60}$$

If N_{new} objects have to be added to this population it must be reduced beforehand to the size $N_{rep}^{new} = N_{rep} - N_{new}$, i.e. a sample of size N_{rep}^{new} has to be selected from N_{rep}. This can be achieved by selecting a partial sample from each strata N_i, i=1, ..., c, so that the total sample size is equal to N_{rep}^{new} and the relation of partial samples is equal to the relation of stratas N_i. Using the rule of proportional allocation the size of partial samples is determined in the following way:

$$N_i^{new} = \left[N_{rep}^{new} \cdot \frac{N_i}{N_{rep}} \right]_{round} \quad , \; i = 1,\ldots,c \tag{4.61}$$

Thus, if N_{new} useful objects are added to the sets of best representatives of clusters, (N_i- N_i^{new}) objects are discarded from each cluster i keeping the relation between the sets of cluster representatives unchanged. Hence, only new, useful objects can influence and change this relation. Note that if there are no new, useful objects for some cluster i, this cluster is not involved in the calculation of the population size (4.60) and no objects are discarded from this cluster. This condition prohibits the degradation of old clusters and guarantees that clusters that are not supported by new objects any more will be preserved anyway (i.e. the set of cluster representatives is 'frozen').

Example 3: Illustration of the updating procedure for the template set.

The example shown in *Figure 48* illustrates the principle of proportional allocation applied to the updating procedure of the template set. Consider the template set in time window T(k) and suppose that the maximum size of this set is equal to 1300. The size of the time window is restricted to 200 objects and the set of the most recent objects keeps objects of the last four time windows. The sets of cluster representatives contain $N_1(T(k))$=260, $N_2(T(k))$=90 and $N_3(T(k))$=150 objects in the current time window.

If 200 new objects arrive during time window T(k+1), then the most recent objects of time window T(k-3) consisting of 30 objects absorbed in Cluster 1, 120 objects absorbed in Cluster 2 and 50 free objects have to be discarded. Suppose that 20 and 80 useful objects are selected as the best representatives for Clusters 1 and 2 respectively and have to be added to the corresponding sets.

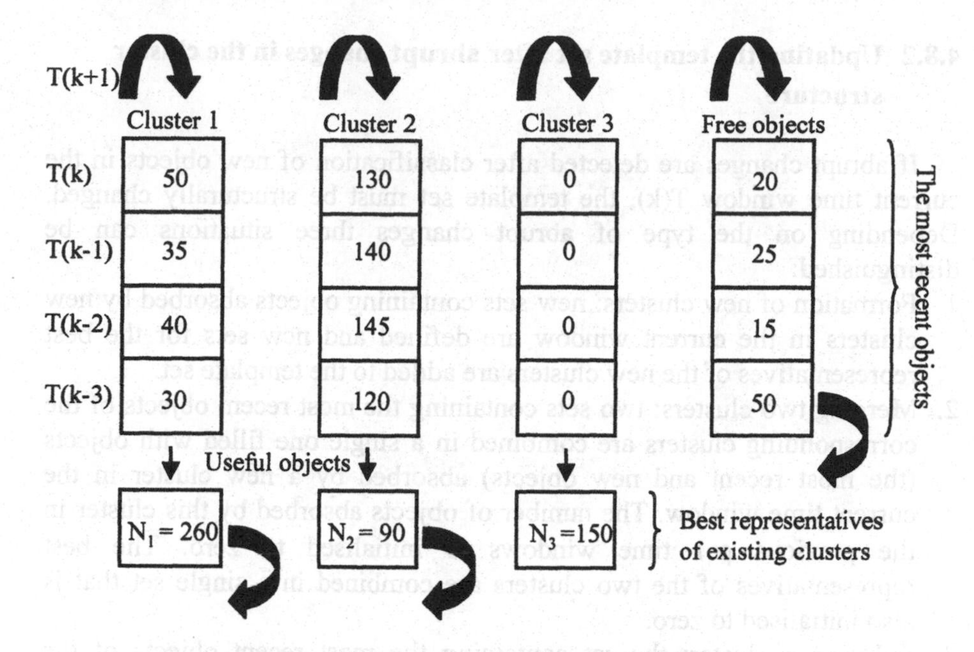

Figure 48. Illustration of the updating procedure of the template set

This means that the number of objects in the sets of the best cluster representatives have to be reduced by N_{new}=100 objects to N_{rep}^{new} =260+90+150-100=400. It must be noted that the set of Cluster 3 is not extended by new useful objects, therefore it remains unchanged. Hence, the size of the population considered is N_{rep}=350 and only sets N_1 and N_2 have to be reduced. The sizes of the partial samples for these two sets are calculated according to equation (4.60):

$$N_1^{new} = \left[250 \cdot \frac{260}{350} = 185.7 \right]_{round} = 186 ,$$

$$N_2^{new} = \left[250 \cdot \frac{90}{350} = 64.2 \right]_{round} = 64 .$$

The relation of the cluster sizes remains unchanged: 260:90=186:64=2.9. After adding new useful objects to the sets of cluster representatives, their sizes are: $N_1(T(k+1))$=206, $N_2(T(k+1))$=144 and $N_3(T(k+1))$=150. As can be seen, the size relation between Clusters 1 and 2 has changed to 206:144=1.4 due to new objects but the total number of objects contained in the sets of cluster representatives remains equal to 500.

4.8.2 Updating the template set after abrupt changes in the cluster structure

If abrupt changes are detected after classification of new objects in the current time window T(k), the template set must be structurally changed. Depending on the type of abrupt changes three situations can be distinguished:

1. Formation of new clusters: new sets containing objects absorbed by new clusters in the current window are defined and new sets for the best representatives of the new clusters are added to the template set.
2. Merging two clusters: two sets containing the most recent objects of the corresponding clusters are combined in a single one filled with objects (the most recent and new objects) absorbed by a new cluster in the current time window. The number of objects absorbed by this cluster in the previous ρ-1 time windows is initialised to zero. The best representatives of the two clusters are combined in a single set that is also initialised to zero.
3. Splitting a cluster: the set containing the most recent objects of the corresponding cluster is split into two sets containing objects (the most recent and new objects) absorbed in two new clusters in the current time window. The number of objects absorbed in these clusters in the previous ρ-1 time windows is initialised to zero. The best representatives of the corresponding cluster are also split into two sets initialised to zero.

The other sets of the most recent objects and cluster representatives are preserved and updated with new objects as described above for the case of gradual changes.

It must be noted that in the case of abrupt changes, after modification of the template set most of the objects previously contained in the sub-sets of clusters that have been changed (merged or split) can be absorbed in the newly formed clusters and concentrated in one sub-set of time window T(k), whereas other sub-sets of these clusters are empty. If new objects arrive during time window T(k+1) a part of the most recent objects have to be discarded and the most useful of them have to be added to the set of the best cluster representatives. In contrast to the updating procedure in the case of gradual changes, objects to be discarded from the sets corresponding to new formed clusters are selected from time window T(k) (instead of time window T(k-ρ)) since they represent the oldest objects in these sets. Thus, it can be generally said that if the template set is getting too large, objects from the bottom of the set of the most recent objects, which may correspond to time windows between T(k-ρ) and T(k), are discarded.

Using the concept of usefulness of objects and an adaptive training set that is updated by including new and useful objects, the time window can be chosen to be of constant length. The choice of this parameter, as well as the number of time windows establishing the set of the most recent objects in the template set, must be defined depending on the application. If the size of the time window is relatively small the number of time windows kept in the template set must be sufficiently large and vice versa. Such a relation should guarantee a sufficient number of new and recent objects needed to recognise abrupt changes in the dynamic cluster structure.

4.9 Cluster Validity Measures for Dynamic Classifiers

During dynamic classifier design the validity measure is used to control the process of adaptation of a classifier to temporal changes in the cluster structure. The goal of a validity measure is to determine whether the adaptation of a classifier leads to an improvement of the partitioning and to make a decision regarding acceptance or rejection of modifications applied to a classifier (such as formation, merging or splitting of clusters). If after having detected abrupt changes a classifier has been re-learned, a validity measure is evaluated for the current partition and compared to the previous value of the validity measure before re-learning. If the validity measure indicates an improvement of the current partitioning compared to the previous one the new classifier is retained, otherwise the previous classifier is restored.

Validity measures are generally used for the evaluation and comparison of the quality of the fuzzy partitions with different numbers of clusters. In other words, validity measures evaluate how good the given partition of data into clusters reflects the actual structure of the data. The requirements for the definition of the optimal partition of data into clusters and the corresponding criteria for the definition of validity measures are usually formulated as follows [Gath, Geva, 1989]:

1. Clear separation between the resulting clusters (separability),
2. Minimum volume of the clusters (compactness),
3. Maximum number of objects concentrated around the cluster centres (density of objects within clusters).

Thus, although fuzzy methods for clustering are used, the aim of clustering fulfilling these requirements is to generate well-defined subgroups of objects leading to a harder partition of the data set.

In the literature a large number of validity measures, which can be separated into three large classes depending on the properties of cluster partitions used, is proposed: measures using properties of the degrees of memberships, measures using properties of the data structure, and measures based on both types of properties. The first class of validity measures is represented by the partition coefficient [Bezdek, 1981, p. 100], classification entropy [Bezdek, 1981, p. 111] and proportion exponent [Windham, 1981]. These measures evaluate the degree of fuzziness of a cluster partition.

The second class of validity measures based only on the data aims at evaluating cluster properties such as compactness and separation. The compactness of clusters characterises the spread or variation of objects belonging to the same cluster. The separation of clusters means the isolation of clusters from each other. Most of the validity measures of this class ([Dunn, 1974], [Gunderson, 1978], [Davies, Bouldin, 1979]) were proposed based on properties of the crisp partition and depend on the topological structure induced by the distance metric used.

The largest class of validity measures includes measures based on both the properties of the degrees of membership and on the data structure. Here one can distinguish between measures based on the criteria of volume and density of fuzzy clusters and those based on compactness and separation. The three best known validity measures of the first group were proposed in [Gath, Geva, 1989]. Suppose that $\mathbf{F}_i$ is the fuzzy covariance matrix of cluster i defined by equation (4.24).

The fuzzy hypervolume criterion is then calculated by

$$v_{HV}(c) = \sum_{i=1}^{c} h_i , \tag{4.62}$$

where the hypervolume of the i-th cluster is defined as $h_i = \sqrt{\det(F_i)}$.

The partition density is calculated by

$$v_{PD}(c) = \frac{\sum_{i=1}^{c} n_i^{good}}{\sum_{i=1}^{c} h_i}, \tag{4.63}$$

where n_i^{good} is the sum of 'good objects' in cluster i defined by (4.25).

The average partition density is calculated by

$$v_{APD}(c) = \frac{1}{c}\sum_{i=1}^{c} \frac{n_i^{good}}{h_i}. \tag{4.64}$$

Density criteria are rather sensitive in cases of substantial overlapping between clusters and large variability in compactness of existing clusters. Since average partition density is calculated as the average of densities of single clusters, it indicates a good partition even if both dense and loose clusters are presented. The partition density corresponds to the common physical definition of density without taking into account the distribution of density over clusters.

Another group of validity measures is based on fuzzy compactness and separation. Based on the ideas introduced in [Xie, Beni, 1991] the global compactness and the fuzzy separation of fuzzy clusters are defined in [Bensaid et al., 1996] in the following way:

$$\pi_i = \frac{\sum_{j=1}^{N} (u_{ij})^2 \left\| x_j - v_i \right\|_A^2}{n_i} \quad i = 1,\ldots,c \tag{4.65}$$

$$S_i = \sum_{r=1}^{c} \left\| v_i - v_r \right\|_A^2 \quad i = 1,\ldots,c \tag{4.66}$$

where $n_i = \sum_{j=1}^{N} u_{ij}$ is the fuzzy cardinality of cluster i and A is an arbitrary M×M symmetric positive definite matrix. The validity index for cluster i is given by the ratio between its fuzzy separation and its compactness and the total validity measure is obtained by summing up this index over all clusters:

$$SC_1(U,V,X) = \sum_{i=1}^{c} \frac{S_i}{\pi_i} = \sum_{i=1}^{c} \frac{n_i \sum_{r=1}^{c} \left\| v_i - v_r \right\|_A^2}{\sum_{j=1}^{N} (u_{ij})^2 \left\| x_j - v_i \right\|_A^2} \tag{4.67}$$

A larger value of SC_1 indicates a better partition, i.e. a fuzzy partition characterised by well-separated and compact fuzzy clusters.

In [Zahid et al., 1999] this definition of the validity measure was extended by considering the ratio between the fuzzy separation and compactness obtained only from the properties of fuzzy membership functions. The fuzzy compactness indicates how well objects are classified, or how close objects are located to cluster centres, by considering the objects' maximum degrees of membership. Compact clusters are obtained if all maximum degrees of membership of objects take high values. The fuzzy separation measures the pairwise intersection of fuzzy clusters by considering the minimum degrees of membership of an object to a pair of clusters. Clusters are well-separated if they do not intersect or their intersection area is minimum.

$$FC = \frac{\sum_{j=1}^{N} (\max_i u_{ij})^2}{n_\cup} \qquad (4.68)$$

$$FS = \sum_{i=1}^{c-1} \sum_{r=1+i}^{c} \frac{\sum_{j=1}^{N} \min (u_{ij}, u_{rj})^2}{n_{ir}} \qquad (4.69)$$

where $n_\cup = \sum_{j=1}^{N} \max_i u_{ij}$ is the cardinality of the set of maximum degrees of membership and $n_{ir} = \sum_{j=1}^{N} \min (u_{ij}, u_{rj})$ is the cardinality of the set of pairwise minimum degrees of membership. The resulting validity measure based on membership functions is obtained as the ratio between the fuzzy separation and compactness:

$$SC_2(U) = \frac{FS}{FC} = \frac{\sum_{i=1}^{c-1} \sum_{r=1+i}^{c} \frac{\sum_{j=1}^{N} \min (u_{ij}, u_{rj})^2}{n_{ir}}}{\frac{\sum_{j=1}^{N} (\max_i u_{ij})^2}{n_\cup}} \qquad (4.70)$$

A compact and well-separated cluster partition corresponds to a low value of SC_2.

The overall validity measure is defined in [Zahid et al., 1999] as the degree of correspondence between the structure of the input data set and the fuzzy partition resulting from the fuzzy clustering algorithm.

$$SC = SC_1(U,V,X) - SC_2(U). \qquad (4.71)$$

A larger value of SC indicates a better fuzzy partition. This validity measure indicates cohesive clusters with a small overlap between pairs of clusters.

All validity measures described above are usually used in combination with clustering algorithms to determine the optimal fuzzy c-partition, i.e. a clustering algorithm is applied to partition a data set into a number of clusters varying between 2 and c_{max} and the partition with the best value of the validity measure is chosen as the optimal one.

Applying validity measures during dynamic classifier design the intention remains unchanged but the comparison is carried out only for two partitions: before and after an adaptation of the classifier. It is assumed that the optimal number of clusters is determined by the monitoring procedure and the task of the validity measure is to confirm the new partition obtained. In order to choose a suitable validity measure it is reasonable to take into account the types of temporal changes that can appear in the cluster structure and lead to the modification of the classifier.

Considering the criteria proposed to detect abrupt temporal changes it can be seen that the decision to split a cluster or form a new one was based on the criteria of density and separation: a classifier must be changed if dense groups well-separated from existing clusters, or within a single cluster, are detected. The decision regarding the merging of clusters was based on the ambiguity property of a fuzzy partition due to the intersection of fuzzy clusters: a classifier must be changed if there are pairs of clusters with a significant overlap. Therefore, the purpose of the adaptation of a classifier is to achieve the most unambiguous hard partition with dense well-separated clusters. It seems reasonable to consider a validity measures based on the average partition density (4.64) as well as on fuzzy separation and compactness.

Instead of evaluating the pairwise intersection of fuzzy clusters, a measure of fuzzy separation can be obtained by considering the difference between the highest and second highest degree of membership of each object to the clusters. This quantity can serve as a much better indicator for separation of clusters since it measures the degree of ambiguity of an object assignment when clusters overlap. A high value of this quantity is a sign for a rather hard assignment of objects. The total value of the validity measure based on the principle of ambiguity can be obtained by aggregating the membership differences over all objects using, for instance, the arithmetic mean operator and considering the ratio between the fuzzy separation and the fuzzy compactness given by (4.68).

$$SC_3(U)=\frac{FSA}{FC}=\frac{\sum_{j=1}^{N}(u_j^{m1}-u_j^{m2})}{\sum_{j=1}^{N}\max_i u_{ij}}, \tag{4.72}$$

where FSA is the fuzzy separation based on ambiguity, $u_j^{m1}=\max_{i\in\{1,\dots,c\}} u_{ij}$ is the highest degree of membership of object j and $u_j^{m2}=\max_{i\in\{1,\dots,c\}\setminus m1} u_{ij}$ is its second highest degree of membership. A larger value of SC3 corresponds to a better partition. The value of FSA equal to 1 corresponds to a hard unambiguous partition, whereas the value equal to zero corresponds to the most ambiguous partition[2].

Using the validity measure as an indicator for adapting the dynamic classifier, it is necessary to compare the value of the validity measure after adaptation with the one before adaptation. A positive difference in the validity measure indicates an improvement of the quality of the dynamic classifier. A negative difference indicates a deterioration of classifier performance in the case of adaptation, thus this must be rejected and the previous classifier must be preserved.

However, a problem may arise in which the validity measure SC_3 does not reflect adequately the improvement of the dynamic classifier. For instance, if during the adaptation of the classifier a cluster was split up, it is to be expected that the fuzzy compactness will increase much more than the fuzzy separation, which can even decrease due to a certain overlapping of new clusters. Moreover, an increase in the average partition density after splitting clusters can be expected. In order to allow the adaptation of the classifier the negative variation of the validity measure can be accepted as a trade-off for the significant increase of compactness and the average partition density. Thus, in order to obtain relevant information about changes in the partition quality, it seems reasonable to consider the variation of three indexes: fuzzy separation, fuzzy compactness and average partition density.

The variation of the fuzzy separation measure between time instants t and t-1 is given by the following equation:

[2] Since the validity measure is used here to evaluate a possibilistic c-partition generated by a dynamic classifier, it cannot be stated that U=[1/c] if the partition is ambiguous as in the case for probabilistic fuzzy c-partition.

$$\Delta FSA_{t,t-1} = \frac{1}{N}\sum_{j=1}^{N}(u_j^1(t)-u_j^2(t)) - \frac{1}{N}\sum_{j=1}^{N}(u_j^1(t-1)-u_j^2(t-1)) =$$
$$\frac{1}{N}\sum_{j=1}^{N}(u_j^1(t)-u_j^2(t))-(u_j^1(t-1)-u_j^2(t-1)) = \qquad (4.73)$$
$$\frac{1}{N}\sum_{j=1}^{N}(u_j^1(t)-u_j^1(t-1))-(u_j^2(t)-u_j^2(t-1))$$

The variation of the fuzzy compactness between time instants t and t-1 is defined as follows:

$$\Delta FC_{t,t-1} = \frac{1}{N}\sum_{j=1}^{N}\max_i u_{ij}(t) - \frac{1}{N}\sum_{j=1}^{N}\max_i u_{ij}(t-1) =$$
$$\frac{1}{N}\sum_{j=1}^{N}(\max_i u_{ij}(t) - \max_i u_{ij}(t-1)) \qquad (4.74)$$

The variation of the average partition density between time instants t and t-1 is obtained in the following way:

$$\Delta v_{t,t-1}^{apd} = \frac{1}{c(t)}\sum_{i=1}^{c(t)}\frac{n_i^{good}(t)}{h_i(t)} - \frac{1}{c(t-1)}\sum_{i=1}^{c(t-1)}\frac{n_i^{good}(t-1)}{h_i(t-1)} \qquad (4.75)$$

If at least one of these three validity measures has a positive variation and the others have a small negative variation, a new classifier can be accepted.

The following examples show that different validity measures can be relevant for adapting the dynamic classifier depending on the changes taking place.

Example 4: Splitting clusters based on the average partition density measure.

Consider the cluster structure at a certain time instant t-1 shown in *Figure 49*. A current classifier is characterised by two clusters with centres at v_1 = (-0.02, -0.01) and v_2=(1.65; 1.99). Due to new objects absorbed by cluster C_2 the monitoring procedure has detected a heterogeneous cluster C_2 which must be split up. The adaptation of the classifier is carried out by re-learning the classifier with a number of clusters c=3 The values of validity measures before and after adaptation are summarised in *Table 4*. As can be seen, the fuzzy separation and compactness of the fuzzy partition have decreased but the average partition density has increased considerably (in more than 2 times). Thus, this result corresponds to the splitting criteria and the new classifier can be accepted.

Table 4. Validity measures for a fuzzy partition before and after splitting clusters

Before splitting	After splitting
FSA=0.68	FSA=0.66
FC=0.69	FC=0.68
SC_3=0.988	SC_3=0.975
v_{apd}=971.68	v_{apd}=2022.5

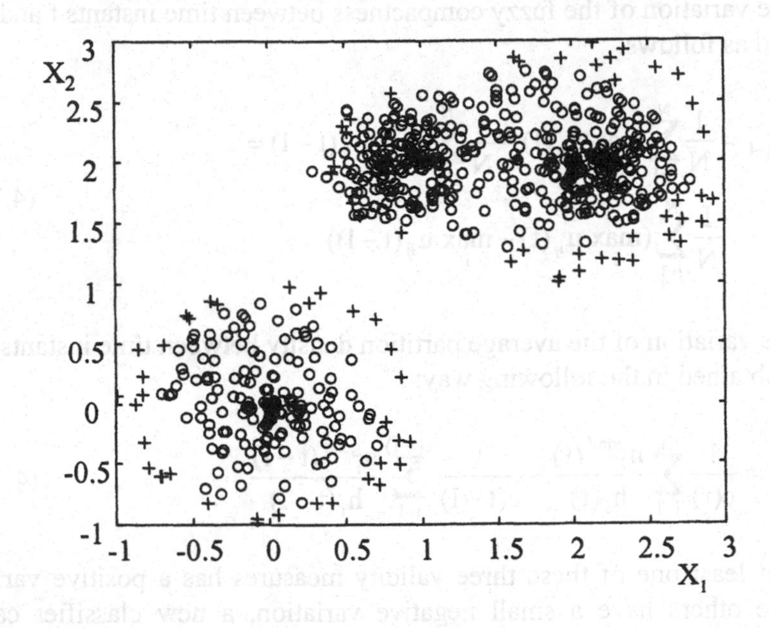

Figure 49. Adaptation of a classifier requires splitting clusters

Example 5: Merging clusters based on the fuzzy separation and average partition density measures.

Figure 50 illustrates a cluster structure containing three clusters with centres at v_1=(-0.02; 0), v_2=(2.09; 1.99) and v_3=(1.09; 2.07). After classification of new objects arriving in time window t, the monitoring procedure has detected two similar clusters, v_2 and v_3, which must be merged. After the adaptation of the classifier by re-learning with a new cluster number c=2 a partition with better separated, more dense but less compact clusters is obtained. The values of validity measures before and after merging clusters are shown in *Table 5*. Since the partition is improved with respect to two measures, a new classifier is accepted.

Table 5. Validity measures for a fuzzy partition before and after merging clusters

Before merging	After merging
FSA=0.607	FSA=0.649
FC=0.683	FC=0.654
SC_3=0.889	SC_3=0.993
v_{apd}=963.09	v_{apd}=1116.0

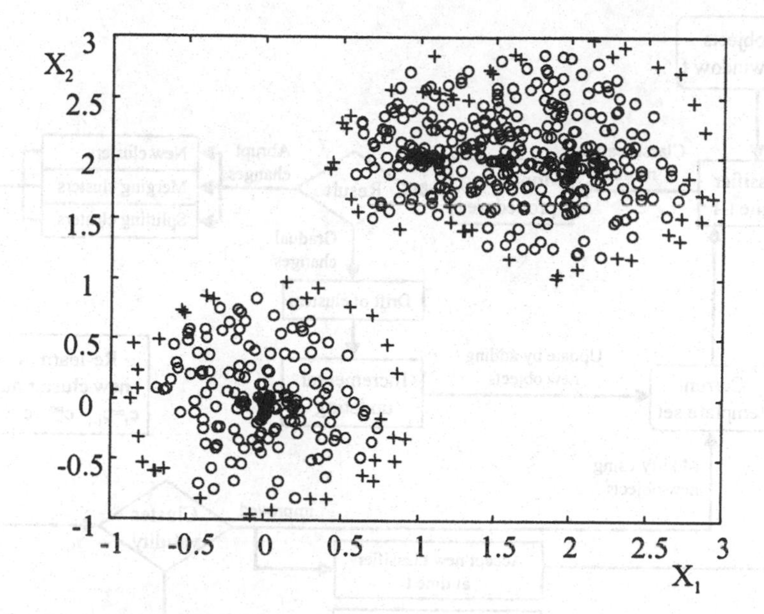

Figure 50. Adaptation of a classifier requires cluster merging

The use of three different validity measures leads to a more complete judgement of changes in a fuzzy partition and establishes more reliable control of the adaptation of a dynamic classifier. The use of validity measures during classifier design and adaptation guarantees the preservation and improvement of classifier performance over time and concludes the learning-and-working cycle of the dynamic pattern recognition system.

4.10 Summary of the Algorithm for Dynamic Fuzzy Classifier Design and Classification

The overall algorithm for dynamic fuzzy classifier design and classification is summarised in *Figure 51.* As can be seen, the design of the dynamic classifier is combined with the classification of new objects in a single learning-and-working cycle in order to keep the classifier up-to-date and to adapt it to the temporal changes in the cluster structure.

Consider a classifier designed in time window t-1. Suppose that new objects arrive in time window t. They are classified by the current classifier and the classification results are introduced into the monitoring procedure for analysis. The aim of the monitoring procedure is to decide whether new objects fit well into the current cluster structure and, if not, to recognise changes. In order to fulfil this task the monitoring procedure has four algorithms at its disposal.

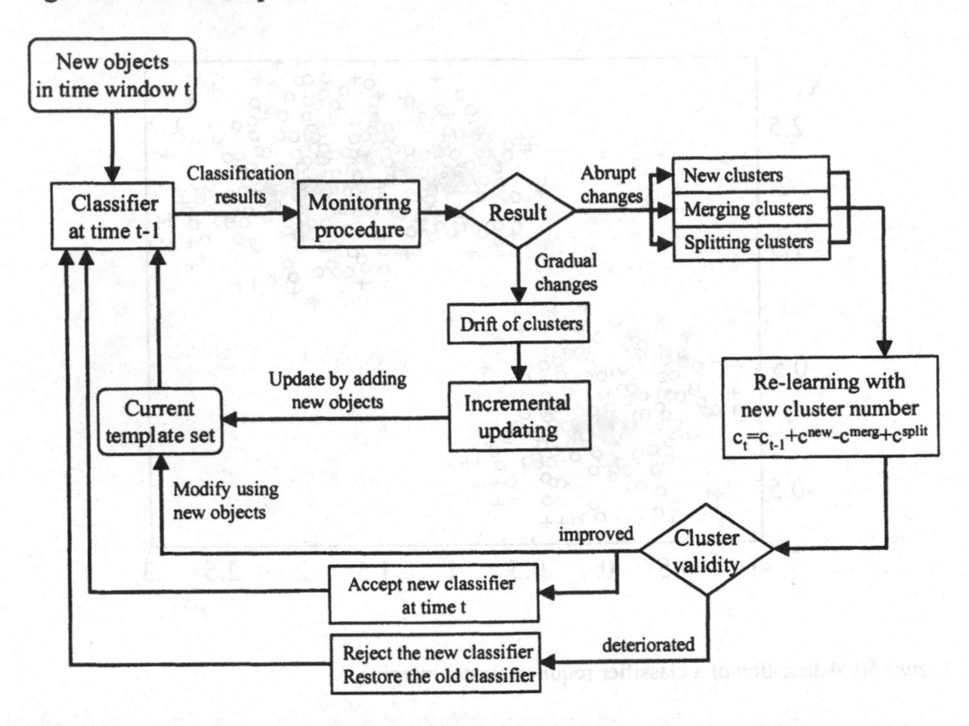

Figure 51. The process of dynamic fuzzy classifier design and classification

Based on the algorithm for evaluating the performance of the classifier applied to new data (Section 4.6), gradual changes such as the drift of cluster centres can be detected. This decision of the monitoring procedure is introduced into the adaptation procedure where incremental updating of corresponding cluster centres is carried out as explained in Section 4.7. Using three algorithms for detection of new, similar and heterogeneous clusters which were presented in Sections 4.3, 4.4 and 4.5, respectively, the monitoring procedure can detect abrupt temporal changes in the cluster structure and results in any combination of three possible decisions: formation of new clusters, merging of clusters, or splitting of clusters. Each of the three algorithms of the monitoring procedure delivers the number of clusters by which the current cluster number must be increased or decreased, as well as estimates of new cluster centres. These results are forwarded to the adaptation procedure where a new number of clusters c_t recognised in time window t is calculated by adding the results of the three algorithms, as described in Section 4.7.

The adaptation of the classifier is achieved by re-learning the classifier with the new cluster number and using the existing template set extended by new data. In order to evaluate the performance of a new classifier obtained in time window t, its performance in compared to the classifier in time window t-1 by applying a set of validity measures. If at least some of these measures indicate an improvement in classifier performance the new classifier at time t is accepted and saved instead of the previous one. If all validity measures have deteriorated this means that the adaptation of the classifier was too early and cannot improve the cluster partition. In this case the new classifier is rejected and the old classifier at time t-1 is restored.

After adapting the classifier the template data set is updated using one of the updating strategies described in Section 4.8. The choice of the relevant strategy depends on the result of the monitoring procedure. If only gradual changes were detected the template set is updated by new data according to the queue principle. If abrupt changes were recognised and the new classifier was accepted, the template set must be structurally modified before extending it with new data. If the new classifier was rejected after adaptation to abrupt changes, the template data set is updated as in the case of gradual changes.

It must be noted that the results of the monitoring procedure can be used to make the diagnosis about the current cluster structure (i.e. current system states). Besides the final results of the monitoring procedure concerning detected changes in the cluster structure, some additional parameters can be provided to the end user. For instance, cluster sizes calculated as the number of objects absorbed by the corresponding clusters (see equation (4.17)) or as the number of 'good' objects (see equation (4.26)) can be interpreted as usefulness of clusters in the current time window. Further parameters, which are calculated during the monitoring procedure and can be relevant for the diagnosis, are the number of free objects, partition densities of existing clusters, new locations of cluster centres that have been moved, the value of the fuzzy separation and the fuzzy compactness for the current partition. Together, all these measures can give a good representation of the current situation.

The algorithm for dynamic fuzzy clustering developed in this chapter provides a possibility to design an adaptive classifier capable of following temporal changes in the cluster structure and adjusting its parameters automatically to detected changes in the course of time.

5 SIMILARITY CONCEPTS FOR DYNAMIC OBJECTS IN PATTERN RECOGNITION

In Sections 2.2.1 and 4.1 it was stated that the task of clustering methods is to partition a number of objects into a small number of homogeneous clusters so that objects belonging to any one of the clusters would be as similar as possible and objects of different clusters as dissimilar as possible. The most important problem arising in this context is the choice of a relevant similarity measure, which is then used for the definition of the clustering criterion. The notion of similarity depends on the mathematical properties of the data set (e.g. mutual distance, angle, curvature, symmetry, connectivity etc.) and on the semantic goal of a specific application.

According to [Bezdek, 1981, p. 44] there is no universally optimal clustering criterion or measure of similarity that can perform well for all possible data structures. This fact is explained by a wide variety of data structures, which can posses different shapes (spherical, elliptical), sizes (intensities, unequal numbers of observations), or geometry (linear, angular, curved) and can be translated, dilated, compressed, rotated, or reordered with respect to some dimensions of the feature space. Therefore, the choice of a particular clustering criterion and a particular measure of similarity is at least partially context dependent. It should be noted that similarity measures represent main components of clustering criteria and can be used either as criteria themselves or in combination with various clustering criteria to obtain different clustering models.

Many clustering methods (e.g. fuzzy c-means [Bezdek, 1981, p. 65], possibilistic c-means [Krishnapuram, Keller, 1993], unsupervised optimal fuzzy clustering [Gath, Geva, 1989], (fuzzy) Kohonen networks [Rumelhart, McClelland, 1988]) use clustering criteria based on the distance between objects x_j and x_l, $x_j, x_l \in X$, which is defined by a symmetric positive-definite function of pairs of elements $d: X \times X \rightarrow \Re^+$. If d additionally satisfies the triangle inequality, then d is a metric on $\Re^M$, a property that is not necessarily required, however. In particular one of the most frequently used measures is the Minkowski metric and its special case the Euclidean metric [Bandemer, Näther, 1992, p. 66-67]. If each feature vector (object) is considered as a point in the

$$s_{jl} = 1 - d^*_{jl} \tag{5.1}$$

$$s_{jl} = \frac{1}{1 + d^*_{jl}} \tag{5.2}$$

$$s_{jl} = e^{-d^*_{jl}} \tag{5.3}$$

It is obvious that similarity measures are also symmetric positive-definite functions defined on the interval [0, 1].

In Section 4.1 it was stated that dynamic objects are represented by a temporal sequence of observations and described by multidimensional trajectories in the feature space, which contain a history of temporal development of each feature and are given by equation (4.7). Since the distance between vector-valued functions is not defined, classical clustering and in general pattern recognition methods are not suited for processing dynamic objects.

There are two possibilities for handling dynamic objects represented by multidimensional trajectories in pattern recognition:

1. to pre-process trajectories in order to transform them into conventional feature vectors;
2. to define a distance or dissimilarity measure for trajectories.

The first approach is used in most applications regarding time series classification and will be described in Section 5.1. The second approach requires a definition of a similarity measure for trajectories that should take into account the dynamic behaviour of trajectories and will be considered in Section 5.2.

When defining a similarity measure for trajectories it is important to determine a specific criterion for similarity. Depending on the application this criterion may require either the best matching of trajectories by minimising the pointwise distance or a similar form of trajectories independent of their relative location to each other. After considering different types of similarity and introducing different similarity models in Section 5.2, a number of definitions of specific similarity measures will be proposed. The use of similarity measures for trajectories for the modification of static pattern recognition methods will be explained in Section 5.3 based on the example of the modification of the fuzzy c-means algorithm.

5.1 Extraction of Characteristic Values from Trajectories

The first approach regarding the use of dynamic objects in pattern recognition is related to the pre-processing of dynamic objects in such a way that they become valid inputs for classical pattern recognition methods. Data pre-processing is one of the obvious steps usually performed before the actual pattern recognition process starts and is concerned with the preparation of data (e.g. scaling and filtering) and complexity reduction. The reasons for performing data pre-processing are the following [Famili et al., 1996]:

1. solving data problems that may prevent the use of any data analysis methods or may lead to unacceptable results (e.g. missing features or feature values, noisy data, different data formats, very large data volumes);
2. understanding the nature of data (through data visualisation and calculation of data characteristics in order to initialise the data analysis process), and
3. extracting more meaningful knowledge from a given data set (e.g. extraction of the most relevant features).

Data pre-processing includes such techniques as signal processing, correlation analysis, regression analysis and discrimination analysis, to name just a few. In some cases, more than one form of data pre-processing is required, therefore the choice of a relevant method for data pre-processing is of crucial importance.

Data pre-processing can be represented as a transformation Z [Famili et al., 1996] that maps a set of the raw real world feature vectors x_{jh} into a set of new feature vectors y_{jk}:

$$y_{jk} = Z(x_{jh}), j = 1,..., N, h = 1,..., M_a, k = 1,...,M, \qquad (5.4)$$

so that Y preserves the valuable information in X, eliminates at least one of the data problems in X and is more useful than X. In (5.4) M_a denotes the number of features before pre-processing (often called attributes) and M the number of features after pre-processing.

Valuable information is characterised by four attributes [Fayyad et al., 1998, p. 6-8]: valid, novel, potentially useful, and ultimately understandable. Valuable information includes components of knowledge that must be discovered by pattern recognition methods and presented in a meaningful way.

Dealing with dynamic objects in pattern recognition, we are confronted with the first data problem: classical methods can not be applied to data in the form of multidimensional trajectories. The goal of pre-processing in this case is to reduce trajectories of features to small sets of real numbers (vectors) by extracting characteristic values from the trajectories. The resulting real-valued vectors are then combined to form a single one and a conventional feature vector is obtained. During this transformation vector-valued features are replaced by one or more real numbers leading to real-valued feature vectors. This idea is illustrated in *Figure 52*.

Let $\mathbf{x}(t) = [\mathbf{x}_1(t), \mathbf{x}_2(t), ..., \mathbf{x}_M(t)]$ denote a vector-valued feature vector whose components are trajectories of single features defined on the same time interval $[t_1, ..., t_p]$. Selecting a set of characteristic values $\{K_1(\mathbf{x}_r), ..., K_{Lr}(\mathbf{x}_r)\}$ from a trajectory of each feature $\mathbf{x}_r(t) = [x_r(t_1), x_r(t_2), ..., x_r(t_p)], r = 1, ..., M$ and placing it in a vector together with other sets, a conventional feature vector is obtained. The number $L_r, r = 1, ..., M$, and the type of characteristic values may vary for single features.

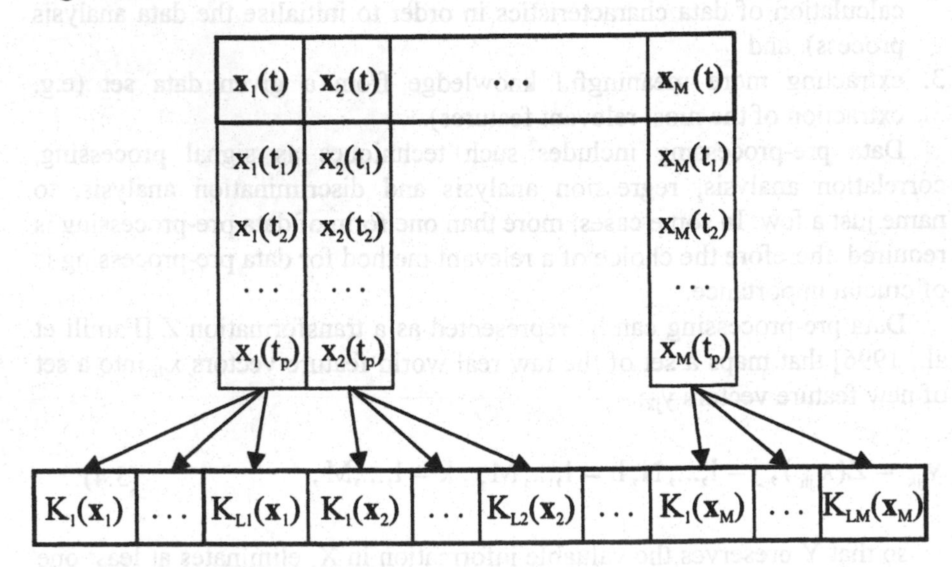

Figure 52. Transformation of a feature vector containing trajectories into a conventional feature vector

This method of data pre-processing produces the following transformation Z:

$$Z : [x_1(t), x_2(t), ..., x_M(t)] \rightarrow [K_1(x_1), ..., K_{LM}(x_M)] = [K_1, ..., K_L], \quad (5.5)$$

where $L = \sum_{r=1}^{M} L_r$ is the total number of characteristic values extracted from a feature vector $\mathbf{x}(t)$.

Since pre-processing of dynamic objects is independent of the pattern recognition method used for analysis, any classical method dealing with objects in the vector space can be applied in conjunction with this method of pre-processing for the purpose of dynamic pattern recognition.

Pre-processing of dynamic objects is used in most applications of time series classification, for instance, in EEG/ECG diagnosis, speech recognition, radar classification, etc. In all these cases static feature vectors are extracted from time series during pre-processing and then used within conventional (static) pattern recognition methods.

One of the best illustrations of the applicability of the method of pre-processing of trajectories is given in [Geva, Kerem, 1998], where the problem of dynamic state recognition and event prediction in biomedical signal processing was considered. The goal of this application was to recognise brain states based on the background electroencephalogram (EEG) activity to form the basis for forecasting a generalised epileptic seizure, which is usually preceded by a preseizure state in the EEG. An EEG state can be defined by a portion of the time series dominated by one rhythm, a particular alternation of rhythms, or a frequent appearance of isolated events. In order to be able to use a conventional clustering algorithm, characteristic features of EEG time series were extracted by the fast wavelet transform [Mallat, Zhong, 1992]. The most relevant features were then selected and introduced into the unsupervised optimal fuzzy clustering (UOFC) algorithm [Gath, Geva, 1989] that was used to detect distinct groups in the data representing EEG states. During dynamic state recognition, feature selection and clustering were periodically applied to segments of the EEG signal obtained in overlapping time windows. In each cycle the classifier was re-learned based on the complete history of data available until the current moment. The results reported in [Geva, Kerem, 1998] show that the UOFC algorithm is suitable for an accurate and reliable identification of bioelectric brain states based on the analysis of the EEG time series.

The disadvantage of this approach for handling dynamic objects is that a part of the important information contained in trajectories and correlations across several time steps gets lost. In order to avoid this drawback, an alternative approach regarding the definition of similarity for trajectories can be considered. The use of this approach for dynamic pattern recognition requires an obvious modification of conventional methods.

5.2 The Similarity Notion for Trajectories

Trajectories, and in particular time series, constitute an important class of complex data. They appear in many financial, medical, technical and scientific applications. Although there is a lot of statistical literature dedicated to time series analysis, which is used for modelling and prediction tasks, the similarity notion for time series, which is of prime importance for data analysis applications, has not been studied enough.

The notion of similarity can be interpreted in different ways depending on the context. In everyday language, the interpretation of similarity is 'having characteristics in common' or 'not different in shape but in size or position' [Setnes et al., 1998, p. 378]. This interpretation can be employed for defining the similarity between trajectories.

Intuitively, two trajectories can be considered similar if they exhibit similar behaviour over a large part of their length. In order to define similarity mathematically, it is essential to determine which mathematical properties of trajectories should be used to describe their behaviour and in what way the comparison (matching) of trajectories should be performed.

When analysing trajectories it is assumed that trajectories can contain [Das et al., 1997]:

- outliers and noise due to measurement errors and random fluctuations;
- different scaling and translating factors and base lines, which can appear due to measurements on different devices or under different conditions.

When comparing trajectories it is necessary to omit outliers and noise from the analysis since they are not representative for describing the behaviour of trajectories. Scaling and translation is often irrelevant when searching for similar trajectories, but there can be applications where such transformations include a useful information about the behaviour of trajectories and could not be ignored.

In Section 2.3.2 it was already stated that depending on the criterion chosen for comparison of trajectories, two types of similarity between trajectories can be distinguished [Joentgen, Mikenina et al., 1999b, p. 83]:

1. *structural similarity*: the better two trajectories match in form / evolution / characteristics, the greater the similarity between these two trajectories;
2. *pointwise similarity*: the smaller pointwise distance between two trajectories in feature space, the greater the similarity between these two trajectories.

In order to determine structural similarity, relevant aspects of the behaviour of trajectories must be specified depending on a concrete application. Based on the chosen aspects, mathematical properties of trajectories (e.g. slope, curvature, position and values of inflection points, smoothness etc.) can be selected, which are then used as comparison criteria. In such a way, structural similarity is suited to situations in which one looks for particular patterns in trajectories that should be well matched.

Pointwise similarity expresses the closeness of trajectories in the feature space. In this case the behaviour of trajectories is not in the foreground and some variations in form are allowed as long as trajectories are spatially close. In contrast to structural similarity, the calculation of pointwise similarity does not require a formulation of characteristic properties of trajectories and is based directly on the values of trajectories.

Thus, the task is to define a similarity measure for trajectories that expresses a degree of matching of trajectories according to some predefined criteria and is invariant to some specific transformations (e.g. scaling, translation, missing values or appearance of incorrect values). In the following sections different definitions of pointwise and structural similarity measures proposed in the literature, which use different similarity models and context-dependent interpretation of the similarity notion, will be considered. Afterwards, a number of definitions for the structural similarity measure, which differ in the underlying similarity criterion and take into account different properties of trajectories, will be proposed.

5.2.1 Pointwise similarity measures

Pointwise similarity determines a degree of closeness of two trajectories with respect to corresponding pairs of points of the trajectories. The closeness of points in the feature space can be determined using the Euclidean distance or some other distance measure, which is interpreted as a dissimilarity of points. The similarity between points is regarded as an inverse of their distance in the feature space. This classical definition of similarity is crisp: as soon as a threshold for the distance or similarity measure is defined, points are separated into two categories of 'similar' and 'non-similar' points. Such a definition in the manner of Boolean logic is very restrictive for a description of the concept of similarity and does not agree with the human perception of similarity, which is coupled with imprecision and uncertainty [Zimmermann, Zysno, 1985, p. 149]. In order to be able to model the inherent fuzziness of the concept of similarity, a similarity measure must allow a gradual transition between 'similar' and 'non-similar' [Binaghi et al., 1993, p. 767]. This property can be fulfilled by modelling the similarity in a fuzzy logic framework.

In [Joentgen, Mikenina et al., 1999b] a method to determine the pointwise similarity between trajectories was presented, which is based on a linguistic description of similar trajectories. This linguistic description represents a subjective evaluation of similarity by a human observer and can vary depending on the context. The idea of this method is to consider the pointwise difference of two trajectories and to calculate a degree of similarity, or proximity, of this sequence of differences to the zero-function (a function equal to zero on the whole universe of discourse). Consider two trajectories $\mathbf{x}(t)$ and $\mathbf{y}(t)$, $t \in T = [t_1, ..., t_p]$, in the feature space $\Re^M$. If the sequence of differences is given by $\mathbf{f}(t) = \mathbf{x}(t) - \mathbf{y}(t) = [\mathbf{f}(t_1), ..., \mathbf{f}(t_p)] = [\mathbf{f}_1, ..., \mathbf{f}_p]$, then the similarity between trajectories $\mathbf{x}(t)$ and $\mathbf{y}(t)$ is defined according to the equivalence equation: $s(\mathbf{x}, \mathbf{y}) = s(\mathbf{x}-\mathbf{y}, 0) = s(\mathbf{f}, 0)$.

For the sake of simplicity the case of one-dimensional trajectories in the feature space $\Re^1$ is considered. A generalisation to the case of multidimensional trajectories will be given afterwards. The following algorithm to determine a measure of pointwise similarity between an arbitrary sequence f(t), $t \in T$, and the zero-function was proposed.

Algorithm 6: Determination of pointwise similarity for trajectories [Joentgen, Mikenina et al., 1999b, p. 85].

1. *A fuzzy set A 'approximately zero' with a membership function u(f) is defined on the universe of discourse F* (Figure 53).
2. *The degree of membership u(f(t)) of the sequence f(t) to the fuzzy set A is calculated for each point $t \in T$. These degrees of membership can be interpreted as (pointwise) similarities $u(f(t_i)) = s_i$, $i = 1, ..., p$, of the sequence f(t) to the zero-function* (Figure 54).
3. *The sequence $u(f(t)) = [s_1, ..., s_p]$ is transformed by using specific transformations (e.g. minimum, maximum, arithmetic mean, γ-operator, integral,) into a real number s(f, 0) expressing the overall degree of f(t) being zero.*

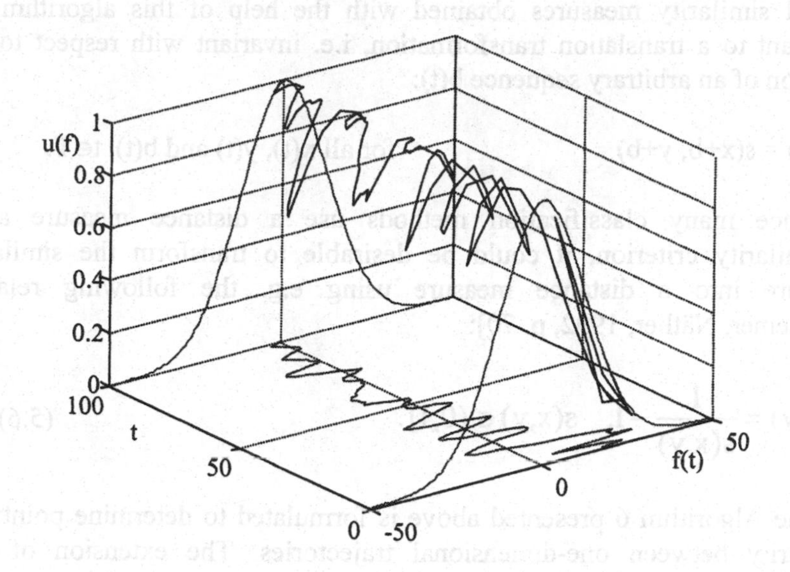

Figure 53. The fuzzy set 'approximately zero' with u(f), the sequence of differences f(t) and the resulting pointwise similarity u(f(t)

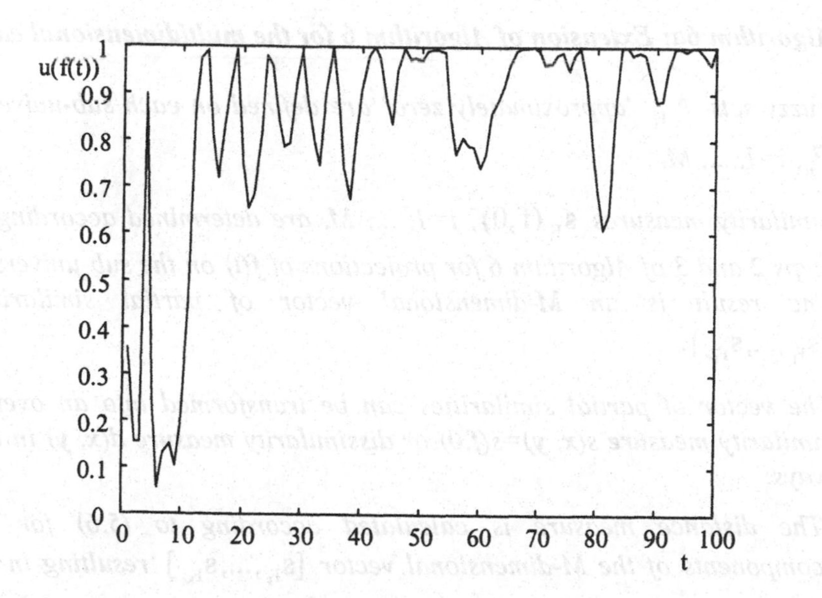

Figure 54. Projection of pointwise similarity between trajectories into the plane (t, u(f(t)))

All similarity measures obtained with the help of this algorithm are invariant to a translation transformation, i.e. invariant with respect to the addition of an arbitrary sequence b(t):

$$s(x, y) = s(x+b, y+b) \qquad \text{for all } x(t), y(t) \text{ and } b(t), t\in T.$$

Since many classification methods use a distance measure as a dissimilarity criterion, it could be desirable to transform the similarity measure into a distance measure using e.g. the following relation [Bandemer, Näther, 1992, p. 70]:

$$d(x,y) = \frac{1}{s(x,y)} - 1, \quad s(x,y) \in (0,1]. \tag{5.6}$$

The Algorithm 6 presented above is formulated to determine pointwise similarity between one-dimensional trajectories. The extension of this algorithm for M-dimensional trajectories $\mathbf{x}(t)$ and $\mathbf{y}(t)$ and for the corresponding M-dimensional sequence of differences $\mathbf{f}(t) \in F_1 \times F_2 \times ... \times F_M$ is straightforward and results in the following two modifications of the algorithm:

Algorithm 6a: Extension of Algorithm 6 for the multidimensional case.

1. *Fuzzy sets A_{F_i} 'approximately zero' are defined on each sub-universe F_i, i=1, ..., M.*
2. *Similarity measures $s_{F_i}(f,0)$, i=1, ..., M, are determined according to steps 2 and 3 of Algorithm 6 for projections of **f**(t) on the sub-universes. The result is an M-dimensional vector of partial similarities $[s_{F_1},...,s_{F_M}]$.*
3. *The vector of partial similarities can be transformed into an overall similarity measure s(**x**, **y**)=s(**f**,0) or dissimilarity measure d(**x**, **y**) in two ways:*

3.1 *The distance measure is calculated according to (5.6) for the components of the M-dimensional vector $[s_{F_1},...,s_{F_M}]$ resulting in the vector $[d_{F_1},...,d_{F_M}]$. The latter is then transformed into an overall distance using e.g. the Euclidean norm:*

$$d(\mathbf{x},\mathbf{y}) = \sqrt{\sum_{i=1,\dots,M} d^2_{Y_i}} \quad . \tag{5.7}$$

3.2 The M-dimensional vector $[s_{F_1},\dots,s_{F_M}]$ *is transformed using specific transformations (e.g. minimum, maximum, arithmetic mean, γ-operator, integral) into an overall similarity s(f, 0). A distance measure is then calculated using (5.6).*

The advantage of Algorithm 6a is that the definition of fuzzy sets A_i for each dimension (feature) allows easy interpretation, is technically simple and can be performed by an expert without any great effort. In the following algorithm a global view on fuzzy set A in a multidimensional space, rather than a consideration of single projections, is preferred.

Algorithm 6b: Extension of Algorithm 6 for the multidimensional case.

1. *The M-dimensional fuzzy set A 'approximately zero' is defined on* $F_1 \times F_2 \times \dots \times F_M$.
2. *The overall similarity measure* $s_{F_1\times\dots\times F_M}(\mathbf{f},0)$ *is obtained for the M-dimensional sequence of differences* ***f****(t) analogously to steps 2 and 3 of Algorithm 6.*
3. *The similarity measure* $s_{F_1\times\dots\times F_M}(\mathbf{f},0)$ *is transformed into a distance measure* $d(\mathbf{x},\mathbf{y})$ *using (5.6).*

The obtained distance measure between the M-dimensional trajectories **x**(t) and **y**(t) can be applied for the modification of classical pattern recognition methods to make them suitable for clustering and classification of multidimensional trajectories. This topic will be discussed in greater detail in Section 5.3.

5.2.1.1 Choice of the membership function for the definition of pointwise similarity

The two design parameters involved in the definition of pointwise similarity in Algorithm 6 are the fuzzy set A 'approximately zero' and the aggregation operator that transforms the vector of pointwise similarities into an overall degree of similarity. Fuzzy set A is defined by its membership function u(f) which represents the meaning of similarity for a certain system variable, i.e. fuzzy set A is context-dependent on the physical domain. The most frequently chosen membership functions are triangular, trapezoidal, and bell-shaped functions [Driankov et al., 1993, p. 116]. The reason for these choices is the easy parametric, functional description of the membership function that can be stored with minimum use of memory and efficiently used during calculations. The different types of membership functions are given by the following equations:

– Triangular function [Driankov et al., 1993, p. 51] shown in *Figure 55* (left):

$$u^{TA}(f,\alpha,\beta)=\begin{cases}0, & f<\alpha, f>\beta\\ (f/\alpha)-1, & \alpha\le f<0\\ 1-f/\beta, & 0\le f\le\beta\end{cases}, \tag{5.8}$$

where the degree of membership equal to 1 (called the peak value of the fuzzy set) corresponds to f=0,

– Trapezoidal function [Driankov et al., 1993, p. 52] shown in *Figure 55* (right):

$$u^{TZ}(f,\alpha,\beta,\gamma,\delta)=\begin{cases}0, & f<\alpha, f>\delta\\ (f-\alpha)/(\beta-\alpha), & \alpha\le f<\beta\\ 1, & \beta\le f\le\gamma\\ (\gamma-f)/(\delta-\gamma), & \gamma<f\le\delta\end{cases} \tag{5.9}$$

where the top of the function is not one point but an interval containing f=0,

– Bell-shaped function defined as the exponential function [Dubois, Prade, 1988, p. 50]:

$$u^{exp}(f,a)=e^{-af^2}, \tag{5.10}$$

which is shown in Figure 56 (left),

– Bell-shaped function defined as the quadratic function [Dubois, Prade, 1988, p. 50]:

$$u^{qdr}(f,a)=\frac{1}{1+af^2}. \tag{5.11}$$

which is shown in Figure 56 (right),

– Bell-shaped function defined as the logistic function [Zimmermann, Zysno, 1985, p. 153]:

$$u^S(f,a,b) = \frac{1}{1+e^{-a(f-b)}}, \tag{5.12}$$

which is the S-shaped function and shown in *Figure 57*.

Parameters a, b, α, β, γ, δ must be defined by an expert depending on the desired meaning of the fuzzy set and similarity. These parameters influence the width of the fuzzy set and in this way determine the sharpness of the similarity. In bell-shaped functions (5.10) and (5.11), parameter a has the following effect on the definition of similarity: the larger the value of parameter a, the narrower the fuzzy set and the stronger the definition of similarity (Figure 56). In logistic function (5.12), parameters a and b have the opposite effect: the larger the value of parameter a, the harder the transition from similar to non-similar, and the larger the value of parameter b, the weaker the definition of similarity (*Figure 57*).

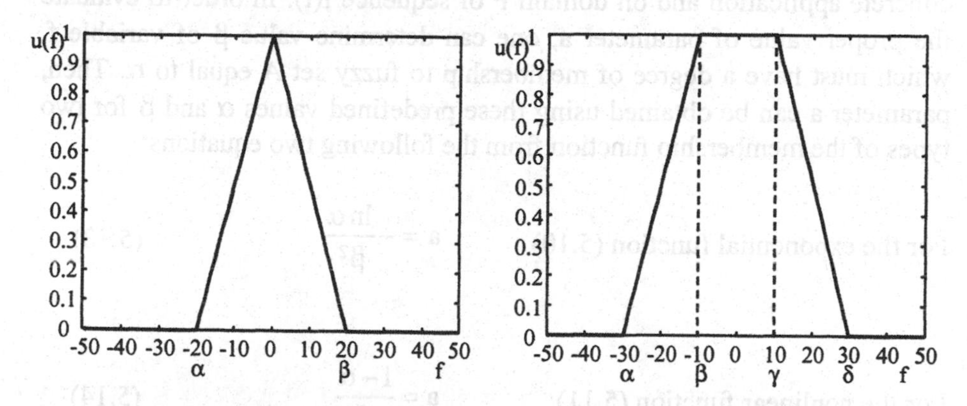

Figure 55. Triangular and trapezoidal membership functions

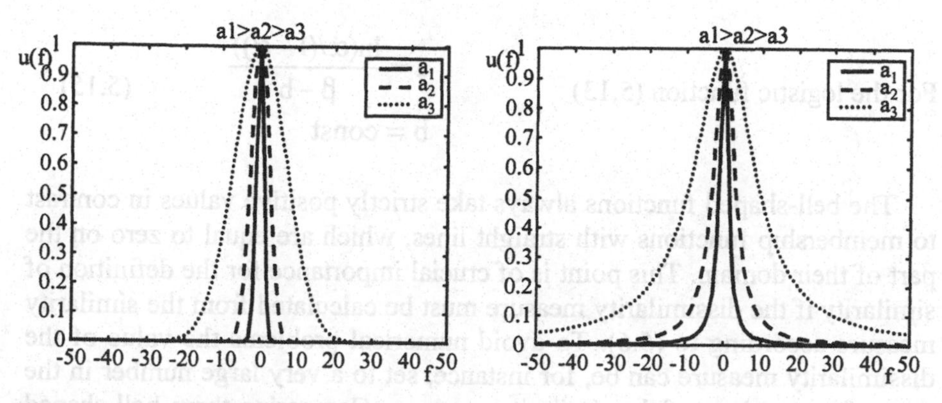

Figure 56. Exponential and non-linear membership functions

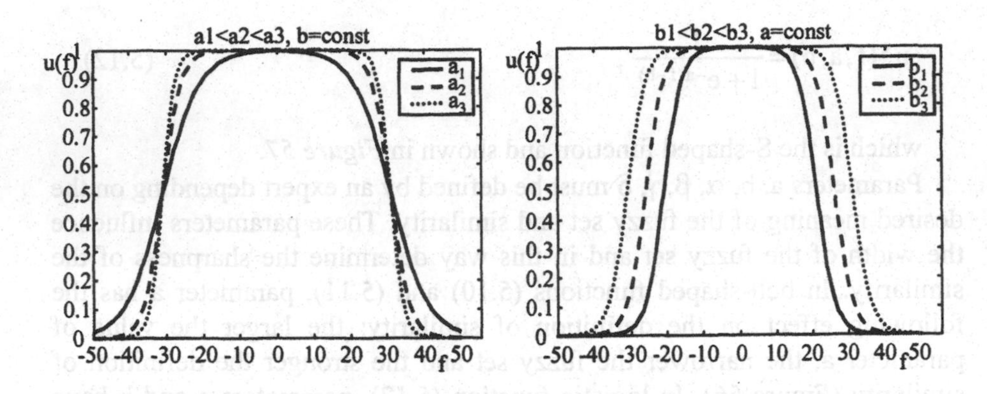

Figure 57. Logistic S-shaped membership function

The choice of parameter a in bell-shaped functions depends on the concrete application and on domain F of sequence f(t). In order to evaluate the proper value of parameter a, one can determine value β of variable f, which must have a degree of membership to fuzzy set A equal to α. Then, parameter a can be obtained using these predefined values α and β for two types of the membership function from the following two equations:

For the exponential function (5.10)

$$a = -\frac{\ln \alpha}{\beta^2} \tag{5.13}$$

For the nonlinear function (5.11)

$$a = \frac{1-\alpha}{\alpha\beta^2} \tag{5.14}$$

For the logistic function (5.13)

$$a = \frac{\ln(\alpha/(1-\alpha))}{\beta - b} \tag{5.15}$$
$$b = \text{const}$$

The bell-shaped functions always take strictly positive values in contrast to membership functions with straight lines, which are equal to zero on the part of their domain. This point is of crucial importance for the definition of similarity if the dissimilarity measure must be calculated from the similarity measure according to (5.6). To avoid numerical problems the value of the dissimilarity measure can be, for instance, set to a very large number in the case of zero values of the similarity measure. Comparing three bell-shaped functions it can be seen that the decrease for the quadratic function (5.11) is not as steep as that for the exponential function and is characterised by higher values on the largest part of its domain. The logistic function has compared to two other functions the largest interval with high values of the membership that allows to define equally high similarity within a certain threshold given by parameter b. Generally, the choice of the membership function depends on the specific application under consideration and on the subjective evaluation of the linguistic meaning of similarity.

5.2.1.2 Choice of the aggregation operator for the definition of pointwise similarity

The choice of the aggregation operator in Step 3 of Algorithm 6 for aggregating the pointwise similarities to the overall similarity measure over the whole time interval depends on the desired interpretation of the similarity between trajectories. In general there are four groups of operators which can be applied for aggregation: t-norms, t-conorms, averaging operators and compensatory operators. Triangular norms (or t-norms) represent a general class of operators for the intersection of fuzzy sets, interpreted as the logical 'and' ([Zimmermann, 1996, p. 29 ff.], [Dubois, Prade, 1988, p. 79, ff.]). The operators belonging to this class are e.g. the min, product, and bounded sum operators. Triangular conorms (or t-conorms) define a general class of aggregation operators for the union of fuzzy sets, interpreted as the logical 'or'. Some typical representatives of this class are the max-operator, algebraic sum, and bounded difference operators. The use of these operators for the aggregation of pointwise similarities leads to the following interpretation of results in both cases considered exemplarily with min and max operators. If the min operator is chosen for aggregation a lower bound of the overall similarity is obtained, since the worst value of pointwise similarity on the given time interval is considered. Choosing the max operator for aggregation yields an upper bound of the overall similarity, which provides the best pointwise similarity value over the given time interval. Both these values can be used if the number of values to be aggregated is rather small. Otherwise these aggregation operators are too restrictive and the information loss during the aggregation is too high. Moreover, if outliers are present in the sequence of pointwise similarities the aggregated value is not representative any more. Therefore, it seems more appropriate to take some average value for the aggregation.

The most frequently used operators in the class of non-parametric averaging operators are the arithmetic and geometric mean values of the sequence $s = [s_1, ..., s_p]$ given by:

$$\bar{s} = \frac{1}{p}\sum_{i=1}^{p} s_i \,, \tag{5.16}$$

$$\bar{s}_G = \sqrt[p]{s_1 \cdot \ldots \cdot s_p} \,. \tag{5.17}$$

In [Thole et al., 1979] it was shown that these operators provide an adequate model for human aggregation procedures in decision environments and perform quite well in empirical research. The geometric mean is usually interpreted as a mean growth rate of a sequence. The arithmetic mean value is considered as the best estimator for the expected value of a sequence in the sense of least mean squares. It provides a representative average value of the sequence in case of small variance. The larger the variance, the less useful the arithmetic mean value. It is used in statistics for the evaluation of unimodal approximately symmetric distributions.

In the case of a large variance in data another averaging operator called the median value is often applied, which in contrast to the arithmetic mean value gives a very robust evaluation. This operator is defined as follows:

$$\tilde{s} = \begin{cases} s_{(r+1)}, & p = 2r+1 \\ \dfrac{s_{(r)} + s_{(r+1)}}{2}, & p = 2r \end{cases} \tag{5.18}$$

where $s_{or} = [s_{(1)}, \ldots, s_{(p)}]$ is the increasingly ordered sequence of values of s. The median value is defined as the mean value of the ordered sequence if the number of elements in the sequence is odd. If the number of elements in the sequence is even there are two mean values, therefore the median is calculated as the arithmetic mean of both mean values [Sachs, 1992, p. 155]. The median value corresponds to the value of a sequence that halves the ordered sequence. It means that 50% of the values of the sequence are above and 50% of the values are below the median value. The median value is frequently used in statistics for the evaluation of asymmetric distributions and when outliers are suspected.

The median is a special case of a more general class of averaging operators in statistics called quantiles. The definition of quantiles is based on similar considerations such as that of the median. The value of the α-quantile defines a value of the increasingly ordered sequence s that $100 \cdot \alpha$% of values are below and $100 \cdot (1-\alpha)$% of values are above the α-quantile value [Sachs, 1992, p. 157]. The ordinal numbers of α-quantiles are given by:

$$Q_\alpha = (p+1)\alpha \tag{5.19}$$

and the corresponding value of the α-quantile of the ordered sequence:

$$s_\alpha = s_{or}(Q_\alpha) = s_{(Q_\alpha)} \tag{5.20}$$

When Q_α is not an integer number it is rounded off to the closest integer. Special cases of quantiles are represented by lower quartile Q_1 with α=0.25, median with α=0.5, upper quartile Q_2 with α =0.75, l-th decile DZ with α = l/10 (l = 1, ..., 9), and r-th percentile PZ with α = r/100 (r = 1, ..., 99).

Applying the quantile values for the aggregation of a sequence $s = [s_1, ..., s_p]$ of pointwise similarities, it seems reasonable to use lower quantiles with $\alpha \in [0.1, 0.5]$ whose value can be determined by an expert. In this case the aggregated value of similarity can be interpreted as the lower bound of similarity for 100·(1-α)% of points of trajectories to be compared. In other words, 100·(1-α)% of pairs of points are similar at least to a degree s_α.

An extension of aforementioned non-weighted averaging operators such as the arithmetic mean and order statistics is represented by a class of the Ordered Weighted Averaging (OWA) operators introduced by [Yager, 1988]. They are defined as a weighted sum with ordered arguments, provided that weights are normalised to one, in the following way. Let $w_1, ..., w_p$ be a set of weights such that $\sum_{i=1}^{p} w_i s_{(i)} = 1$. The OWA operator on $[0, 1]^p$ is defined as

$$OWA_{w_1,...,w_p}(s_1,...,s_p) = \sum_{i=1}^{p} w_i s_{(i)}, \tag{5.21}$$

where $s_{(i)}$ is the increasingly ordered sequence of values of s.

OWA operators are often applied in multicriteria decision theory with weights expressing importance on criteria. Considering OWA operators for the aggregation of partial values of pointwise similarity, it seems difficult to define weights for different elements of the sequence, or different points of time. Generally it is assumed that the values of similarity at all time instants are equally important, though some special situation are conceivable.

The averaging operators provide a 'fix' compensation between the logical 'and' and the logical 'or'. In order to be able to vary the degree of compensation, a more general operator called 'compensatory and' was introduced and empirically tested in [Zimmermann, Zysno, 1980], where the degree of compensation is expressed by a parameter γ. The 'compensatory and' or γ-operator is defined as follows:

$$s_\gamma = \left(\prod_{i=1}^{p} s_i\right)^{(1-\gamma)} \left(1-\prod_{i=1}^{p}(1-s_i)\right)^{\gamma}, \quad \gamma \in [0,1] \tag{5.22}$$

The γ-operator represents a combination of the algebraic product and sum. The parameter γ indicates where the operator is located between the logical 'and' and 'or', i.e. it reflects the weighting of both operators. For $\gamma = 0$ the γ-operator is reduced to the algebraic product, for $\gamma = 1$ the algebraic sum is obtained. Using the γ-operator for aggregating pointwise similarities, numerical problems may appear due to a product operation in (5.22). If the number of elements in sequence s is rather large each additional multiplication with a new element $s_i \in [0, 1]$ decreases the resulting aggregated value that tends towards zero. Since it is often necessary to transform the overall similarity value into the distance measure, this behaviour of the aggregation operator seems undesirable. Therefore, the γ-operator is better suited for aggregation of pointwise similarities over short time intervals.

The results of aggregation with different operators are illustrated in the following example. Consider two one-dimensional trajectories x(t) and y(t) defined on the time interval [1, 100] and shown in *Figure 58*. The sequence of pointwise differences f(t) of these trajectories is illustrated in *Figure 59*. It can be seen that these trajectories are relatively close to each other on the whole time interval with the exception of one point t = 2 where the difference f(2) is equal to 120. For the calculation of pointwise similarity, fuzzy set A 'approximately zero' is defined with the exponential membership function (5.11) and with parameter $a = 0.0026$. The value of a is obtained using equation (5.13), where it was defined that the value $\beta = 30$ of the difference f(t) has a degree of membership to fuzzy set A equal to $\alpha = 0.1$. The resulting sequence of pointwise similarities $u(f) = [s_1, ..., s_{100}]$ is shown in *Figure 60*.

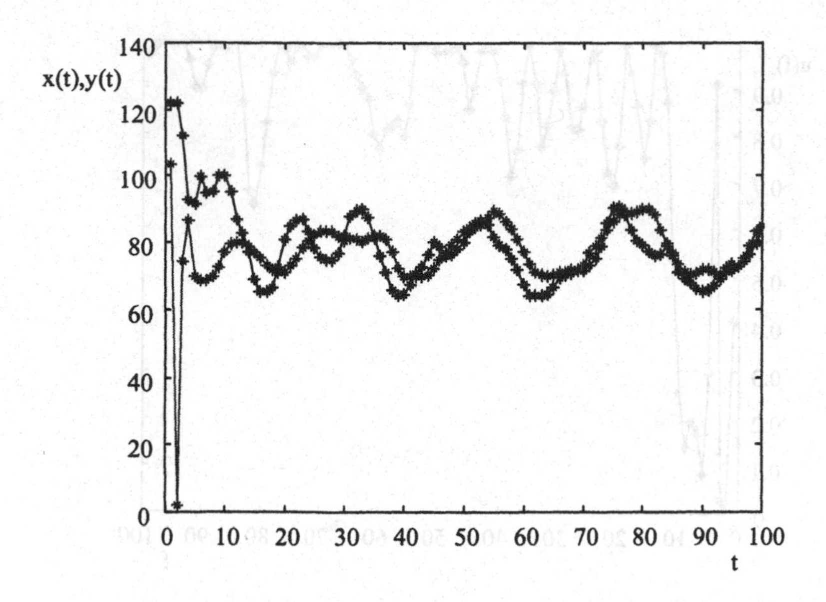

Figure 58. An example of two trajectories x(t) and y(t)

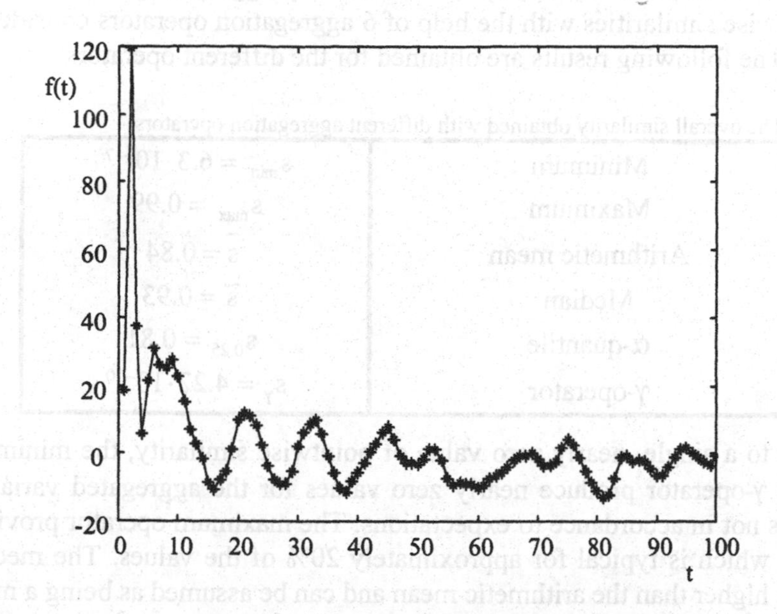

Figure 59. The sequence of differences between trajectories x(t) and y(t)

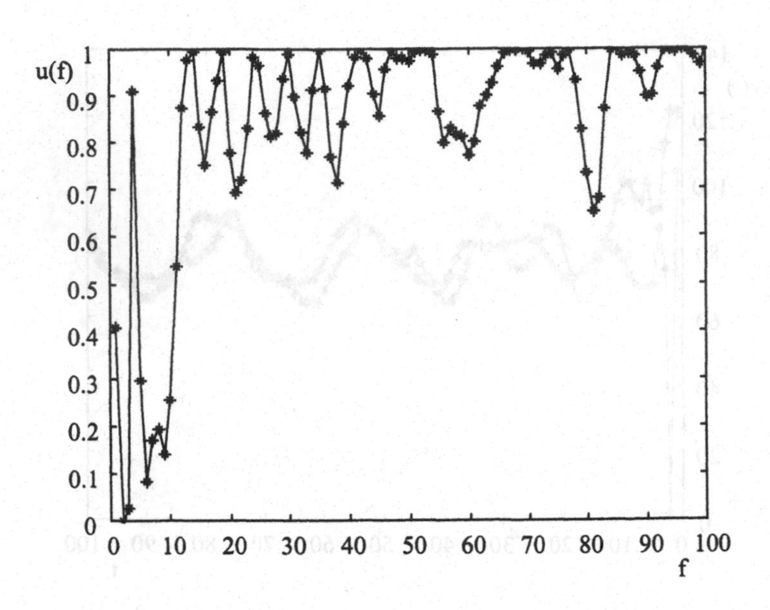

Figure 60. The sequence of pointwise similarities of trajectories x(t) and y(t)

The overall similarity s(f, 0) is calculated by aggregating sequence u(f) of pointwise similarities with the help of 6 aggregation operators considered above. The following results are obtained for the different operators:

Table 6. The overall similarity obtained with different aggregation operators

Minimum	$s_{min} = 6.3 \cdot 10^{-17}$
Maximum	$s_{max} = 0.99$
Arithmetic mean	$\bar{s} = 0.84$
Median	$\tilde{s} = 0.93$
α-quantile	$s_{0.25} = 0.81$
γ-operator	$s_{\gamma} = 4.22 \cdot 10^{-10}$

Due to a single, nearly zero value of pointwise similarity, the minimum and the γ-operator produce nearly zero values for the aggregated variable, which is not in accordance to expectations. The maximum operator provides a value which is typical for approximately 20% of the values. The median value is higher than the arithmetic mean and can be assumed as being a more reliable evaluation since the variance of data is rather large ($\sigma^2 = 0.76$). The value of 0.25-quantil equal to 0.81 provides a lower bound of similarity for 75% of the values, which is still pretty high. It can be seen that the final choice of the aggregation operator is a matter of interpretation and desired results.

5.2.2 Structural similarity measures

Searching for structural similarity between trajectories means searching for measures of similar behaviour of trajectories over time. The crucial point is, however, to determine the meaning of similarity. It obviously depends on the underlying dynamic process and the intended usage of trajectories. Sometimes the global statistical behaviour is important, whereas in other cases local changes in behaviour are relevant. It is rather difficult to define one specific measure of similarity that would be useful in all applications. The different nature of trajectories requires a problem specific selection of a suitable similarity measure for trajectories. Thus, it is more favourable to define a general framework for similarity measures and to determine in each specific case which aspects are important for the analysis of trajectories in order to formulate a precise criterion of similarity.

Thus, structural similarity can be viewed as a degree of proximity of trajectories with respect to such aspects as form, evolution trend, size or orientation in $\Re^M$ representing their behaviour. Depending on the chosen aspect different mathematical properties and characteristics of trajectories may be relevant to describe similarity, e.g. slope, curvature, position and values of extreme points or other information like smoothness or monotony. A set of relevant characteristics has to be chosen in such a way that trajectories that are close to each other with respect to these characteristics would exhibit similar behaviour. In the following sections, different approaches to determine a degree of similarity presented in the literature will be discussed. Afterwards, a general algorithm for determination of structural similarity between trajectories will be considered. Finally, a number of specific definitions of structural similarity based on relevant characteristics of trajectories will be formulated and their properties will be discussed.

5.2.2.1 Similarity model using transformation functions

A general similarity model for time series was proposed in [Das et al., 1997], which is based on local transformations of time series and can avoid the influence of outliers, different scaling and translating factors. This model can be generalised for trajectories.

Let *G* be a set of transformation functions mapping integers to integers. Set *G* could consist of either all linear or quadratic or monotone functions or the identity function. Consider two trajectories $\mathbf{x}(t) = [\mathbf{x}(t_1),\dots,\mathbf{x}(t_p)]$ and $\mathbf{y}(t) = [\mathbf{y}(t_1),\dots,\mathbf{y}(t_p)]$, which will be denoted for simplicity by sequences $\mathbf{x} = [\mathbf{x}_1,\dots,\mathbf{x}_p]$ and $\mathbf{y} = [\mathbf{y}_1,\dots,\mathbf{y}_p]$, where the time index is dropped.

According to an intuitive understanding of similar behaviour, two sequences **x** and **y** are considered *G*-similar if there is a function $g \in G$ such that a long sub-sequence **x'** of **x** can be approximately mapped into a long sub-sequence **y'** using the transformation g. It is not required that sub-sequences **x'** and **y'** consist of consecutive points of **x** and **y**, respectively. By contrast, sequences **x** and **y** are allowed to contain some missing or additional values that are not matched by sub-sequences. This means that points of **x'** and **y'** should only preserve the same relative order as in **x** and **y**, respectively.

The idea of using the longest common sub-sequence for the definition of the similarity measure between sequences of objects was first introduced in [Yazdani, Ozsoyoglu, 1996], but the model proposed could not deal with sequences modified by scaling and translation. A similar model was also introduced in [Agrawal et al., 1995] whose main disadvantage is that it does not allow outliers within time windows of a specified length and the choice of linear functions used for mapping is restricted. Hence, the similarity model of [Das et al., 1997] will be discussed below.

The notion of similar sequences can be mathematically formulated as follows.

Definition 1 [Das et al., 1997]. Given two trajectories $\mathbf{x} = [\mathbf{x}_1,...,\mathbf{x}_p]$ and $\mathbf{y} = [\mathbf{y}_1,...,\mathbf{y}_p]$ and numbers $0 < \gamma, \varepsilon < 1$. Sequences x and y are (G, γ, ε)-similar if and only if there exists a function $g \in G$ and sub-sequences $\mathbf{x}_g = [\mathbf{x}_{i_1},...,\mathbf{x}_{i_{\gamma p}}]$ and $\mathbf{y}_g = [\mathbf{y}_{j_1},...,\mathbf{y}_{j_{\gamma p}}]$, where $i_k \le i_{k+1}$ and $j_k \le j_{k+1}$ for all $k = 1,...,\gamma n - 1$, such that

$$\frac{y_{j_k}}{1+\varepsilon} \le g(x_{i_k}) \le y_{j_k}(1+\varepsilon), \quad 1 \le k \le \gamma n . \tag{5.23}$$

Parameter γ denotes the length of the sub-sequence of **x** which can be mapped to **y**. Parameter ε defines the precision of mapping, i.e. how close sub-sequences should be matched. Based on this definition the similarity measure for trajectories can be defined as a coefficient that maximises the length of sub-sequences satisfying condition (5.23).

Definition 2 [Das et al., 1997]. For given trajectories x and y, a set *G* and numbers $0 < \gamma, \varepsilon < 1$, the similarity of x and y is defined by:

$$s_{G,\varepsilon}(x, y) = \{\max \gamma \mid x, y \text{ are } (G, \gamma, \varepsilon) - \text{similar}\} . \tag{5.24}$$

This similarity measure takes its values in the interval [0, 1], higher values meaning greater similarity.

In [Das et al., 1997] only the set of linear functions is considered:

$$G_{lin} = \{g = ax + b \mid a, b \in \Re, a \neq 0\}, \tag{5.25}$$

which allows to determine similarity between trajectories with different scaling and translation factors.

If the transformation function $g \in G$ is known, then the problem of calculating the similarity measure is reduced to searching for the longest common sub-sequence between g(x) and y. This problem can be solved by dynamic programming in $O(p^2)$ time. If the length of the longest common sub-sequence is at least p-h, $h \in \aleph$, then the computation time is reduced to O(hp).

In order to find the linear transformation g that maximises the length of the longest common sub-sequence between g(x) and y (with tolerance ε), it is necessary to identify all different linear functions specified by pairs (a, b) and check each one. Since this procedure of finding the longest common sub-sequence for each pair (a, b) is very time consuming (computation time is $O(p^6)$) and is practically unfeasible, Das et al. have proposed an approximation algorithm based on the use of methods from computational geometry.

The main idea of this algorithm is to reduce the number of candidate pairs (a, b) representing different linear functions. This is done by computing bounds for possible values of a and b and by defining a grid to sample the area restricted by the bounds. For each point of the grid, the longest common sub-sequence between g(x) and y is then calculated and the best solution is chosen. The computational time of this algorithm is $O(z(1-\gamma)p^2)$, where z is the number of sampling points on the grid. The accuracy of the algorithm depends on the size of the sampling grid.

The algorithm presented above provides a clear intuitive model for measuring similarity between trajectories. However, the problem of searching for a transformation that best matches two trajectories is restricted to the class of linear functions because of its complexity. The underlying similarity model uses the shape of trajectories as a basic criterion for comparison allowing a number of local transformations such as scaling, translation and a time difference between similar sub-sequences of trajectories. For this reason, the model can not deal with global transformations of trajectories, for instance, a variation in the trend or orientation of trajectories in $\Re^M$.

5.2.2.2 Similarity measures based on wavelet decomposition

An alternative approach to using a similarity model is concerned with the consideration of statistical properties of time series and the underlying dynamic. In this case the similarity is defined using these statistical properties rather than the shape of time series.

In [Struzik, Siebes, 1998] a similarity framework based on wavelet decomposition is introduced, which provides a variety of criteria for the evaluation of similarity between time series. Using a hierarchical representation of time series (up to a certain resolution) in this framework two classes of similarity measures are considered: global similarity based on scaling properties of time series and local similarity using the scale-position bifurcation representation of time series.

To define the global statistical similarity measure it is essential that this remains unchanged for any arbitrary part of the time series, provided that the characteristics of the time series do not change in time (stationarity) or with the length of a considered part of the time series. A suitable parameter for indicating global similarity of the time series with its parts is the exponent that has to be used as a re-scaling factor for the height of the (sliding) time window (through which a part of the time series is observed) in order to obtain a new time series similar to the original one. It means that time series can be considered to be statistically similar if they possess similar values of scaling parameters such as the exponent. Struzik and Siebes propose the use of the Hurst exponent for the comparison of time series, which was developed within the domain of fractal geometry and is broadly applicable in time series analysis. The values of the Hurst exponent [Falconer, 1990] give evidence about three possible types of behaviour of the time series: a long range positive correlation in the time series expressed visually by moderate jumps, a negative correlation (so-called anti-correlation) displayed by more 'wild' behaviour and numerous intensive jumps, and the Brownian motion shown as random noise. Using the Hurst exponent as a criterion for similarity it is possible to classify time series by their scaling behaviour, whereas the form of trajectories is irrelevant.

The scaling parameters of functions can be successfully estimated by the wavelet transform as shown in [Arneodo et al., 1995]. In particular, the Hurst exponent can be derived from the wavelet transform modulus maxima (WTMM) representation of time series introduced by [Mallat, Zhong, 1992].

In order to evaluate the local similarity measure between time series, Struzik and Siebes use the bifurcation representations of two time series and estimate the degree of similarity of these representations. Bifurcations [Struzik, 1995] form a set of points reflecting the landscape of the wavelet transform tree and capture the intricate structure of the time series. Each bifurcation can be represented by its position and scale co-ordinates and the corresponding value of the wavelet transform in the bifurcation point. The bifurcations as well as wavelet transform itself can be evaluated for the time series up to a certain resolution so that only coarse features of time series are taken into account.

In order to determine a degree of similarity of two bifurcation representations, the correlation function between them is estimated as a normalised sum of correlation measures over all pairs of bifurcation points. The correlation measure is parameterised by the additional scale and position shift of representations with respect to each other in order to find a better match between representations.

The two discussed types of similarity measures are able to recognise statistically similar time series or similar parts within the time series in the presence of scaling, translation and polynomial bias and to discover regularities in the time series. Using other scale-position localised features of the time series dependent on the applied wavelet instead of bifurcations, it is possible to design a variety of local similarity measures. For their practical use, efficient techniques for increasing the accuracy of the representation with compactly supported wavelets and for optimising algorithms are required. However, it must be noted that these measures take into account only certain statistical properties of the time series, which can be relevant for some specific applications but ignore the form of the time series and their temporal behaviour. Thus, this similarity framework considers only one very special aspect of similarity, which corresponds to a somewhat narrow viewpoint on the general similarity notion between trajectories.

5.2.2.3 Statistical measures of similarity

In many applications the purpose of time series classification is to partition a collection of time series into groups, or clusters, of series with similar dynamics. It means that the notion of similarity is used to quantify the closeness of dynamic systems and their attractors rather than individual time series. For dynamic systems with M degrees of freedom, attractors are defined as a subset of M-dimensional phase space towards which almost all sufficiently close trajectories get 'attracted' asymptotically [Grassberger, Procaccia, 1983].

One of the measures characterising the local structure of attractors is the correlation integral describing the interdependence between points (observations) of time series. It measures the degree of randomness between subsequent points on the attractor by calculating the spatial correlation between pairs of points on the attractor.

Consider two sets of points on the attractor $\mathbf{x} = [\mathbf{x}_1, \ldots, \mathbf{x}_p]$ and $\mathbf{y} = [\mathbf{y}_1, \ldots, \mathbf{y}_p]$ obtained from two time series

$$x(t) = [x(t_1 + \tau), ..., x(t_1 + p\tau)] = [x(t_1 + k\tau)]_{k=1}^{p}$$

$$y(t) = [y(t_1 + \tau), ..., y(t_1 + p\tau)] = [y(t_1 + k\tau)]_{k=1}^{p}$$

with a fixed time increment τ between successive observations. The correlation integral is defined according to [Grassberger, Procaccia, 1983] by

$$C(\varepsilon) = \lim_{p\to\infty} \frac{1}{p^2} \left\{(i, j) \middle| \left\| \mathbf{x}_i - \mathbf{y}_j \right\| < \varepsilon \right\} = \lim_{p\to\infty} \frac{1}{p^2} \sum_{i,j=1}^{p} \Theta\left(\left\| \mathbf{x}_i - \mathbf{y}_j \right\| - \varepsilon\right) = \int_0^{\varepsilon} c(\varepsilon') d^M \varepsilon' \tag{5.26}$$

where $\Theta(x)$ is the Heaviside function given by

$$\Theta(x - x_0) = \begin{cases} 1 & \text{for } x \geq x_0 \\ 0 & \text{for } x < x_0 \ (x_0 > 0) \end{cases}, \tag{5.27}$$

and $c(\varepsilon)$ is the standard correlation function.

The correlation integral measures a degree to which points are grouped in the phase space by calculating the number of such pairs $(\mathbf{x}_i, \mathbf{y}_j)$, i, j = 1, ..., N, for which the distance between $\mathbf{x}_i$ and $\mathbf{y}_j$ is less than ε.

The correlation integral can be generalised to the cross-correlation sum [Kantz, 1994]:

$$C_{xy}(\varepsilon) = \frac{1}{p^2} \sum_{i,j=1}^{p} \Theta\left(\left\| x_i - y_j \right\| - \varepsilon\right) \tag{5.28}$$

and can be used directly or after normalisation

$$s(x, y) = \frac{C_{xy}}{\sqrt{C_{xx} C_{yy}}} \tag{5.29}$$

as a similarity measure between time series ([Manuca, Savit, 1996], [Schreiber, Schmitz, 1997]).

Another possibility to define the similarity measure is to use the cross-prediction error based on some time series models [Schreiber, 1997].

The usefulness and applicability of similarity measures based on cross-correlation sum and cross prediction error was illustrated by some examples in [Schreiber, Schmitz, 1997], where clustering of time series was based on dissimilarities between time series defined as d(x(t),y(t)) = 1 - s(x(t),y(t)). It was shown that time series generated with different parameter settings can be clearly separated into distinct clusters. In order to visualise clustering results, an abstract space of dynamic properties of time series was introduced. Once c clusters are formed, they can be represented in the space of mutual dissimilarities of dynamic objects, i.e. each co-ordinate i, i=1, ..., c, is the average dissimilarity of each object to the objects in cluster i. This representation allows a visual judgement on the presented structure in dynamic objects based on their mutual dissimilarities.

5.2.2.4 Smoothing of trajectories before the analysis of their temporal behaviour

As stated in 5.2, trajectories may contain undesired deviations due to random fluctuations or measurement errors. In order to reduce their influence on the analysis results and to filter the specific behaviour of trajectories, smoothing techniques are often applied before the actual analysis starts. One of the most frequently used methods for smoothing is a median filter of length r. Consider a sequence of p measurements $\mathbf{x} = [\mathbf{x}_1, \ldots, \mathbf{x}_p]$. The values of a new sequence $\bar{\mathbf{x}}$ are calculated as mean values over r values in the following way:

if r is an uneven number, (r-1)/2 left and (r-1)/2 right neighbouring values are considered

$$\bar{x}_k = \frac{x_{i-(r-1)/2} + \ldots + x_{i-1} + x_i + x_{i+1} + \ldots + x_{i+(r-1)/2}}{r} = \frac{1}{r} \sum_{j=i-(r-1)/2}^{i+(r-1)/2} x_j \qquad (5.30)$$

where $k = 1 + \frac{r-1}{2}, \ldots, p - \frac{r-1}{2}$,

if r is an even number, r/2 left and r/2 right neighbouring values are considered

$$\bar{x}_k = \frac{x_{i-r/2} + \ldots + x_{i-1} + x_i + x_{i+1} + \ldots + x_{i+r-1/2}}{r+1} = \frac{1}{r} \sum_{j=i-r/2}^{i+r/2} x_j \qquad (5.31)$$

,

where $k = 1 + \frac{r}{2}, \ldots, p - \frac{r}{2}$.

The example in *Figure 61* illustrates the effect of smoothing of trajectories. The original trajectory x(t) contains a significant part of random fluctuations. Applying a median filter of length 250, a smooth trajectory preserving the temporal behaviour but filtered from noise is obtained.

Smoothing with equations (5.30) and (5.31) can be used for the off-line analysis of trajectories obtained during a period of time from different dynamical systems. If trajectories are monitored over time and parts of them are compared with each other during the on-line analysis, these equations are not applicable in this form since future values of the trajectories are not available. Therefore, in order to smooth a trajectory the mean value is calculated over the last r values and is assigned to the current value of a trajectory. Mean values are updated along with new incoming values of a trajectory. A smoothed trajectory is characterised by a particular property, this being that it is situated above an original trajectory if a trajectory exhibits a decreasing trend, and under a trajectory if a trajectory exhibits an increasing trend. This modified smoothing technique is broadly used in applications of time series analysis such as the technical analysis of share prices [Welcker, 1991, p. 48].

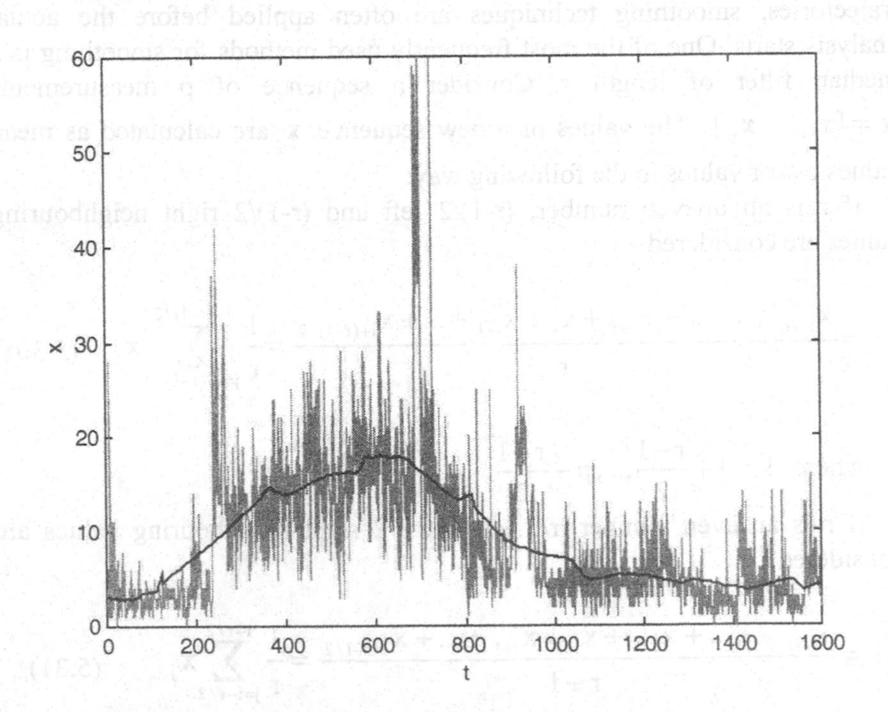

Figure 61. A trajectory x(t) before and after smoothing

The length r of the interval for smoothing is chosen depending on the length of sequence x and on the desired accuracy. For instance, for the analysis and forecasting of share prices, trajectories (time series) obtained over 6 or 12 months on the basis of daily measurements are usually smoothed over the interval of 100 or 200 measurements, respectively [Welcker, 1991, p. 48]. For traffic control and traffic forecasts [Engels, Chadenas, 1997], the data is usually collected every minute during one day and then aggregated over five, ten or fifteen minute intervals. In the fault diagnosis in anaesthesia [Vesterinen et al., 1997], measurement data of gas flow signals are gathered each second and sequences of, for example, 200 measurements are sent to the fault diagnosis software for analysis. To eliminate deviations in gas flow signals a median filter of length 3 is sufficient.

If a trajectory has a wavy form it can be represented by trigonometric polynoms. In order to obtain them, a trajectory is first smoothed by some median filter and then the discrete Fourier transform (DFT) is applied. The resulting function is a sum of waves with different periods weighted by the Fourier coefficients. Half of the coefficients are responsible for waves with higher frequencies and can be neglected. A smooth trajectory is obtained by applying the inverse Fourier transform to the remaining coefficients.

5.2.2.5 Similarity measures based on characteristics of trajectories

The matching approaches based on distance measures in the feature space or functional transformations and the statistical approaches based on cross-correlation measures or on specific parameters of time series presented in previous sections, yield a quantitative evaluation of similarity between trajectories. In many cases the evaluation of similarity should, however, reproduce the judgement of an expert based mainly on qualitative and possibly subjective features. Since in such kind of problems similarity can not be inferred by quantitative analysis, the modelling of similarity as a cognitive process simulating human decision making seems to be appropriate [Binaghi et al., 1993].

Among various knowledge representation formalisms proposed for reasoning in the presence of uncertainty and imperfect knowledge typical for human cognitive processes [Graham, Jones, 1988], fuzzy set theory provides the most plausible tool for modelling cognitive processes, in particular a recognition process [Pedrycz, 1990]. Since the evaluation of similarity is one of the steps of the recognition process, it presents itself to take advantage of the fuzzy framework in this context as well. With the help of fuzzy set theory, it is possible to model a gradual representation of similarity according to human judgement and in this way to fuzzify the difference between similar and non-similar dynamic objects.

One of the first implementations of the structural similarity was presented in [OMRON ELECTRONICS GmbH, 1991] in the area of signal analysis. The purpose of the designed system was to distinguish between different types of balls based on the analysis of oscillation spectra after the impact of balls on the base. The identification of a specific oscillation pattern allows for the recognition of a certain ball type. For the definition of structural similarity between balls four characteristics concerning the number of high and low frequency impacts, pulse frequency and oscillation duration were considered and modelled as linguistic variables. Using the fuzzy rule-based system it was possible to recognise slight differences between impact sequences and to classify balls into three groups. This application example shows the advantages of using fuzzy structural similarity in the pattern recognition process which is carried out in accordance with human subjective evaluation.

A general algorithm to determine a measure of structural fuzzy similarity between trajectories based on relevant characteristics of these trajectories is proposed in [Joentgen, Mikenina et al., 1999b, p. 84]. The idea of this algorithm is to represent the behaviour of trajectories by a finite number of temporal characteristics and to measure the similarity of trajectories with respect to each of these characteristics using a fuzzy definition of similarity. In this algorithm expert knowledge is incorporated at two points: by the choice of characteristics that must adequately describe the temporal structure of trajectories and by the subjective evaluation of the admissible difference of a characteristic's values for two trajectories (from the viewpoint of their similarity). The subjective evaluations for each characteristic are given in the form of fuzzy sets, which are used as a basis of the similarity definition. By choosing different sets of characteristics a number of specific definitions of similarity can be derived.

For the sake of simplicity, the algorithm is formulated for the case of one-dimensional trajectories x(t) and y(t) and will be generalised to a multidimensional case afterwards.

Algorithm 7: Determination of structural similarity between trajectories [Joentgen, Mikenina et al., 1999b, p. 84].

1. *A set of relevant characteristics* $\{K_1,\ldots,K_L\}$ *used for description of structural similarity is chosen.*
2. *For each characteristic* K_i, $i=1, \ldots, L$, *a fuzzy set* A_i *labelled 'admissible difference for characteristic* K_i*' with membership function* u_i *is defined.*
3. *All characteristics' values* $K_i(x)$ *for the trajectory* $x(t)$ *and* $K_i(y)$ *for the trajectory* $y(t)$ *are calculated.*

4. *For each characteristic K_i, i=1, ..., L, the difference* $\Delta K_i = |K_i(x) - K_i(y)|$ *is calculated.*

5. *The degree of membership* $s_i = u_i(\Delta K_i)$ *of the difference* ΔK_i *to the fuzzy set A_i is calculated for each characteristic K_i. These membership values can be interpreted as similarities between trajectories x(t) and y(t) with respect to the chosen characteristics.*

6. *Finally the vector* $[s_1,...,s_L]$ *of partial similarities is transformed using specific transformations (e.g. γ-operator, fuzzy integral, minimum, maximum) into a real number s(x, y) expressing the overall degree of similarity:* $s(x,y) = aggr(s_1,...,s_L)$.

Each value of a membership function u_i of a fuzzy set A_i, i=1, ..., L, expresses a degree to which a certain difference of two values of characteristic K_i can be considered admissible and two trajectories can be considered similar with respect to this characteristic. Obviously the maximum degree of the membership function corresponds to the difference value equal to zero, i.e. trajectories are definitely similar if their characteristics' values are equal. The shape and the support of the membership function should be defined context dependently by an expert. Thus, fuzzy sets represent an expert's understanding and his/her subjective judgement as regards the meaning of similarity.

The extension of the algorithm to the case of M-dimensional trajectories **x**(t) and **y**(t), **x**, **y** $\in Y_1 \times Y_2 \times ... \times Y_M$, is straightforward and leads to the following two modifications of the algorithm:

Algorithm 7a: Extension of Algorithm 7 to the multidimensional case.

1. *A set of relevant characteristics* $\{K_1,...,K_L\}$ *used for description of structural similarity is chosen.*

2. *For each characteristic K_i fuzzy sets* $A_i^{Y_j}$ *'admissible difference for characteristic K_i' are defined on each sub-universe Y_j, j=1, ..., M.*

3. *Characteristics values $K_i(\mathbf{x})$ for the trajectory **x**(t) and $K_i(\mathbf{y})$ for the trajectory **y**(t) are calculated with respect to each of M dimensions resulting in vectors* $K_i(\mathbf{x}) = [K_i(x_1),...,K_i(x_m)]$ *and* $K_i(\mathbf{y}) = [K_i(y_1),...,K_i(y_m)]$.

4. *For each characteristic K_i an M-dimensional difference vector* $\Delta K_i = |K_i(\mathbf{x}) - K_i(\mathbf{y})|$ *is calculated.*

5. *Partial similarity measures* $s_i^{Y_j}$, *j=1, ..., M, with respect to characteristic* K_i *are determined according to Algorithm 7 for each dimension. The result is the M-dimensional vector of partial similarities* $[s_i^{Y_1},...,s_i^{Y_M}]$. *This vector can be aggregated to a real number* s_i *expressing a partial similarity measure with respect to characteristic* K_i *in three ways:*

5.1 *The vector* $[s_i^{Y_1},...,s_i^{Y_M}]$ *is transformed using specific transformations (e.g. minimum, maximum, arithmetic mean, γ-operator, integral,) into a real number* $s_i = \mathrm{aggr}(s_i^{Y_1},...,s_i^{Y_M})$. *The resulting vector* $[s_1,...,s_L]$ *of partial similarities is transformed using one of the given specific transformations into a real number s(x, y) expressing the overall degree of similarity:* $s_{Y_1 \times Y_2 .. \times X_M}(\mathbf{x},\mathbf{y}) = \mathrm{aggr}(s_1,...,s_L)$.

5.2 *Components of the vector* $[s_i^{Y_1},...,s_i^{Y_M}]$ *are transformed into distance measures expressing partial dissimilarities according to equation (5.6). The resulting vector* $[d_i^{Y_1},...,d_i^{Y_M}]$ *is then transformed into the distance measure* d_i *with respect to characteristic* K_i *using e.g. the Euclidean norm (5.7). The aggregation of partial distance measures* d_i *to an overall distance* $d_{Y_1 \times Y_2 .. \times X_M}(\mathbf{x},\mathbf{y})$ *is performed with the help of the Euclidean norm as well.*

5.3 *The vector* $[s_i^{Y_1},...,s_i^{Y_M}]$ *is transformed using specific transformations into a real number* $s_i = \mathrm{aggr}(s_i^{Y_1},...,s_i^{Y_M})$ *Components of the resulting vector* $[s_1,...,s_L]$ *are transformed into distance measures expressing partial dissimilarities according to equation (5.6). The resulting vector* $[d_1, ..., d_L]$ *is aggregated to an overall distance measure* $d_{Y_1 \times Y_2 .. \times X_M}(\mathbf{x},\mathbf{y})$ *using the Euclidean norm (5.7).*

The advantage of Algorithm 7a is the separate definition of fuzzy sets A_i for each dimension, which due to the better interpretability can simplify the task for an expert. However, the algorithm requires an additional aggregation step for the transformation of M-dimensional vectors into real numbers.

Algorithm 7b: Extension of Algorithm 7 for the multidimensional case.

1. *A set of relevant characteristics* $\{K_1,...,K_L\}$ *used for describing structural similarity is chosen.*

2. *For each characteristic K_i the M-dimensional fuzzy set Ai 'admissible difference for characteristic K_i' is defined on* $Y_1 \times Y_2 \times ... \times Y_M$.

3. *Characteristics values $K_i(x)$ for the trajectory x(t) and $K_i(y)$ for the trajectory y(t) are calculated with respect to each of M dimensions resulting in vectors* $K_i(x) = [K_i(x_1),...,K_i(x_m)]$ *and* $K_i(y) = [K_i(y_1),...,K_i(y_m)]$.

4. *For each characteristic K_i an M-dimensional difference vector* $\Delta K_i = |K_i(x) - K_i(y)|$ *is calculated.*

5. *Partial similarity measures* $s_i = u_i(\Delta K_i), i = 1,..., L,$ *are calculated as degrees of membership of M-dimensional difference vectors* ΔK_i *to the M-dimensional fuzzy set A_i.*

6. *Partial degrees of similarity are transformed, analogously to the one-dimensional case, into the overall degree of similarity* $s_{Y_1 \times Y_2 .. \times X_M}(x, y)$.

In contrast to Algorithm 7a, all calculations in Algorithm 7b are performed for M-dimensional vectors and not for components of these vectors. This means that one has to deal with multidimensional fuzzy sets whose definition is usually more sophisticated compared to the one-dimensional case.

The general definition of structural similarity according to Algorithm 7 can be used to derive a set of specific definitions that take into account different properties of trajectories. It should be emphasised that each particular definition of structural similarity considers a certain aspect of the similarity notion and is suitable only for a certain number of problems. In the following section a number of definitions of structural similarity with respect to different properties of trajectories is introduced. This list can be extended by combinations of these definitions and by introducing other relevant properties of trajectories.

Definition of structural similarity based on the trend of trajectories. The trend of a trajectory is defined as a slow but consistent unidirectional change of a trajectory [Vesterinen et al., 1997]. It describes a simple general behaviour of a trajectory. The form of a trend can be e.g. a slope, a step or a damped step. Trends that are shaped like a slope can be estimated by fitting a straight line to a given set of measurements (constituting a trajectory) with the least squares approximation. Given a set of measurements $(\mathbf{x}_1,t_1)$, $(\mathbf{x}_2,t_2)$, ..., $(\mathbf{x}_p,t_p)$, where $\mathbf{x}_k$, k = 1, ..., p, are known to be subject to measurement errors, the task of approximation is to estimate a regression line $F(t)=a_1t+a_2$. Regression coefficients a_1 and a_2 are determined so that the values

$\hat{\mathbf{x}}_k = a_1 t_k + a_2, k = 1,...,p$ are as close as possible to the values $\mathbf{x}_1$, $\mathbf{x}_2$, ..., $\mathbf{x}_p$ in a least square sense. This requires the minimisation of the sum of squared errors:

$$e^2 = \sum_{k=1}^{p} (\mathbf{x}_k - \hat{\mathbf{x}}_k)^2 . \tag{5.32}$$

Regression coefficients can also be estimated using the covariance of $\mathbf{x}(t)$ and t [Rüger, 1989, p. 104]:

$$a_1 = \frac{\mathrm{Cov}(\mathbf{x}, t)}{\mathrm{Var}(t)}, \qquad a_2 = \bar{\mathbf{x}} - a_1 \bar{t}, \tag{5.33}$$

where

$$\mathrm{Cov}(\mathbf{x}, t) = \frac{1}{p} \sum_{k=1}^{p} (\mathbf{x}_k - \bar{\mathbf{x}})(t_k - \bar{t}) \tag{5.34}$$

$$\mathrm{Var}(t) = \sum_{k=1}^{p} (t_k - \bar{t})^2 = \sum_{k=1}^{p} t_k^2 - \bar{t}^2 \tag{5.35}$$

$\bar{t}$ and $\bar{\mathbf{x}}$ are the mean values of t and $\mathbf{x}(t)$, respectively, over the time interval $[t_1, t_p]$.

Definition 3. Two trajectories x(t) and y(t) are considered similar with respect to their temporal trend if they are characterised by similar values of parameter a_1.

This definition of structural similarity is illustrated in *Figure 62* where trajectories represent three classes of behaviour: increasing, constant and decreasing. Although these trajectories are rather similar with respect to their form, they are not considered to be similar with respect to their temporal trend.

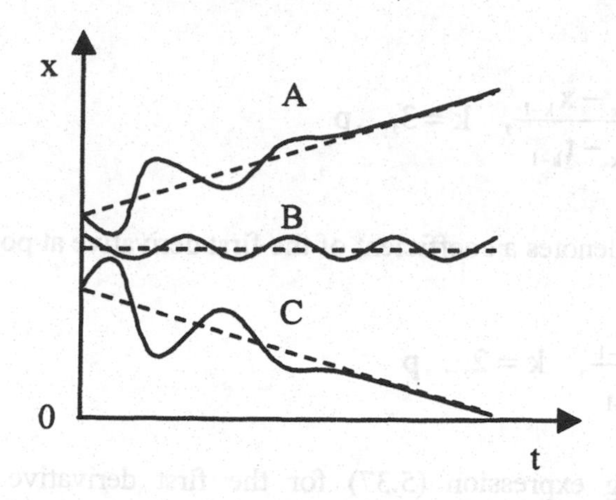

Figure 62. Structural similarity based on the temporal trend of trajectories

The corresponding similarity measure s(**x**, **y**) can be calculated with Algorithm 7, where parameter a_1 is used as characteristic K of trajectories and fuzzy set A denotes the 'admissible difference for the trend'. This structural similarity can be applied to find clusters of trajectories with a similar temporal trend, where the translation of trajectories in the feature space along M dimensions or translation over time and a degree of their fluctuation are irrelevant.

If the translation of trajectories is a relevant criterion for analysis, i.e. the location of trajectories in the feature space must be considered together with the trend, then Definition 3 of the similarity measure can be presented as follows.

Definition 4. Two trajectories x(t) and y(t) are considered similar with respect to their temporal trend and location if they are characterised by similar values of parameter a_1 and a_2.

Definition of structural similarity based on the curvature of trajectories.

The curvature of a trajectory at each point describes the degree to which a trajectory is bent at this point. It is evaluated by the coefficients of the second derivative of a trajectory at each point that can be defined by the following equation[3] (for one-dimensional trajectory):

---

[3] Equations (5.36) and (5.37) represent only one of the possibilities to calculate numerically the first and the second derivatives.

$$cv_k = x_k'' = \frac{x_k' - x_{k-1}'}{t_k - t_{k-1}}, \quad k = 3,...,p \tag{5.36}$$

where x_k' denotes a coefficient of the first derivative at point x_k given by

$$x_k' = \frac{x_k - x_{k-1}}{t_k - t_{k-1}}, \quad k = 2,...,p \tag{5.37}$$

Substituting expression (5.37) for the first derivative into equation (5.36), the following equation based on the values of the original trajectory is obtained for the coefficients of the second derivative:

$$cv_k = x_k'' = \frac{x_k - 2x_{k-1} + x_{k-2}}{(t_k - t_{k-1})^2}, \quad k = 3,...,p \tag{5.38}$$

If trajectories possess local minima and maxima, which can be detected by looking for a sign change in the values of the first derivative, then it can be sufficient to consider the coefficients of the second derivative only in these specific points where the curvature is maximum. The most distinctive feature when considering the curvature is the sign of coefficients of the second derivative. If the coefficient is positive over a certain time period, then a trajectory is convex on this interval (open to the top). If the coefficient is negative over a certain time period a trajectory is concave (open to the bottom). If the coefficient is equal to zero at some point which is then called an inflection point, there is no curvature at this point. Inflection points appear in oscillating trajectories and indicate the change of curvature from convex to concave or vice versa. All linear functions are characterised by zero curvature at all points.

Definition 5. Two trajectories x(t) and y(t) are considered similar with respect to their curvature if they are characterised by similar coefficients cv_k of the second derivative.

This definition of structural similarity is illustrated in *Figure 63* where three types of trajectories are represented: concave trajectory B, convex trajectory C and trajectory A with oscillating behaviour changing from concave to convex.

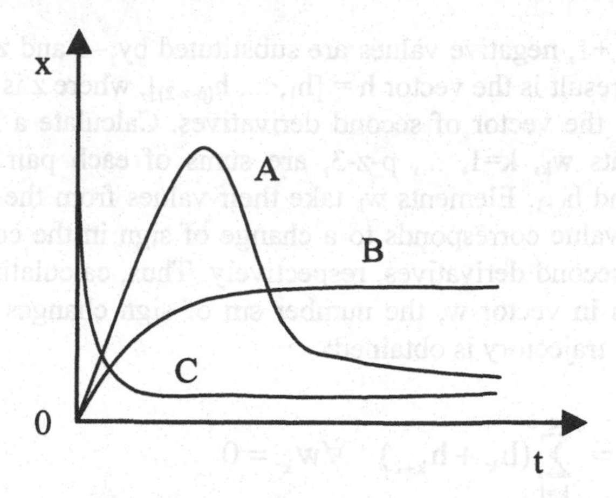

Figure 63. Structural similarity based on curvature of trajectories

The similarity measure s(**x**, **y**) based on the curvature can be calculated using Algorithm 7 or one of its extensions, where characteristic K is a vector of coefficients of the second derivative $K=[cv_3,...,cv_p]$. For multidimensional trajectories, components of vectors K(**x**) and K(**y**) are also vectors consisting of partial second derivatives. Fuzzy set A denotes the 'admissible difference for the curvature'.

If a temporal translation is irrelevant to the process of recognising similar patterns in trajectories, then vectors K(**x**) and K(**y**) of characteristic K obtained for two trajectories can be cyclically shifted with respect to each other and the similarity measure is defined for each combination. In this way the maximum similarity corresponding to the best match of trajectories with respect to their curvature can be found.

The similarity measure based on the curvature is particularly suitable for trajectories with a low number of fluctuations and a wavy form. This measure is, however, sensitive to scaling, i.e. a trajectory transformed by a scaling factor has a different curvature than the initial one.

Definition of structural similarity based on the smoothness of trajectories.

The smoothness of a trajectory describes the degree of oscillations in its behaviour, and can be characterised by the number of sign changes in its second derivative. The higher this number, the more oscillations the trajectory contains and the less smooth it is.

In order to estimate smoothness of a trajectory $x(t) = [x_1, ..., x_p]$, consider a vector of coefficients of the second derivative of a trajectory and transform it to a binary vector h using the following coding rule: positive values are substituted by +1, negative values are substituted by –1 and zero values are dropped. The result is the vector $h = [h_1, ..., h_{(p-z-2)}]$, where z is the number of zero values in the vector of second derivatives. Calculate a new vector w, whose elements w_k, k=1, ..., p-z-3, are sums of each pair of successive elements h_k and h_{k+1}. Elements w_k take their values from the set {-2, 0, 2}, where a zero value corresponds to a change of sign in the code vector and the vector of second derivatives, respectively. Thus, calculating the number of zero values in vector w, the number sm of sign changes of the second derivative of a trajectory is obtained:

$$sm = \sum_{k=1}^{p-z-3} w_k = \sum_{k=1}^{p-z-3} (h_k + h_{k+1}) \quad \forall w_k = 0 \tag{5.39}$$

Definition 6. Two trajectories x(t) and y(t) are considered similar with respect to their smoothness if they are characterised by similar values of parameter sm.

In *Figure 64* trajectories with different degrees of smoothness are shown. Trajectories A and B are both characterised by oscillating behaviour, where A (sm = 6) is more oscillating than B (sm = 4). In contrast, trajectory C exhibits very smooth behaviour (sm = 0).

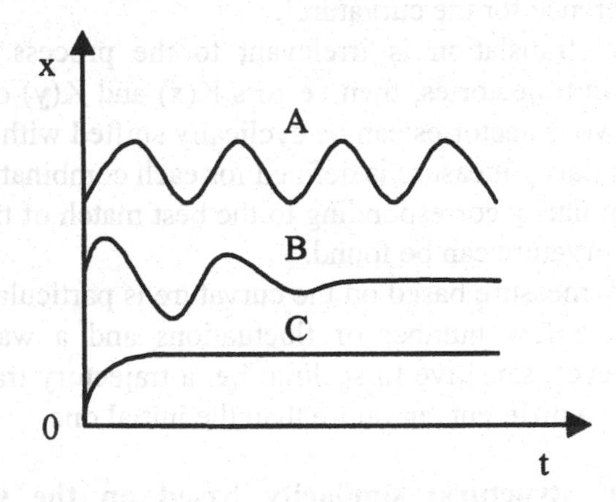

Figure 64. Structural similarity based on the smoothness of trajectories

The similarity measure s(**x**, **y**) based on the smoothness of trajectories can be obtained using Algorithm 7, or one of its extensions, where characteristic K is chosen to be parameter sm of smoothness and fuzzy set A denotes the 'admissible difference for smoothness'.

This similarity measure can be applied to the comparison of trajectories, where oscillating behaviour is assumed. It is more general than the similarity measure based on the pointwise curvature of trajectories since it considers only the total value of oscillations but not the degree of curvature of each separate wave. This similarity measure is suitable for trajectories for which scaling and translation transformations are irrelevant, but it is sensitive to outliers. To avoid the influence of outliers on the value of similarity, trajectories must be pre-processed and smoothed by a median filter according to (5.30) and (5.31) or using some other smoothing technique.

Definition of structural similarity based on specific temporal parameters of trajectories.

Similarity measures based on curvature and smoothness characterise the general behaviour of trajectories with respect to their form and oscillating character. For some problems, it may be important to consider concrete parameters of single waves appearing in a trajectory in order to identify similar temporal patterns in trajectories.

Consider the trajectory shown in *Figure 65*. This temporal pattern in the trajectory can be decomposed into segments indicating local trends in such a way that each segment is bounded by inflection points or an inflection point and an extreme value ([Bakshi et al., 1994], [Angstenberger et.al, 1998]).

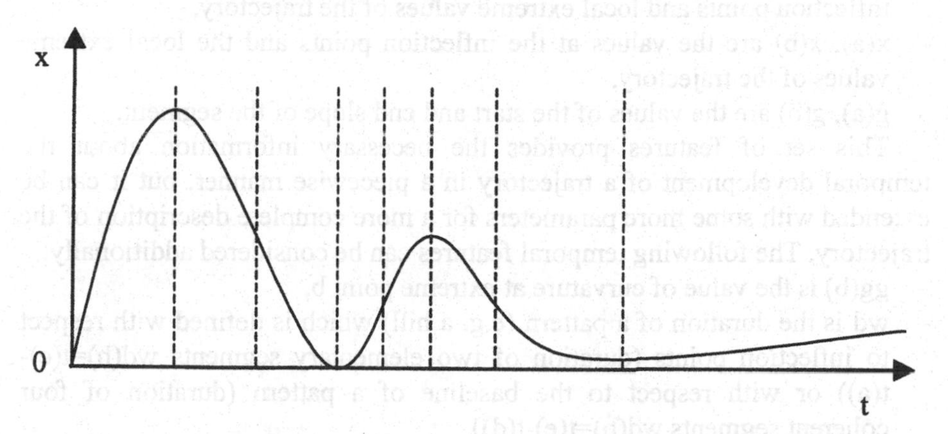

Figure 65. **Segmentation of a temporal pattern of a trajectory according to elementary trends**

Seven types of elementary segments (trends) can be distinguished in a trajectory, each of which is characterised by a constant sign of the first and second derivatives (*Figure 66*).

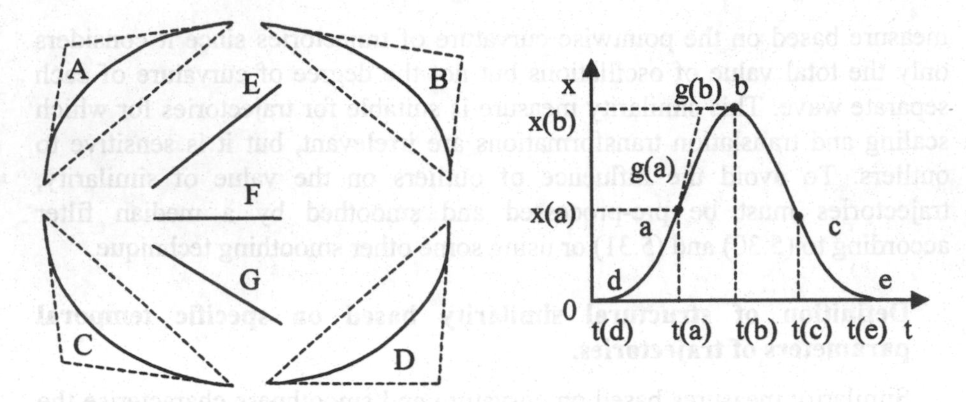

Figure 66. Qualitative (left) and quantitative (right) temporal features obtained by segmentation

Such a triangular representation of trends provides qualitative features for a description of the segments. In order to derive quantitative information from the segments, these are described by the following set of temporal characteristics (see segment a-b in *Figure 66*):

- t(a), t(b) are the start and end time of the segment, i.e. time instants of inflection points and local extreme values of the trajectory,
- x(a), x(b) are the values at the inflection points and the local extreme values of the trajectory,
- g(a), g(b) are the values of the start and end slope of the segment.

This set of features provides the necessary information about the temporal development of a trajectory in a piecewise manner, but it can be extended with some more parameters for a more complete description of the trajectory. The following temporal features can be considered additionally:

- gg(b) is the value of curvature at extreme point b,
- wd is the duration of a pattern (e.g. a hill) which is defined with respect to inflection points (duration of two elementary segments wd(b)=t(c)-t(a)) or with respect to the baseline of a pattern (duration of four coherent segments wd(b)=t(e)-t(d)),
- tn_1 is the time interval until the first zero value of a trajectory (tn_1 = t(e)),
- tn_2 is the time interval until the second zero value of a trajectory,
- I is an integral of the part of a trajectory until its first zero value,
- CG is the centre of gravity of the part of a trajectory until its first zero value,
- Med is the median of a trajectory's values,
- RV is the range of values of a trajectory,
- lv is a plateau or the limiting value of a trajectory,
- statistical measures (e.g. mean value, standard deviation, correlation coefficients between segments of a trajectory).

All these temporal parameters, or characteristics, may be calculated for the original trajectory as well as for any derived trajectory (e.g. derivatives, transformations, etc.). Furthermore, the definition of these parameters may be limited to parts of its time domain (e.g. centre of gravity of a trajectory on the domain $t \in [t_1, t_2]$). The set of temporal parameters describing a trajectory can be determined by an expert taking into consideration specific properties of the dynamic system under consideration. The choice of the set of relevant temporal parameters can be simplified if the shape of specific patterns in a trajectory is known. For instance, for the analysis of ECG waveforms for medical diagnosis, each pattern (ECG cycle) can be represented by five characteristic points including the starting and the termination points and three top points of the waves. In order to describe the shape of each pattern four parameters are proposed in [Nemirko et al., 1994]: the duration of the pattern, the amplitude of the pattern as a difference between extreme values, the area limited by the pattern (integral), and the mean of the amplitude range with respect to the baseline. These parameters are sufficient for the comparison of ECG patterns and thus can be used for the definition of similar trajectories.

Definition 7. Given a set of relevant temporal parameters, trajectories x(t) and y(t) can be considered similar if the values of temporal parameters describing elementary patterns in trajectories are similar.

In order to determine the structural similarity measure based on specific temporal parameters of trajectories, the parameters are summarised to a vector of relevant characteristics K and fuzzy sets A_i 'admissible difference for a parameter K_i' are defined for each parameter K_i, $i=1, ..., L$. Similarity measure $s(\mathbf{x}, \mathbf{y})$ is calculated according to Algorithm 7 for one-dimensional trajectories or Algorithm 7a or Algorithm 7b for multidimensional ones.

The parameters listed above allow a precise description of the shape of temporal patterns present in a trajectory. They take into account the number and size of hills, their slope and curvature, the moments of their appearance and their duration, where scaling and translation factors have an effect on the parameter values. This similarity measure is suitable for the recognition and comparison of specific patterns in trajectories.

Definition of structural similarity based on the peaks of trajectories.

In a number of applications, the behaviour of trajectories is characterised by significant increases in the values of a measured characteristic x during a short period of time. These abrupt increases are referred to as peaks. They contain information about the state and development of the underlying dynamic system and can be used as a criterion for the comparison of trajectories.

In peak analysis the following parameters of peaks are usually used: height of a peak, position of a peak (time instant of peak appearance), peak limits, and the peak area [Eckardt et al., 1995]. If the number of peaks in a trajectory is too large, only those peaks that are higher than a predefined threshold are considered as being relevant.

Definition 8. Trajectories x(t) and y(t) are considered similar if they possess similar peaks in similar positions (*Figure 67*).

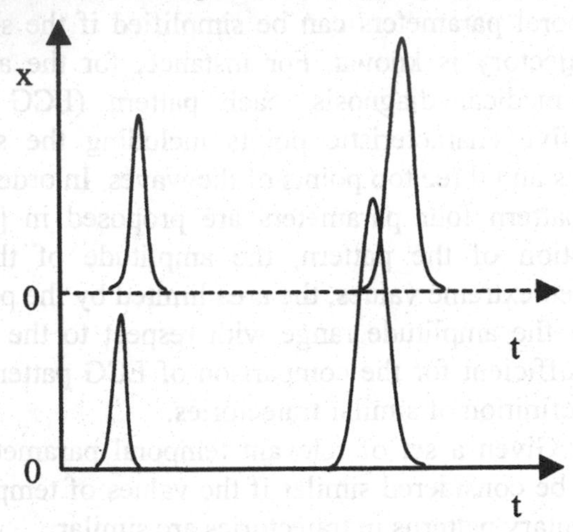

Figure 67. Similarity measure based on peaks

In order to calculate the similarity measure with respect to the peaks of trajectories, peaks are detected in each trajectory and are used as relevant characteristics K_i, i=1, ..., L, where L is the number of detected peaks. Parameters of each peak h_i are summarised to a vector of characteristic values $K_i = [K_{i,1},\ldots,K_{i,pn}]$, where pn is the number of peak parameters. In this context, peaks can be interpreted as pn-dimensional characteristics of trajectories. The resulting pn×L dimensional matrix $K=[K_{ij}]$, , i=1, ..., L, j=1, ..., pn, contains all characteristic values of a trajectory. Fuzzy sets A_j, j=1, ..., pn, the 'admissible difference of a peak parameter j' have to be defined for each peak parameter. Similarity measures are first determined for each peak with respect to peak parameters by applying Algorithm 7 or Algorithm 7a or Algorithm 7b, and aggregated over peak parameters to partial similarities. The vector of partial similarities with respect to each peak is transformed into the overall degree of similarity s(**x**, **y**) between trajectories.

The structural similarity measure with respect to peaks is suitable for the comparison of trajectories exhibiting a specific impulse behaviour. This similarity measure expresses a degree of matching of peaks. If the position of peaks is not considered as a parameter, then the obtained similarity measure is indifferent to temporal translations of peaks.

The above five definitions of structural similarity measures can be used in different combinations to obtain a more complete evaluation of the similarity of trajectories.

In some cases structural similarity can be reduced to pointwise similarity. For instance, considering pointwise similarity for the first derivatives of the trajectories, the comparison of trajectories is performed with respect to their pointwise slope whereas scaling parameters are ignored. If pointwise similarity is determined for the second derivatives of trajectories, then the pointwise curvature of trajectories is considered as a relevant aspect for a comparison whereas scaling parameters and the slope of trajectories are irrelevant.

5.3 Extension of Fuzzy Pattern Recognition Methods by Applying Similarity Measures for Trajectories

As already stated, many pattern recognition methods (e.g. fuzzy c-means [Bezdek, 1981, p.65], possibilistic c-means [Krishnapuram, Keller, 1993], (fuzzy) Kohonen networks [Rumelhart, McClelland, 1988]) use the distance between pairs of feature vectors describing objects as a measure of dissimilarity between these objects. Using one of the equations (5.1), (5.2) or (5.3) a distance d(**x**, **y**) between objects **x** and **y** can be transformed into a similarity measure s(**x**, **y**). Conversely, each strictly positive similarity measure defines a distance measure.

All aforementioned pattern recognition methods use the distance between objects and cluster prototypes as a clustering criterion in order to calculate degrees of membership of objects to clusters, whereas the locations of cluster centres are obtained based on the locations of objects in the feature space weighted by their degrees of membership. Therefore, it is sufficient to provide a distance for pairs of objects and / or class representatives to be able to calculate degrees of cluster membership. These considerations were used to develop a modified version of the fuzzy c-means algorithm, which is called the functional fuzzy c-means (FFCM) [Joentgen, Mikenina et al., 1999b, p. 88]. The main advantage of the FFCM algorithm is its ability to partition dynamic objects described by multi-dimensional trajectories. Instead of the Euclidean distance measure for real-valued feature vectors the distance measure generated from pointwise similarity for trajectories was integrated into the calculation procedure of the FCM algorithm. Obviously the cluster centres obtained with the FFCM algorithm are multi-dimensional trajectories in the feature space, since the algorithm is applied to cluster objects whose features are represented by trajectories.

Pointwise and structural similarity measures for trajectories described in this Chapter can generally be integrated into an arbitrary static or dynamic pattern recognition method using the distance or similarity measure as a clustering criterion. The main principle of this combination is illustrated on *Figure 68*. In the next Chapter similarity measures for trajectories will be used in the algorithm of Gath and Geva applied for dynamic fuzzy classifier design and classification.

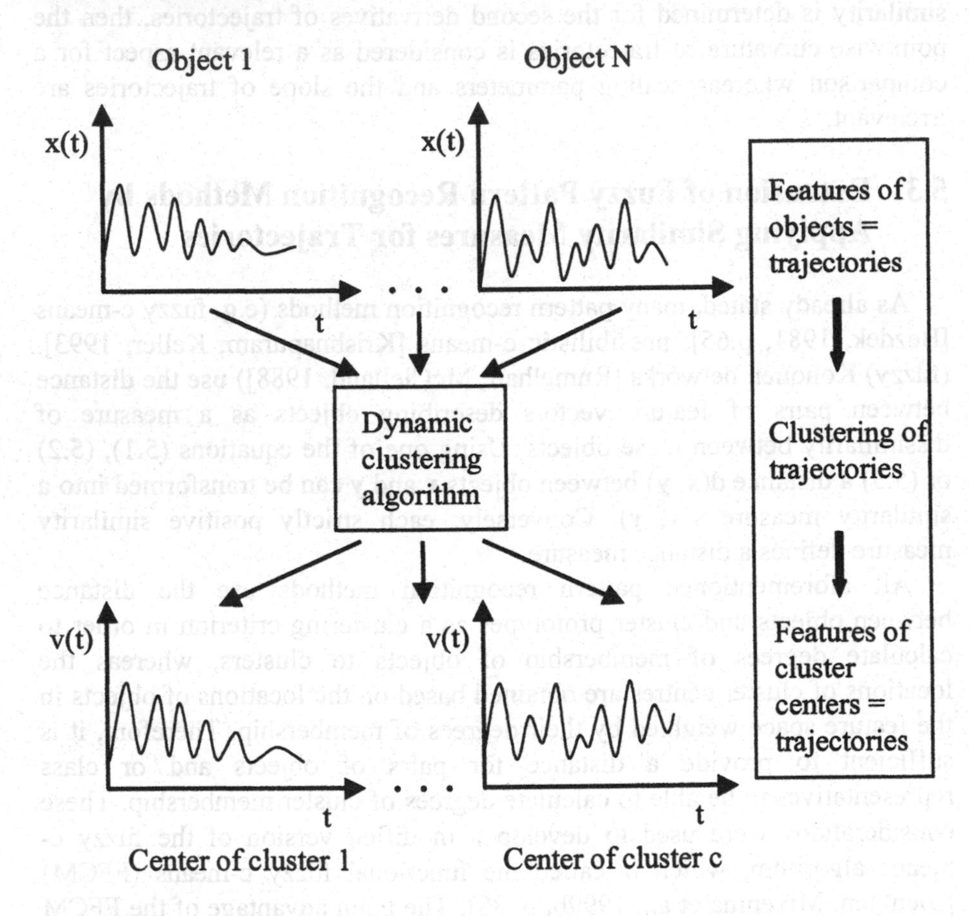

Figure 68. The structure of dynamic clustering algorithms based on similarity measures for trajectories

6 APPLICATIONS OF DYNAMIC PATTERN RECOGNITION METHODS

In order to demonstrate the practical relevance of the new method for dynamic fuzzy clustering developed in this book, two application examples are considered in this chapter. The first example taken from credit industry and presented in Section 6.1 is concerned with the problem of bank customer segmentation based on customers' behavioural data. After a description of the credit data of bank customers and the formulation of the goals of the analysis, two types of customer segmentations based in the first case on the whole temporal history covering two years and in the second case on a partial temporal history of half a year will be carried out. The clustering results obtained in the first case represent the structure within the customer portfolio related to long-term payment behaviour whereas the results generated in the second case provide customer segments based on short-term behaviour and the information about temporal changes in customer behaviour. This section contains a detailed description of the customer segments obtained during both types of analysis, an evaluation of the quality of the fuzzy partitions and a comparison of the different clustering results. The second application example, presented in Section 6.2, is related to the analysis of data traffic in computer networks and allows the optimisation of the network load based on on-line monitoring and dynamic recognition of typical states of data traffic. Dynamic fuzzy classifier design and classification will be performed based on pointwise as well as structural similarity measures for trajectories and the clustering results will be compared and evaluated.

Both applications are carried out using the implementation of the method developed in this book with the software package MATLAB 5.1. The list of implemented algorithms is given in Appendix.

6.1 Bank Customer Segmentation based on Customer Behaviour

Customer segmentation is one of the most important applications of data mining methodologies used in marketing and customer relationship management. Clustering customers based on their behavioural data helps to recognise their buying behaviours and purchase patterns, to derive strategic marketing initiatives and to increase the response rates by addressing only the customers most likely to buy a specific product. Knowing the best customers is the key to developing targeted, personal and relevant marketing campaigns, allocating resources by segment profitability and retaining the most valuable customers. Creating customer segments enables companies to understand the customer portfolio and highlights obvious marketing opportunities.

In finance the success of business activities depends to a large degree on credit services. Banks are interested, on the one hand, in maximising the credit volumes and, on the other hand, in minimising the insolvency rate, i.e. the number of customers unable to pay back their loan. By conducting a segmentation of bank customers based on their payment behaviour it is possible to distinguish between customers representing a 'good risk' for the bank, to which special services can be offered, and customers representing a 'bad risk,' which according to their bad payment behaviour have a high probability of becoming insolvent. Identifying typical segments within a customer portfolio allows a bank to recognise 'bad risk' cases in time, to react by applying corresponding measures and to increase the profitability of a 'good risk' segment. Another goal of bank customer segmentation can be the recognition of typical groups of users of specific bank services. Usually there are at least two segments corresponding to 'active users' using the provided bank services to a full degree and 'passive users' or even 'non-users,' using these services to a low degree. A knowledge of these segments can support a bank in developing a goal oriented marketing strategy for its private and commercial customers.

It must be noted that the goal and results of the analysis depend on the provided data set. In this section the second goal of bank customer segmentation will be discussed, that is, to distinguish between different user groups among bank customers.

6.1.1 Description of the credit data of bank customers

In the application under consideration, 24,267 commercial bank customers ('objects') with revolving credit were observed over two years and the measurements of customer features were carried out monthly (i.e. the length of the temporal history is equal to 24 months). The bank customers are described by two static features, seven dynamic features and one categorical feature. The first two features are used as unique identification numbers of customers such as the customer number and account number. The seven dynamic features characterise the state of an account each month and are represented by sequences of 24 measurements. They are summarised in the following table:

Table 7. Dynamic features describing bank customers

Feature	Description
1	Overdraw limit on account
2	Current end-of-month balance
3	Maximum balance this month
4	Minimum balance this month
5	Average credit utilisation this month
6	Credit turnover this month
7	Number of bank-initiated payment reversals, i.e. returned cheques / cancelled direct debits in the current month

The first feature 'Overdraw limit on account' is defined for each customer individually by the bank and its value usually remains constant for each customer over a long period of time or is changed on rare occasions. This feature can be used to evaluate a customer, but it does not describe his/her behaviour. Therefore, it will not be considered during customer segmentation, but only for the analysis of clustering results afterwards. The second feature 'Current month-end balance' is defined as a sum of all transactions in an account during the current month. The definition of the third feature 'Maximum balance this month' is formulated by the bank in the following way: if there is a negative balance in an account during a given month, then the maximum balance is the largest absolute value of all negative account balances during this month; otherwise the maximum balance is the lowest positive account balance during this month. Vice versa, the fourth feature 'Minimum balance this month' is defined by the bank as follows: if there is a positive balance in an account during a given month, then the minimum balance is the largest positive account balance during this month; otherwise the minimum balance is the lowest absolute value of all negative account balances during this month. The fifth feature 'Average credit utilisation this month' expresses how much of the credit was used on average by the customer during a given month. The sixth feature 'Credit turnover this month' corresponds to the sum of all positive entries in an account during a given month. The seventh feature 'Number of bank-initiated payment reversals' contains the number of failed debits since an account was overdrawn. As will be shown below, positive values of this feature are very rare among customers, thus it is better suited to the characterisation of customers rather than their segmentation. Therefore, only features 2 to 6 will be used for clustering.

One of the categorical features provided for bank customers determines special account properties which can be savings / time deposits / depots and can take two values such as 'yes' or 'no'. According to bank experts, customers with or without these account properties must be treated separately since they may exhibit different payment behaviour. Therefore, the data set will be separated into two subsets according to this categorical feature. The first set of customers characterised by feature value 'yes' and possessing a savings account or depots consists of 4,688 customers, while the other set includes 19,579 customers without the said properties of their accounts. These two sets of customers will be denoted hereinafter as groups 'Y' and 'N', respectively, and the analysis of the customer structure will be performed separately for each group.

After a preliminary analysis of data sets including the calculation of the mean, minimum and maximum values of trajectories and their variances, it can be seen that the value ranges of the seven features are very large and different. The value ranges, some quantiles s_α and the main statistical characteristics of the data group 'Y' are summarised in *Table 8* and *Table 9*. The values of the chosen quantiles show that a large percentage of the data lies in a range of feature values considerably smaller than the entire value range. More information concerning the value range is provided by values $\mu \pm 3\sigma$, where μ is the mean value and σ is a standard deviation. According to statistics, these values, given in *Table 9*, define the limits between which any observed value of a feature falls with a probability of 0.9974.

Table 8. The value range and main quantiles of each feature of Data Group 'Y'

Features	Value range	s_α (α=0.1)	s_α (α=0.2)	s_α (α=0.8)	s_α (α=0.9)	s_α (α=0.95)
1	[0, 6500000]	0	0	25000	50000	100000
2	[-4535483, 119436927]	-51506	-19962	80457	171319	311408
3	[-60309196, 4900000]	-78646	-37000	45911	89362	164654
4	[-4383894, 119436927]	-34509	-8495	132405	287197	521230
5	[0, 4493026]	0	0	18020	46903	82326
6	[0, 223273926]	0	0	179193	402423	870054
7	[0, 4]	0	0	0	0	1

Table 9. Main statistics of each feature of the Data Group 'Y'

Features	Mean value μ	Standard deviation σ	$\mu-3\sigma$	$\mu+3\sigma$
1	14510.06	48097.47	-129782.34	158802.46
2	16347.19	134581.82	-387398.26	420092.63
3	7.77	100780.11	-302332.56	302348.10
4	37655.46	273818.07	-783798.76	859109.67
5	7946.71	36071.66	-100268.26	116161.68
6	81002.74	636081.44	-1827241.60	1989247.10
7	0.01	0.00	-0.11	0.12

The value ranges, some quantiles and the principal statistical characteristics of data Group 'N' are given in *Table 10* and *Table 11.*

Table 10. The value range and main quantiles of each feature of Data Group 'N'

Features	Value range	s_α (α=0.1)	s_α (α=0.2)	s_α (α=0.8)	s_α (α=0.9)	s. (α=0.95)
1	[0, 4000000]	0	0	40000	95000	150000
2	[-14038790, 9594900]	-86414	-35227	42160	92119	166458
3	[-154474415, 5374429]	-104723	-46757	20118	48698	90323
4	[-13963346, 90026464]	-66107	-24025	63976	139351	251797
5	[0, 13984126]	0	0	31896	80186	149686
6	[0, 154506043]	0	0	103729	206334	368865
7	[0, 6]	0	0	0	0	1

Table 11. Main statistics of each feature of Data Group 'N'

Features	Mean value μ	Standard deviation σ	$\mu-3\sigma$	$\mu+3\sigma$
1	20811.16	63966.97	-171089.75	212712.06
2	-4278.04	145423.24	-440547.76	431991.68
3	-13393.86	153139.98	-472813.81	446026.08
4	7360.17	157022.60	-463707.63	478427.98
5	17426.61	123226.32	-352252.36	387105.59
6	35406.50	318825.34	-921069.54	991882.53
7	0.01	0.00	-0.10	0.11

As can be seen in the four tables shown above, only 5% of the data takes values of Feature 7 larger than zero. A feature with such a skewed distribution is not very informative and, therefore, can be dropped.

Due to the low number of data with extreme outlying values and a considerable amount of data within a similar value range, the features are characterised by a highly exponential distribution. *Figure 69*, a, shows the distribution of data with respect to Feature 6, where only mean values of each trajectory over the time interval [0, 24] are considered. In order to reduce the effect of outliers and to improve the performance of clustering algorithms, it is necessary to pre-process the data using some standardisation and normalisation techniques. The most appropriate technique for an exponential distribution is a logarithmic transform. The data transformation is carried out in two steps. Firstly, the minimum value over all trajectories is determined for each feature, and if it is negative it is subtracted pointwise from all trajectories of this feature. Then the values of these trajectories are increased by one in order to achieve only positive values after the transformation. In the second step, all values of the trajectories are transformed logarithmically. As a result, the logarithmically transformed data lies in the value range [0, 20] and the distribution of data for all features is roughly normal. It should be noted that fuzzy clustering methods do not strictly require the data to be normally distributed, but the algorithms work better in many cases if this criterion is satisfied. *Figure 69*, b, illustrates the distribution of the mean values of trajectories of Feature 6 after logarithmic transformation.

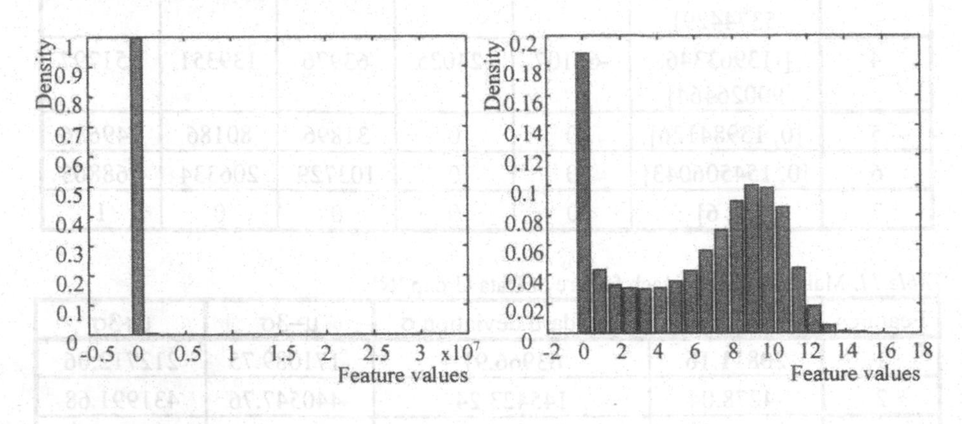

Figure 69. Distribution of data with respect to Feature 6

a) Before logarithmic transformation, b) After logarithmic transformation

After pre-processing the data on bank customers using standardisation and logarithmic transformation, the analysis of the data can be started.

6.1.2 Goals of bank customer analysis

The goals of the dynamic analysis of bank customers can be formulated as follows:

1. To find segments of customers with similar payment behaviour based on the whole temporal history covering two years,
2. To find segments of customers with similar payment behaviour based on the temporal history of half a year, and to follow changes in the cluster structure and in the assignment of customers to the clusters over time.

The first goal can be achieved by clustering customers represented by trajectories of their features on the time interval of two years. The clustering results provide information about the structure within the customer portfolio appearing during this time interval until the current moment. These results are suitable for distinguishing between 'good' and 'bad' customers according to their long-term payment behaviour. The analysis of a long history is often carried out by banks to achieve reliable results, particularly for recognising 'bad' credit customers. The drawback of this analysis is, however, that the classifier cannot be used to classify new observations of existing customers or observations of new customers for the next two years, since the cluster prototypes are described by trajectories with a length of 24 months and thus cannot be compared with shorter sequences of observations. Thus, the classification of new observations and updating the classifier (if necessary) can be repeated every two years. In this case the design of the classifier is static, but the classifier is dynamic in nature since it is applied to dynamic objects.

A more applicable classifier can be designed by clustering sequences of observations over half a year, which is the second goal of the analysis conducted. This analysis allows one to recognise customer segments based on the short-term payment behaviour of customers and to detect temporal changes in the customer behaviour. The classification of new observations of existing or new customers can be repeated every six months providing up-to-date information about the customers' states and their development. If changes in the customer structure are detected the classifier is adapted according to the detected changes, which corresponds to an update of the customer segments and their descriptions. Therefore, this type of analysis is based on dynamic classifier design and classification applied to dynamic objects.

There are in general two viewpoints on the classifier design for customer segmentation. Some companies are interested in static classifier design, if they are sure that they know their typical segments and do not want to change their description. In this case the classifier is designed just once (supervised or unsupervised), the cluster prototypes are 'frozen' and the classification of new observations is repeated in the course of time in order to detect changes of customers with respect to existing (static) clusters. Another viewpoint requires the adaptation of the classifier to temporal changes. Since conditions and properties of bank services can change in the course of time, a change of customer behaviour can also be expected. Thus, there is a need to update the prototypes of customer segments in order to represent appropriately a new environment.

According to the aforementioned goals of customer segmentation and two types of customers, the analysis will include the following steps. Two types of customers will be clustered based on two different lengths of temporal history leading to four clustering problems. Regarding the choice of the appropriate type of the similarity measure for trajectories the following preliminary analysis has been carried out.

Considering some of the trajectories of bank customers it can be seen that their temporal behaviour has either a fluctuating or almost constant character with occasional steps. Since the trajectories are relatively short (24 points of time if the whole temporal history is considered or just 6 time points in a time window) it is difficult to recognise a specific behavioural pattern in trajectories. The analysis of temporal characteristics of trajectories such as temporal trend, smoothness and pointwise derivatives has shown that the values of these characteristics for most of the trajectories are quite similar. For instance, the temporal trend of the trajectories of customers of group 'Y' takes values in the interval [-0.1, 0.1] and more than 95% of the values are around zero. An analogous situation can be observed calculating the average of pointwise derivatives over each time interval of six months. The smoothness for the most of these trajectories takes values between 20 and 23 indicating a strong fluctuating behaviour. Since the consideration of fluctuating behaviour of the trajectories using the structural similarity measure is not very promising, it seems reasonable to analyse the structure of customer portfolio based on the absolute values of trajectories instead of some temporal characteristics describing the behaviour of these trajectories. Fluctuating behaviour usually requires extensive smoothing, which would lead to a loss of relevant information, taking into account that the maximum length of a trajectory is only 24. Therefore, segmentation of bank customers will be carried out based on pointwise similarity between trajectories.

The tasks of analysis of bank customers and the corresponding clustering problems which will be solved in the following sections are summarised in the table below.

Table 12. Scope of the analysis of bank customers

I. Length of temporal history	II. Type of customers	III. Type of similarity measure for trajectories
The whole temporal history t=[1, 24]	Customers of group 'Y'	Pointwise similarity
Time windows equal to half a year	Customers of group 'N'	

6.1.3 Parameter settings for dynamic classifier design and bank customer classification

Customer segmentation will be carried out using the algorithm for dynamic classifier design and classification developed in Chapter 4. For clustering customers in a certain time window the clustering algorithm of Gath and Geva is applied, which uses the fuzzy c-means algorithm for initialisation. In order to be able to deal with dynamic objects represented by 5-dimensional trajectories, the distance measure generated from the pointwise similarity measure for trajectories is integrated into the FCM and the Gath-Geva algorithm.

The pointwise similarity measure is obtained according to Algorithm 6a. For its definition the quadratic membership function given by equation (5.11) is chosen to represent the fuzzy set 'approximately zero', which is defined with respect to each feature f_r, $r=1,...,M=5$. Parameter a(r) of each membership function is determined by equation (5.14), which requires the setting of a certain feature value $\beta(r)$ and its membership degree α to the fuzzy set 'approximately zero'. In order to take into account the value range of each feature and the range of maximal differences between trajectories of each feature, respectively, parameter $\beta(r)$ can be evaluated as the average value on the domain where trajectories of feature r take their values (the values of pairwise differences between trajectories also lie in this domain). The value of parameter $\beta(r)$ is calculated as the mean value of the maximal values of the trajectories of the corresponding feature:

$$\beta(r) = \frac{1}{N}\sum_{j=1}^{N} \max_t \, [x_{jr}(t)], \quad r = 1,...,M \tag{6.1}$$

where $x_{jr}(t)$ is the j-th trajectory of feature r and N is the number of trajectories considered.

The membership degree of feature value $\beta(r)$ to the fuzzy set 'approximately zero' can be chosen between 0 and 1, but it seems reasonable to choose this value between 0.1 and 0.6. The smaller the value of α, the larger the value of parameter a(r), the narrower the fuzzy set and the stronger the definition of similarity between trajectories. Empirical research has shown that the value α=0.5 is best suited for this application. Thus, parameter a(r) for the definition of membership functions is obtained as follows:

$$a(r) = \frac{1-0.5}{0.5 \cdot \beta(r)^2} = \frac{1}{\beta(r)^2}. \tag{6.2}$$

The second parameter needed for the definition of the pointwise similarity measure is the aggregation operator applied to transform the vector of pointwise similarities to the overall similarity measure on the time interval. The arithmetic mean value is chosen using the consideration in Section 5.2.1.2.

The distance measure obtained from the pointwise similarity measure is used instead of the Euclidean distance measure in the calculation schemes of the FCM and the Gath-Geva algorithms. In both algorithms the distance between objects and cluster centres is involved in the calculation of degrees of membership of objects to clusters (see Equation (4.5)). It is also used in the Gath-Geva algorithm to calculate the fuzzy covariance matrices of clusters given by (7.4) and an exponential distance function (7.2), which is finally applied for evaluating degrees of membership.

It is known that the use of an exponential function in the Gath-Geva algorithm can lead to numerical problems in many practical cases, since the distances take either extremely low or extremely high values. This problem was also observed during some test clustering runs applied to the data of bank customers, which are characterised by a considerable variance of values. In order to avoid these numerical problems, the exponential distance function is substituted by the quadratic function and modified in such a way that it provides only constant values above a certain high value. The loss of precision due to such a modification is not significant because of the very low values of the resulting degrees of membership.

For dynamic classifier design it is necessary to define a set of thresholds used within the monitoring procedure. For the algorithms for detecting new clusters or similar clusters to be merged the following parameter settings are chosen:

Table 13. Parameter settings for the detection of new clusters during customer segmentation

Parameter	Value
Absorption threshold	$u^o = 0.5$
Share of the average cluster size	$\alpha^{cs} = 0.7$
Share of the average cluster density	$\alpha^{dens} = 0.25$
Threshold for the choice of 'good' free objects	$\alpha^{good} = u^o = 0.5$
Merging threshold	$\lambda = 0.6$

6.1.4 Clustering of bank customers in Group 'Y' based on the whole temporal history of 24 months and using the pointwise similarity measure

Clustering of bank customers belonging to Group 'Y,' based on their temporal behaviour over the time period of 24 months, is performed with the modified algorithm of Gath and Geva. The distance measure used in the algorithm is generated from the pointwise similarity measure for trajectories defined according to Algorithm 6a. Parameter a for the definition of the pointwise similarity measure is calculated according to equation (6.2) and yields the following values:

	Feature 1	Feature 2	Feature 3	Feature 4	Feature 5	Feature 6
a	12.8612	235.4318	320.9688	234.7001	28.8568	106.0196

Clustering starts with the number of clusters c=2 and the optimal number of clusters is determined by applying the algorithms of the monitoring procedure described in Sections 4.3, 4.4 and 4.5. After possibilistic classification and absorption of customers into two detected clusters according an absorption threshold of 0.5, the sizes of clusters are equal to 3,081 and 1,367 and the number of free objects is equal to 240. For a better judgement of the cluster structure the number of objects absorbed into the clusters with different absorption thresholds is summarised in *Table 14*. The densities of clusters, the average partition density and the density of the group of free objects are represented in *Table 15*, which also includes the values of validity measures such as fuzzy separation and compactness. According to the criterion of the minimal cluster size and the criterion of compactness of a group of free objects, no new cluster can be assumed and free objects are considered as stray data.

Table 14. Number of absorbed and free objects of Data Group 'Y'

	Absorbed		Stray
	C_1	C_2	
u^o=0.3	3114	1419	155
u^o=0.4	3099	1396	193
u^o=0.5	3081	1367	240
u^o=0.6	3065	1308	315

Validity measures for the generated fuzzy partition are given in the following table:

Table 15. Validity measures for fuzzy partition with two clusters for Data Group 'Y'

Partition density of Cluster 1	$PD_1 = 841.3$
Partition density of Cluster 2	$PD_2 = 35.76$
Average partition density	$v_{APD} = 438.53$
Fuzzy separation	FSA = 0.825
Fuzzy compactness	FC = 0.919
Partition density of the group of free objects	$PD_{free} = 0.109 \cdot 10^{-4}$

In order to verify whether the number c=2 is really the correct number for the given group of customers, clustering was performed with the number of clusters equal to 3 and 4. The validity measures for the three generated fuzzy partitions are represented in *Table 16*. As can be seen, the number of stray objects and the average partition density for partitions with 3 and 4 clusters exceeds the corresponding values for the partition with 2 clusters. However, the values of fuzzy separation and compactness obtained with 3 and 4 clusters lie below the values obtained with 2 clusters.

Table 16. Number of stray objects and validity measures for different fuzzy partitions of Data Group 'Y'

	c = 2	c = 3	c = 4
N_{free}	240	692	560
v_{apd}	438.53	9525.7	3754.4
FSA	0.825	0.432	0.419
FC	0.919	0.849	0.688

When applying the algorithm for detection of similar clusters to be merged to partitions with 3 and 4 clusters, it can be stated that some pairs of clusters are highly overlapping. In the case of 3 clusters the similarity measure between clusters 1 and 2 yields $s(C_1, C_2)=0.77$, which indicates that Clusters 1 and 2 can be merged if the merging threshold $\lambda=0.6$ is chosen. In the case of 4 clusters the following similarity measures between pairs of Clusters 1 and 3 and 2 and 4 are obtained: $s(C_1, C_3)=0.82$ and $s(C_2, C_4)=0.97$. These pairs of clusters can be merged as well.

Thus, it can be assumed that the optimal number of clusters for segmentation of customers in Group 'Y' is equal to 2. Centres of 2 customer segments obtained for this group of bank customers with respect to each feature using the modified Gath-Geva algorithm, based on the pointwise similarity measure for trajectories, are given below.

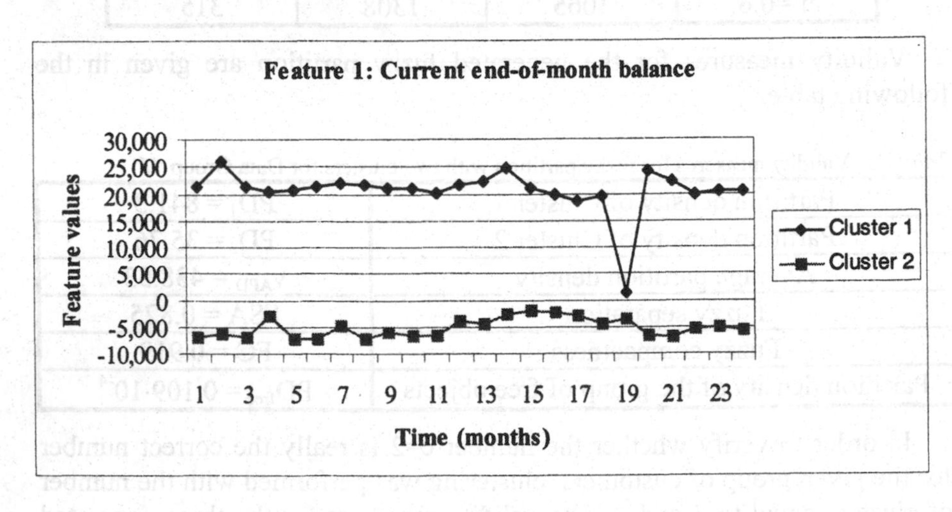

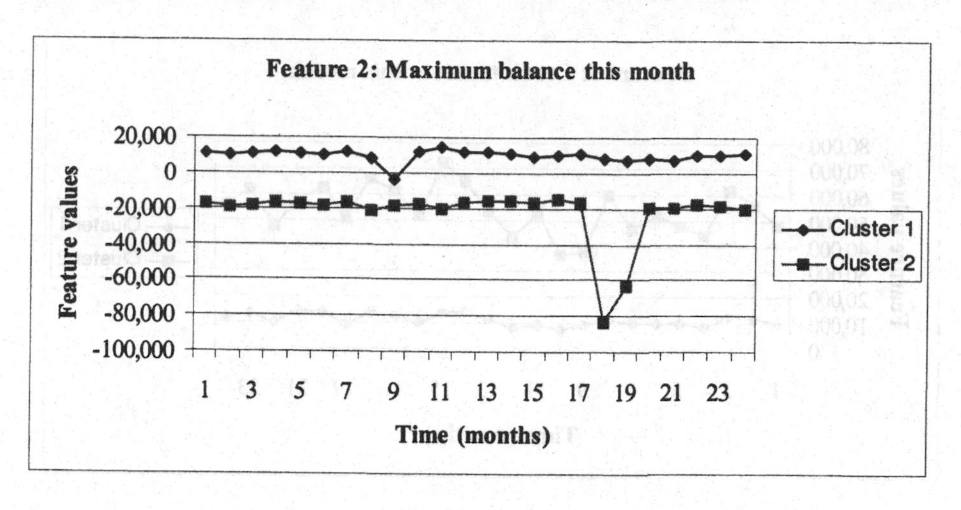
Feature 2: Maximum balance this month
20,000
0
-20,000
-40,000
-60,000
-80,000
-100,000
Feature values
1 3 5 7 9 11 13 15 17 19 21 23
Time (months)
Cluster 1
Cluster 2

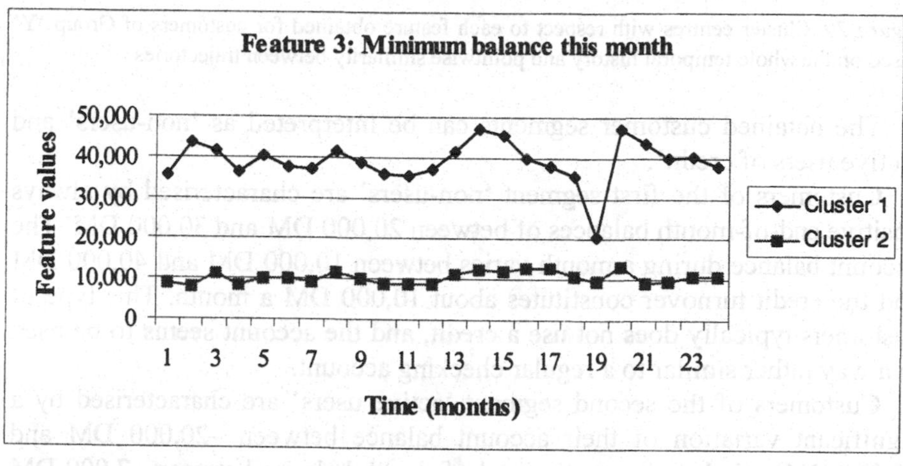
Feature 3: Minimum balance this month
50,000
40,000
30,000
20,000
10,000
0
Feature values
1 3 5 7 9 11 13 15 17 19 21 23
Time (months)
Cluster 1
Cluster 2

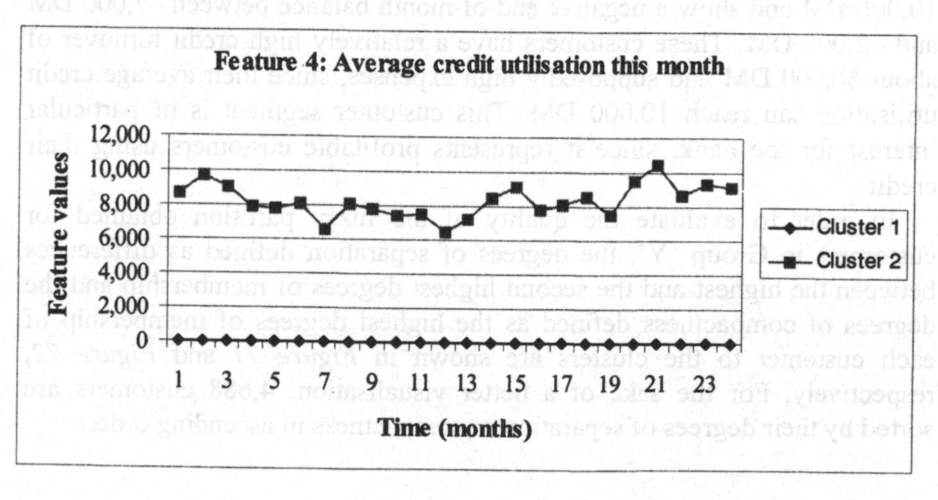
Feature 4: Average credit utilisation this month
12,000
10,000
8,000
6,000
4,000
2,000
0
Feature values
1 3 5 7 9 11 13 15 17 19 21 23
Time (months)
Cluster 1
Cluster 2

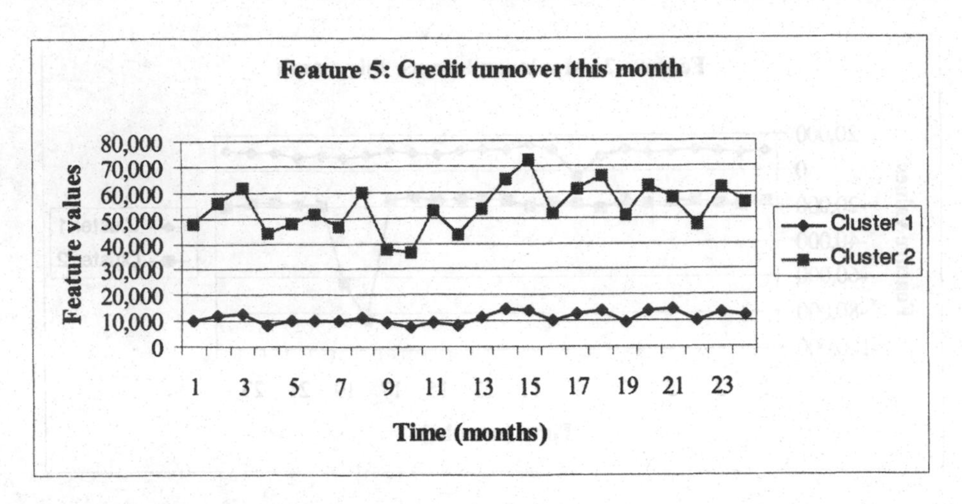

Figure 70. Cluster centres with respect to each feature obtained for customers of Group 'Y' based on the whole temporal history and pointwise similarity between trajectories

The obtained customer segments can be interpreted as 'non-users' and 'active users of credit'.

Customers of the first segment 'non-users' are characterised by always positive end-of-month balances of between 20,000 DM and 30,000 DM. The account balance during a month varies between 10,000 DM and 40,000 DM and the credit turnover constitutes about 10,000 DM a month. This type of customers typically does not use a credit, and the account seems to be used in a way rather similar to a regular checking account.

Customers of the second segment 'active users' are characterised by a significant variation of their account balance between –20,000 DM and 10,000 DM and show a negative end-of-month balance between –7,000 DM and –2,000 DM. These customers have a relatively high credit turnover of about 50,000 DM and supposedly high expenses, since their average credit utilisation can reach 10,000 DM. This customer segment is of particular interest for the bank, since it represents profitable customers using their credit.

In order to evaluate the quality of the fuzzy partition obtained for customers in Group 'Y', the degrees of separation defined as differences between the highest and the second highest degrees of membership and the degrees of compactness defined as the highest degrees of membership of each customer to the clusters are shown in *Figure 71* and *Figure 72*, respectively. For the sake of a better visualisation, 4,688 customers are sorted by their degrees of separation or compactness in ascending order.

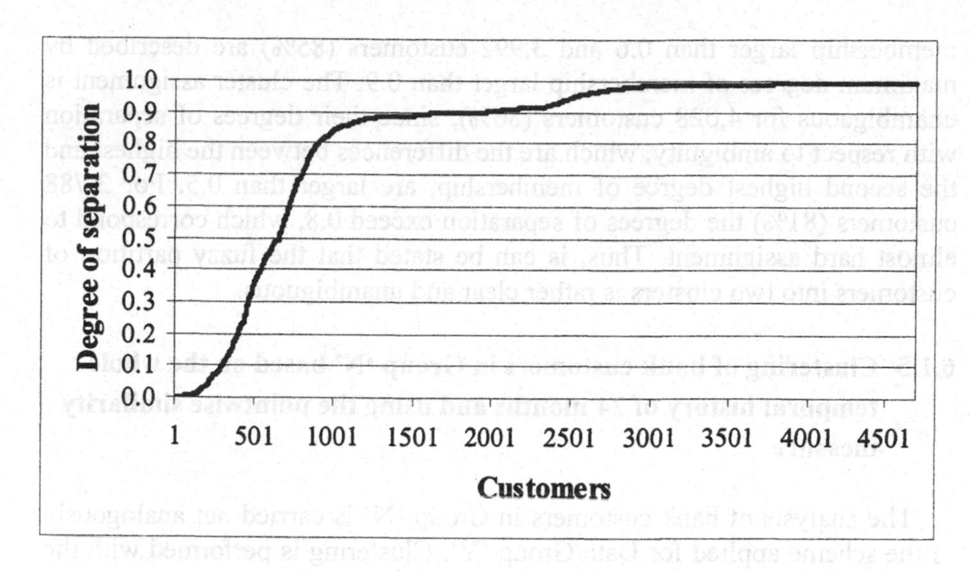

Figure 71. **Degrees of separation between clusters obtained for customers in Group 'Y' based on the whole temporal history**

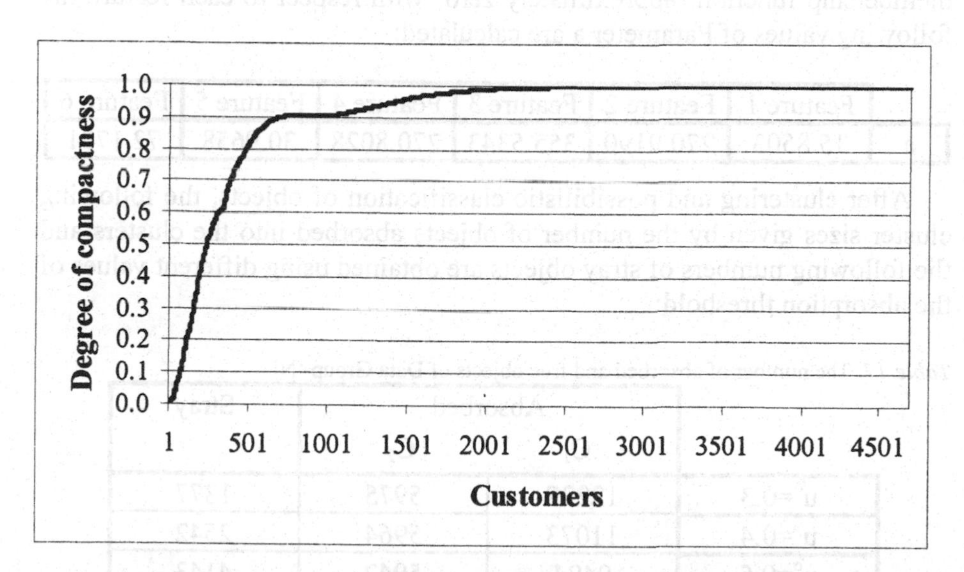

Figure 72. **Degrees of compactness of clusters obtained for customers in Group 'Y' based on the whole temporal history**

The analysis of the figures shown above leads to the following considerations concerning the fuzzy partition obtained. Most of the customers are characterised by high values of maximum degrees of membership, which allows a clear assignment of customers to one of the clusters. In particular, 4,370 customers (93%) have maximum degrees of membership larger than 0.6 and 3,992 customers (85%) are described by maximum degrees of membership larger than 0.9. The cluster assignment is unambiguous for 4,028 customers (86%), since their degrees of separation with respect to ambiguity, which are the differences between the highest and the second highest degree of membership, are larger than 0.5. For 3,788 customers (81%) the degrees of separation exceed 0.8, which correspond to almost hard assignment. Thus, is can be stated that the fuzzy partition of customers into two clusters is rather clear and unambiguous.

6.1.5 Clustering of bank customers in Group 'N' based on the whole temporal history of 24 months and using the pointwise similarity measure

The analysis of bank customers in Group 'N' is carried out analogously to the scheme applied for Data Group 'Y'. Clustering is performed with the modified version of Gath-Geva algorithm based on the pointwise similarity measure for trajectories starting with c=2. For the definition of the membership function 'approximately zero' with respect to each feature the following values of Parameter a are calculated:

	Feature 1	Feature 2	Feature 3	Feature 4	Feature 5	Feature 6
a	15.8503	270.9190	355.5343	270.8028	30.9658	72.1701

After clustering and possibilistic classification of objects, the following cluster sizes given by the number of objects absorbed into the clusters and the following numbers of stray objects are obtained using different values of the absorption threshold:

Table 17. The number of absorbed and free objects of Data Group 'N'

	Absorbed		Stray
	C_1	C_2	
u^o=0.3	12227	5975	1377
u^o=0.4	11073	5964	2542
u^o=0.5	9494	5942	4143
u^o=0.6	6479	5801	7299

The values of different validity measures for the generated fuzzy partition and the density for the group of free objects are presented in the following table.

Table 18. Validity measures for fuzzy partition with two clusters for Data Group 'N'

Partition density of Cluster 1	PD_1=3493
Partition density of Cluster 2	PD_2=22720
Average partition density	v_{apd}= 13107
Fuzzy separation	FSA= 0.5087
Fuzzy compactness	FC= 0.6802
Partition density of the group of free objects	0.0031

Since the number of free objects is not sufficient to declare a new cluster (4143<0.7·5942=4180) and the density of the group of free objects does not exceed the density threshold (0.0031<0.25·13107=3276), no new cluster can be assumed and free data are considered as stray.

In order to verify the result of the monitoring procedure providing the partition with 2 clusters, clustering is performed with the number of clusters equal to 3 and 4. The quality of these fuzzy partitions is evaluated using three validity measures whose values are given in *Table 19.* It can be seen that the number of stray objects grows as the number of clusters increases, which is a sign for a deterioration of the fuzzy partition. The average partition density for a partition with 4 clusters exceeds considerably the corresponding values for the partitions with 2 and 3 clusters. The increase of density can be explained by the fact that the number of objects absorbed in the four clusters is much smaller than the corresponding numbers for two other partitions, which results in extremely small cluster volumes. However, a high value of density can not be considered as an improvement of the fuzzy partition, since the majority of objects can not be absorbed into clusters. Considering the values of fuzzy compactness, it can be seen that the partition with 3 clusters has the best value but the value of fuzzy separation is much lower than the one for 2 clusters. This means that although objects have high degrees of membership, some of the three clusters have an important overlap and can not be clearly distinguished. The values of fuzzy separation and compactness obtained for a partition with 4 clusters lie below the corresponding values obtained for two other partitions and also indicate the intersection of clusters.

Table 19. Number of stray objects and validity measures for different fuzzy partitions of Data Group 'N'

	c = 2	c = 3	c = 4
n_{free}	3052	4563	14270
v_{apd}	13107	1528.6	2373162
FSA	0.5087	0.3046	0.2662
FC	0.6802	0.9395	0.4488

When the algorithm for detection of similar clusters to partitions with 3 and 4 clusters is applied, the fact of highly overlapping clusters is confirmed. In the case of 3 clusters the similarity measure between Clusters 2 and 3 is equal to $s(C_2, C_3)=0.699$, which indicates that these can be merged if the merging threshold $\lambda=0.6$ is chosen. In the case of 4 clusters the following similarity measures for some pairs of clusters are obtained: $s(C_1, C_4)=0.67$ and $s(C_2, C_3)=0.62$. According to the chosen merging threshold, these pairs of clusters can be merged.

Thus, it can be assumed that the optimal number of clusters for fuzzy partitioning is equal to 2. The centres of the 2 customer segments identified for bank customers of group 'N' with respect to each feature using the modified Gath-Geva algorithm based on the pointwise similarity measure for trajectories are presented on *Figure 73*.

The obtained customer segments can again be interpreted as 'non-users' and 'active users of credit'.

Customers belonging to the first segment, 'non-users,' are characterised by always positive end-of-month balances of between 10,000 DM and 15,000 DM. The account balance during a month varies between 10,000 DM and 20,000 DM and the credit turnover constitutes about 10,000 DM per month. As can be seen from the low values of the average credit utilisation, this type of customers does not use a credit from the bank. These customers use their accounts in a way rather similar to a regular checking account.

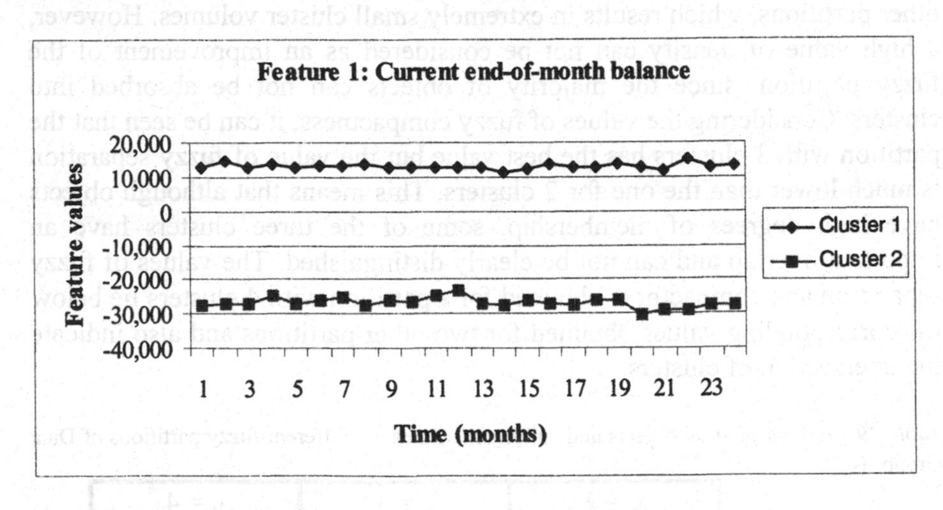

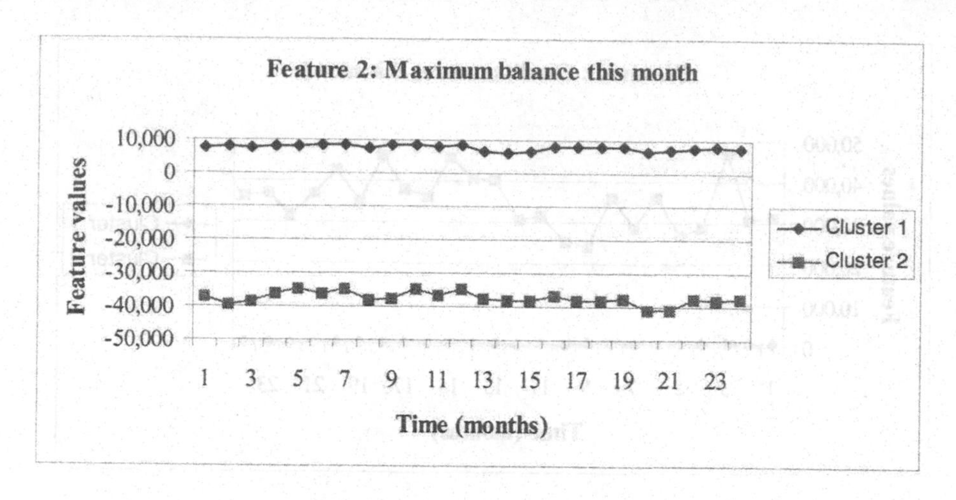
Feature 2: Maximum balance this month
10,000
0
-10,000
-20,000
-30,000
-40,000
-50,000
Feature values
1 3 5 7 9 11 13 15 17 19 21 23
Time (months)
Cluster 1
Cluster 2

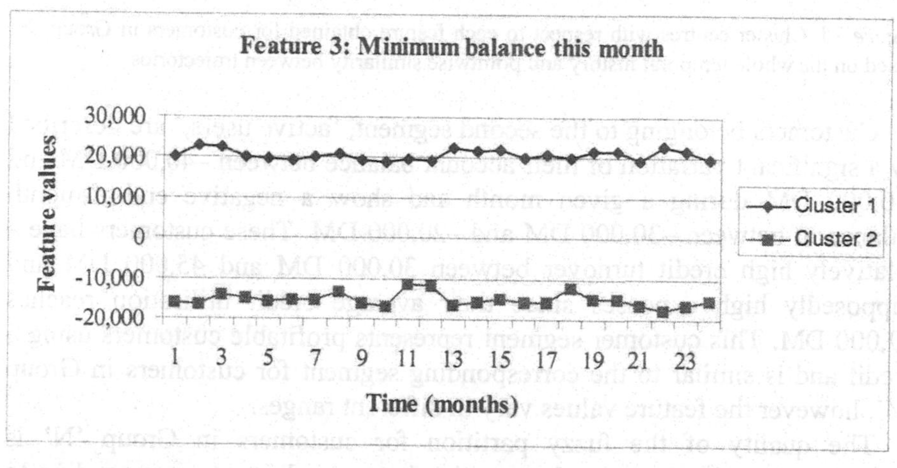
Feature 3: Minimum balance this month
30,000
20,000
10,000
0
-10,000
-20,000
Feature values
1 3 5 7 9 11 13 15 17 19 21 23
Time (months)
Cluster 1
Cluster 2

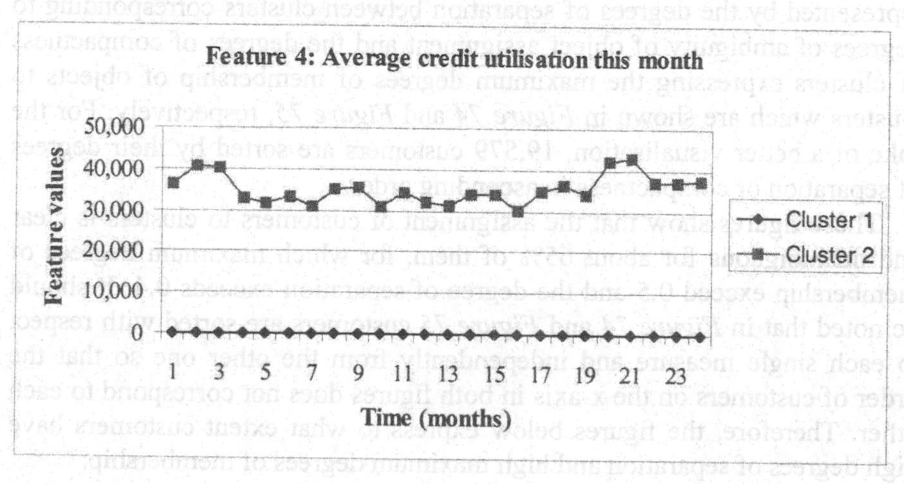
Feature 4: Average credit utilisation this month
50,000
40,000
30,000
20,000
10,000
0
Feature values
1 3 5 7 9 11 13 15 17 19 21 23
Time (months)
Cluster 1
Cluster 2

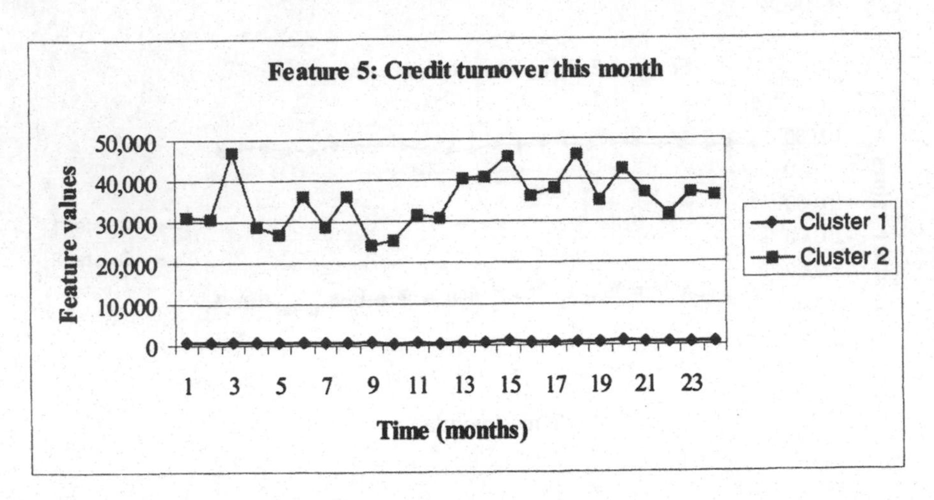

Figure 73. Cluster centres with respect to each feature obtained for customers in Group 'N' based on the whole temporal history and pointwise similarity between trajectories

Customers belonging to the second segment, 'active users,' are described by a significant variation of their account balance between –40,000 DM and -10,000 DM during a given month and show a negative end-of-month balance of between –30,000 DM and –20,000 DM. These customers have a relatively high credit turnover between 30,000 DM and 45,000 DM and supposedly high expenses since their average credit utilisation reaches 40,000 DM. This customer segment represents profitable customers using a credit and is similar to the corresponding segment for customers in Group 'Y', however the feature values vary in different ranges.

The quality of the fuzzy partition for customers in Group 'N' is represented by the degrees of separation between clusters corresponding to degrees of ambiguity of object assignment and the degrees of compactness of clusters expressing the maximum degrees of membership of objects to clusters which are shown in *Figure 74* and *Figure 75*, respectively. For the sake of a better visualisation, 19,579 customers are sorted by their degrees of separation or compactness in ascending order.

These figures show that the assignment of customers to clusters is clear and unambiguous for about 65% of them, for which maximum degrees of membership exceed 0.5 and the degree of separation exceeds 0.4. It should be noted that in *Figure 74* and *Figure 75* customers are sorted with respect to each single measure and independently from the other one so that the order of customers on the x-axis in both figures does not correspond to each other. Therefore, the figures below express to what extent customers have high degrees of separation and high maximum degrees of membership.

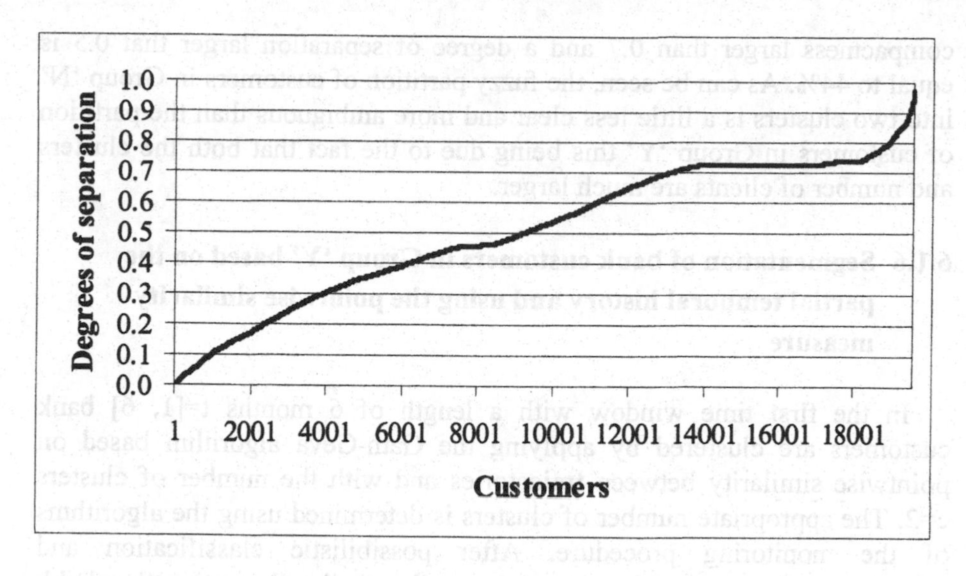

Figure 74. **Degrees of separation between clusters obtained for each customer in Group 'N' based on the whole temporal history**

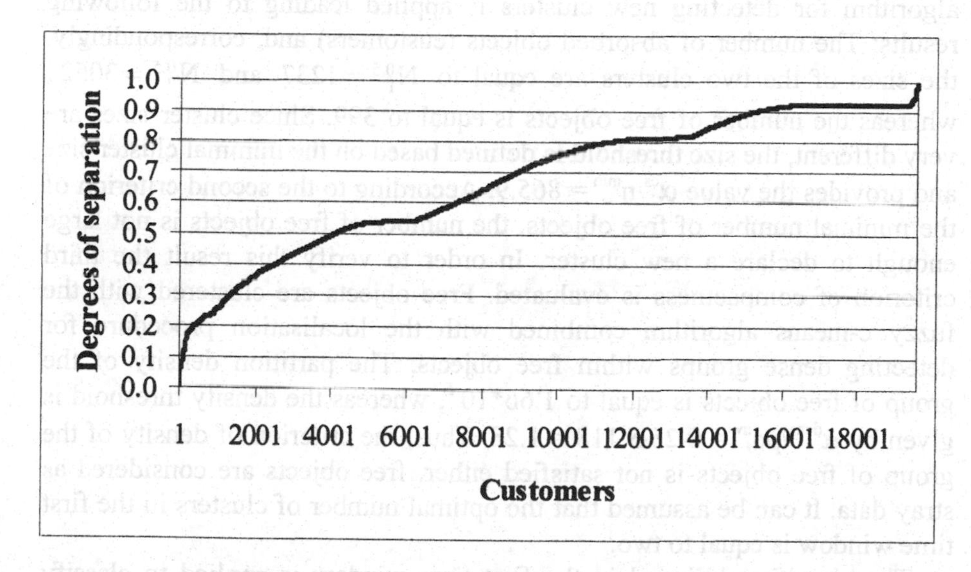

Figure 75. **Degrees of compactness of clusters for each customer in Group 'N' based on the whole temporal history**

The number of customers whose degrees of separation and compactness exceed certain thresholds can be calculated from the original sequences sorted by the customer number by choosing customers satisfying both conditions. For instance, the number of customers with a degree of compactness larger than 0.7 and a degree of separation larger that 0.5 is equal to 44%. As can be seen, the fuzzy partition of customers in Group 'N' into two clusters is a little less clear and more ambiguous than the partition of customers in Group 'Y' this being due to the fact that both the clusters and number of clients are much larger.

6.1.6 Segmentation of bank customers in Group 'Y' based on the partial temporal history and using the pointwise similarity measure

In the first time window with a length of 6 months t=[1, 6] bank customers are clustered by applying the Gath-Geva algorithm based on pointwise similarity between trajectories and with the number of clusters c=2. The appropriate number of clusters is determined using the algorithms of the monitoring procedure. After possibilistic classification and assignment of customers to clusters according to the absorption threshold, whether or not the current cluster structure represents the customer structure in the best way is verified during the monitoring procedure. Firstly, the algorithm for detecting new clusters is applied leading to the following results: The number of absorbed objects (customers) and, correspondingly, the sizes of the two clusters are equal to $N_1^{0.5} = 1237$ and $N_2^{0.5} = 3052$, whereas the number of free objects is equal to 399. Since cluster sizes are very different, the size threshold is defined based on the minimal cluster size and provides the value $\alpha^{cs} \cdot n^{min} = 865.9$. According to the second criterion of the minimal number of free objects, the number of free objects is not large enough to declare a new cluster. In order to verify this result the third criterion of compactness is evaluated. Free objects are clustered with the fuzzy c-means algorithm combined with the localisation procedure for detecting dense groups within free objects. The partition density of the group of free objects is equal to $1.66 * 10^{-5}$, whereas the density threshold is given by $\alpha^{dens} \cdot pd^{av} = 0.25 \cdot 5.118 = 1.28$. Thus, the criterion of density of the group of free objects is not satisfied either, free objects are considered as stray data. It can be assumed that the optimal number of clusters in the first time window is equal to two.

The classifier designed in the first time window is applied to classify new observations of customers in Time Windows 2 to 4. The monitoring procedure cannot detect any abrupt changes in the cluster structure over these three periods of time. The classifier fits the data structure well, which seems to be similar for all time windows. The results of clustering and classification for all four time windows are summarised in *Table 20*, providing the number of customers absorbed into each cluster and the number of free objects rejected for absorption, which are obtained with different values of the absorption threshold. These results provide information about the size of customer segments detected.

Table 20. Number of Customers 'Y' assigned to two clusters in four time windows

	Time Window 1			Time Window 2			Time Window 3			Time Window 4		
	Absorbed		Stray	Absorbed		Stray	Absorbed		Stray	Absorbed		Stray
	C_1	C_2		C_1	C_2		C_1	C_2		C_1	C_2	
u^o=0.3	1481	3093	114	1435	3144	109	1466	3153	69	1459	3137	92
u^o=0.4	1340	3070	278	1308	3119	261	1410	3148	130	1328	3114	246
u^o=0.5	1237	3052	399	1214	3105	369	1240	3143	305	1225	3092	371
u^o=0.6	1030	3032	626	1022	3082	584	1047	3138	503	1051	3070	567

Table 21. Partition densities of clusters, fuzzy separation and compactness indexes obtained for Customers 'Y' in four time windows

	Time Window 1	Time Window 2	Time Window 3	Time Window 4
PD_1	1.191	1.170	1.195	1.184
PD_2	9.044	9.228	9.384	9.215
v_{apd}	5.118	5.199	5.289	5.199
FSA	0.595	0.594	0.583	0.604
FC	0.840	0.845	0.857	0.848
PD_{free}	$1.66*10^{-5}$	$3.94*10^{-7}$	$4.85*10^{-6}$	$2.11*10^{-5}$

The centres of the customer segments obtained for customers in Group 'Y' in the first time window (the first half a year) are shown in *Figure 76.*

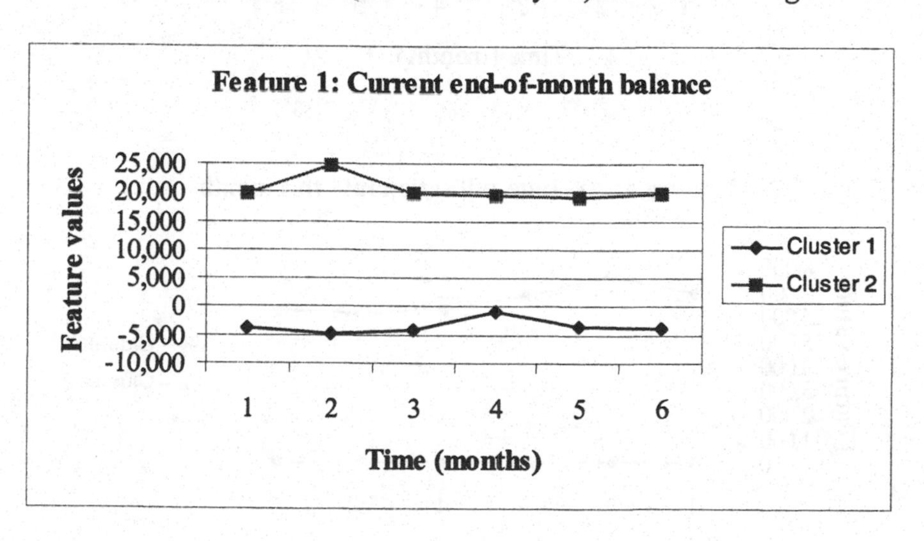

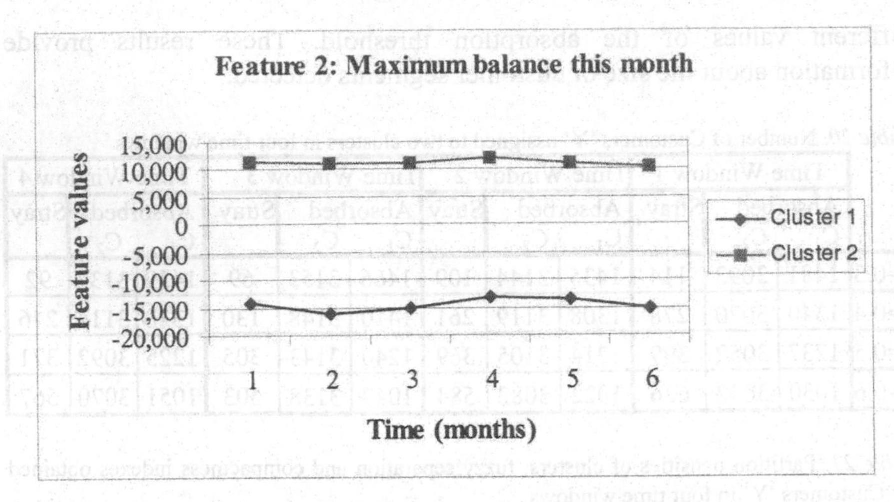

Feature 2: Maximum balance this month
15,000
10,000
5,000
0
-5,000
-10,000
-15,000
-20,000
Feature values
1
2
3
4
5
6
Time (months)
Cluster 1
Cluster 2

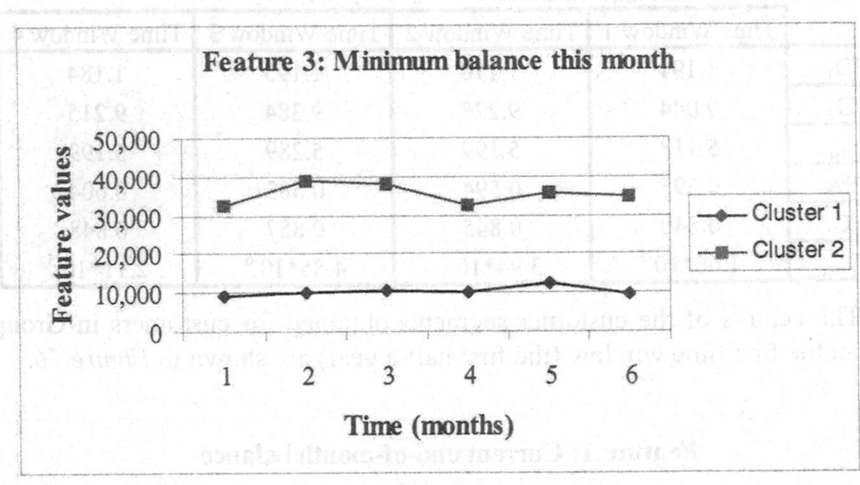

Feature 3: Minimum balance this month
50,000
40,000
30,000
20,000
10,000
0
Feature values
1
2
3
4
5
6
Time (months)
Cluster 1
Cluster 2

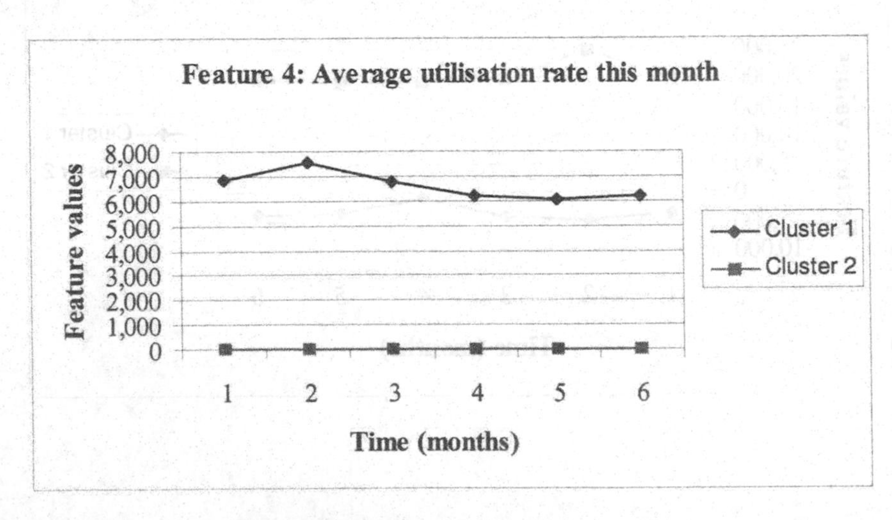

Feature 4: Average utilisation rate this month
8,000
7,000
6,000
5,000
4,000
3,000
2,000
1,000
0
Feature values
1
2
3
4
5
6
Time (months)
Cluster 1
Cluster 2

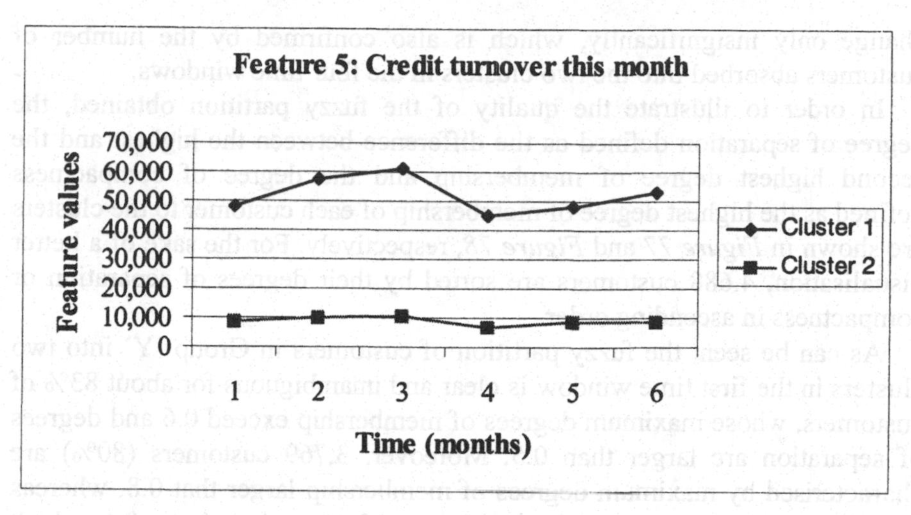

Figure 76. Cluster centres with respect to each feature obtained for customers in Group 'Y' in the first time window and based on pointwise similarity between trajectories

The customer segments obtained in the first time window can be interpreted as 'non-users' and 'active users of credit' similar to customer segments recognised for customers in Group 'Y' based on the whole temporal history of 24 months.

Customers of the first segment, 'non-users,' are described by always positive end-of-month balances of between 20,000 DM and 25,000 DM and, correspondingly, a zero value of the average credit utilisation. The account balance during a given month varies between 10,000 DM and 40,000 DM and the credit turnover constitutes about 10,000 DM per month. It is obvious that this type of customer does not use a credit from the bank, but uses his/her account rather like a regular checking account.

The account balances of customers in the second segment, 'active users,' exhibit a significant variation of between –15,000 DM and 10,000 DM during a given month and possess a negative end-of-month balance of about –5,000 DM. The credit turnover of these customers is relatively high varying between 50,000 DM and 60,000 DM, but at the same time the average monthly credit utilisation by these customers reaches 8,000 DM. Thus, the accounts of these customers are characterised by a large number of transactions. This customer segment makes use of an available credit providing the bank with a desirable profit.

As can be seen, the description and the properties of these two segments are very similar to the ones of segments obtained for customers in Group 'Y,' based on the whole temporal history. This result shows that the two customer segments detected are rather stable over time and their properties change only insignificantly, which is also confirmed by the number of customers absorbed into the two clusters in the four time windows.

In order to illustrate the quality of the fuzzy partition obtained, the degree of separation defined as the difference between the highest and the second highest degree of membership and the degree of compactness defined as the highest degree of membership of each customer to the clusters are shown in *Figure 77* and *Figure 78*, respectively. For the sake of a better visualisation, 4,688 customers are sorted by their degrees of separation or compactness in ascending order.

As can be seen, the fuzzy partition of customers in Group 'Y' into two clusters in the first time window is clear and unambiguous for about 83% of customers, whose maximum degrees of membership exceed 0.6 and degrees of separation are larger than 0.5. Moreover, 3,769 customers (80%) are characterised by maximum degrees of membership larger that 0.8, whereas 2,369 of these customers possess a degree of separation above 0.6, which corresponds to rather sharp assignment.

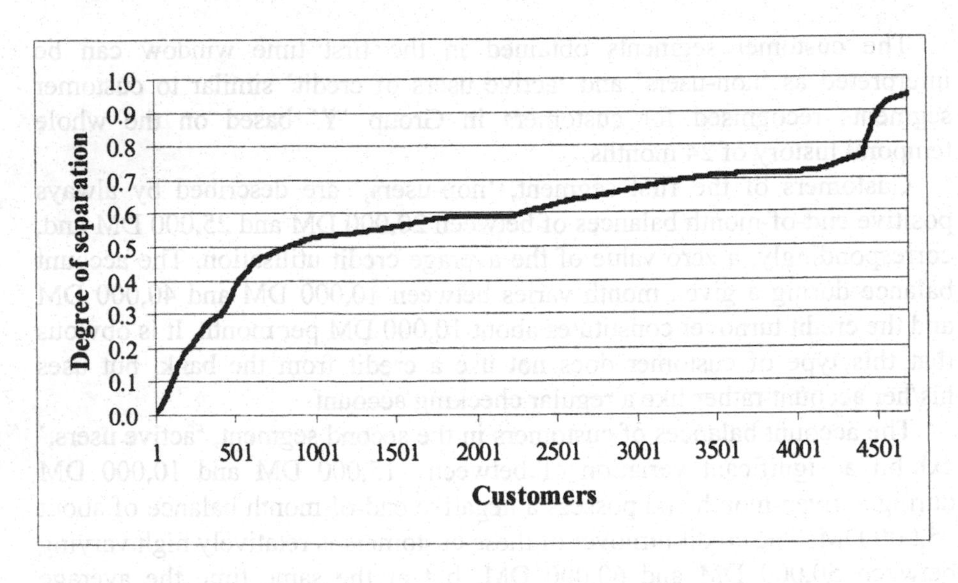

Figure 77. Degrees of separation between clusters obtained for customers in Group 'Y' in the first time window

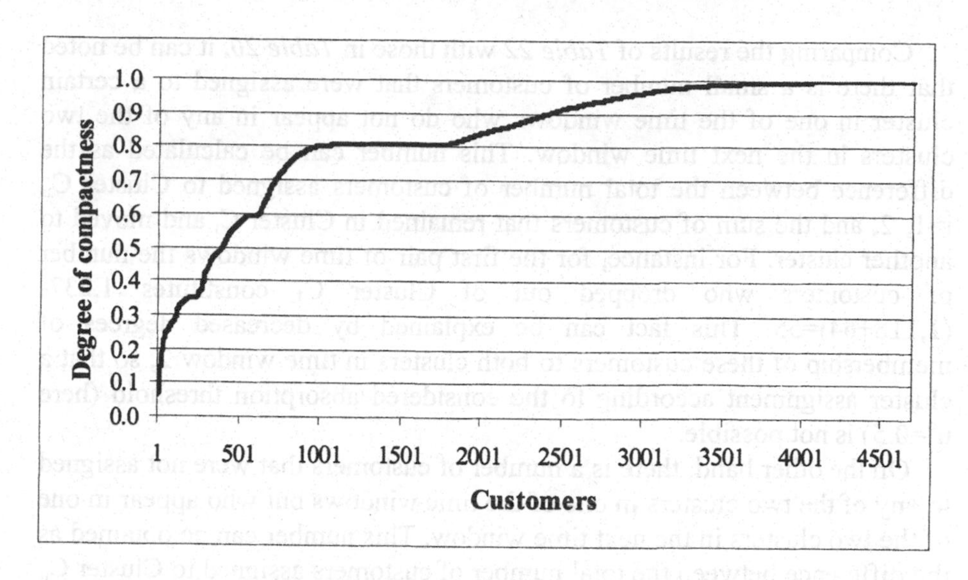

Figure 78. Degrees of compactness of clusters calculated for customers in Group 'Y' in the first time window

During the next step of analysis, the temporal change of customer assignment to clusters achieved using absorption threshold u^o=0.5 (the number of customers assigned to each cluster is given in *Table 20*) is considered. The task is to determine for each pair of subsequent time windows whether customers have remained in the same cluster to which they were assigned in the previous time window or whether they have moved from one cluster into another due to changing properties of their new observations. The results of this analysis are summarised in *Table 22*, where the i-th time window is denoted as tw_i, i=1,..., 4. As can be seen, most of the customers assigned to Cluster (segment) C_1 or C_2 in the first time window remain in these clusters in the following time windows, and just a small percentage of customers has moved from one cluster to another.

Table 22. Temporal change of assignment of customers in Group 'Y' to clusters

	From tw_1 to tw_2		From tw_2 to tw_3		From tw_3 to tw_4	
Number of customers	C_1	C_2	C_1	C_2	C_1	C_2
Remained in C_i	1118	2947	1046	2921	1120	2997
Moved from C_1 into C_2	64		91		45	
Moved from C_2 into C_1	31		102		40	
Dropped out of C_i	55	74	77	82	75	106
Appeared in C_i	65	94	92	131	65	50

Comparing the results of *Table 22* with those in *Table 20*, it can be noted that there is a small number of customers that were assigned to a certain cluster in one of the time windows who do not appear in any of the two clusters in the next time window. This number can be calculated as the difference between the total number of customers assigned to Cluster C_i, i=1, 2, and the sum of customers that remained in Cluster C_i and moved to another cluster. For instance, for the first pair of time windows the number of customers who dropped out of Cluster C_1 constitutes 1,237-(1,118+64)=55. This fact can be explained by decreased degrees of membership of these customers to both clusters in time window 2, so that a cluster assignment according to the considered absorption threshold (here u^o=0.5) is not possible.

On the other hand, there is a number of customers that were not assigned to any of the two clusters in one of the time windows but who appear in one of the two clusters in the next time window. This number can be obtained as the difference between the total number of customers assigned to Cluster C_i, i=1, 2, in the second of two time windows and the sum of customers that remained in Cluster C_i and moved to this cluster from another one. For instance, for the first pair of time windows the number of customers assigned to Cluster C_1 in the second time window but not present in any of the clusters in the first time window is equal to 1,214-(1,118+31)=65. This change in a cluster assignment is due to an increase in degrees of membership of some customers in the second time window compared to those in the first time window.

In short, it can be stated that the fuzzy partition with two clusters represents a good segmentation of customers in Group 'Y' and fits well the natural data structure.

6.1.7 Clustering of bank customers in Group 'N' based on partial temporal history and using the pointwise similarity measure

After clustering with a modified version of the Gath-Geva algorithm based on the pointwise similarity for trajectories and possibilistic classification of objects, the following cluster sizes given by the number of objects absorbed into clusters and the following numbers of free (stray) objects are obtained for each time window using different values of the absorption threshold:

Table 23. Number of customers 'N' assigned to two clusters in four time windows

	Time Window 1			Time Window 2			Time Window 3			Time Window 4		
	Absorbed		Stray	Absorbed		Stray	Absorbed		Stray	Absorbed		Stray
	C_1	C_2		C_1	C_2		C_1	C_2		C_1	C_2	
u^o=0.3	11397	6902	1280	11410	6901	1268	11172	6971	1436	11233	6987	1359
u^o=0.4	11326	6029	2224	11321	6209	2049	11092	6161	2326	11146	6186	2247
u^o=0.5	11227	4987	3365	11198	5075	3306	10966	5015	3598	11045	5058	3476
u^o=0.6	11068	4599	3912	11049	4656	3874	10801	4631	4147	10869	4638	4072

Table 24. Partition densities of clusters, fuzzy separation and compactness indexes obtained for customers in Group 'N' in four time windows

	Time Window 1	Time Window 2	Time Window 3	Time Window 4
PD_1	23611.49	23555.05	22977.04	23190.53
PD_2	3682.93	3744.26	3708.01	3728.95
V_{apd}	13647.21	13649.66	13342.53	13459.75
FSA	0.533	0.523	0.515	0.518
FC	0.717	0.720	0.709	0.713
PD_{free}	$2.40*10^{-7}$	$2.68*10^{-7}$	$4.88*10^{-7}$	$2.43*10^{-7}$

The centres of the customer segments obtained for customers in Group 'N' in the first time window (the first half a year) are shown in *Figure 79.* The customer segments obtained in the first time window can be interpreted as 'non-users' and 'active users of credit' similar to the customer segments recognised for Customer Group 'Y' based on the partial temporal history.

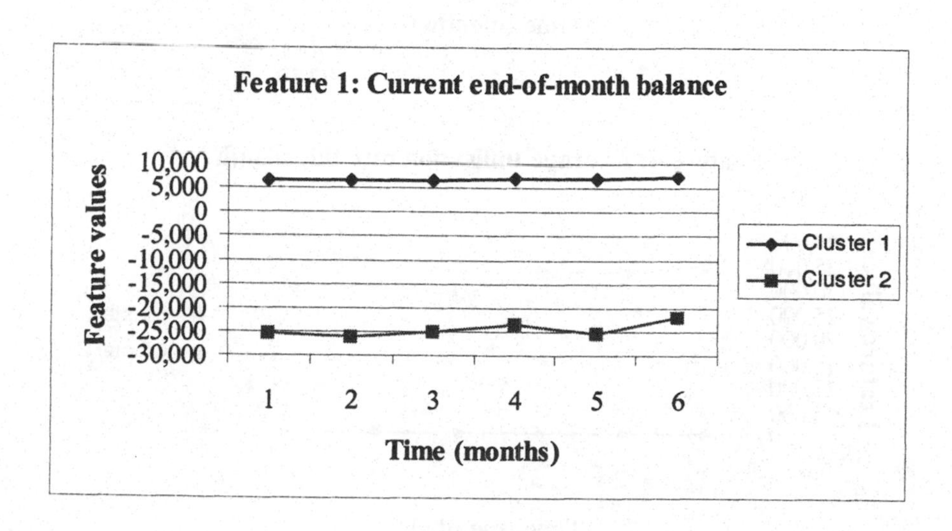

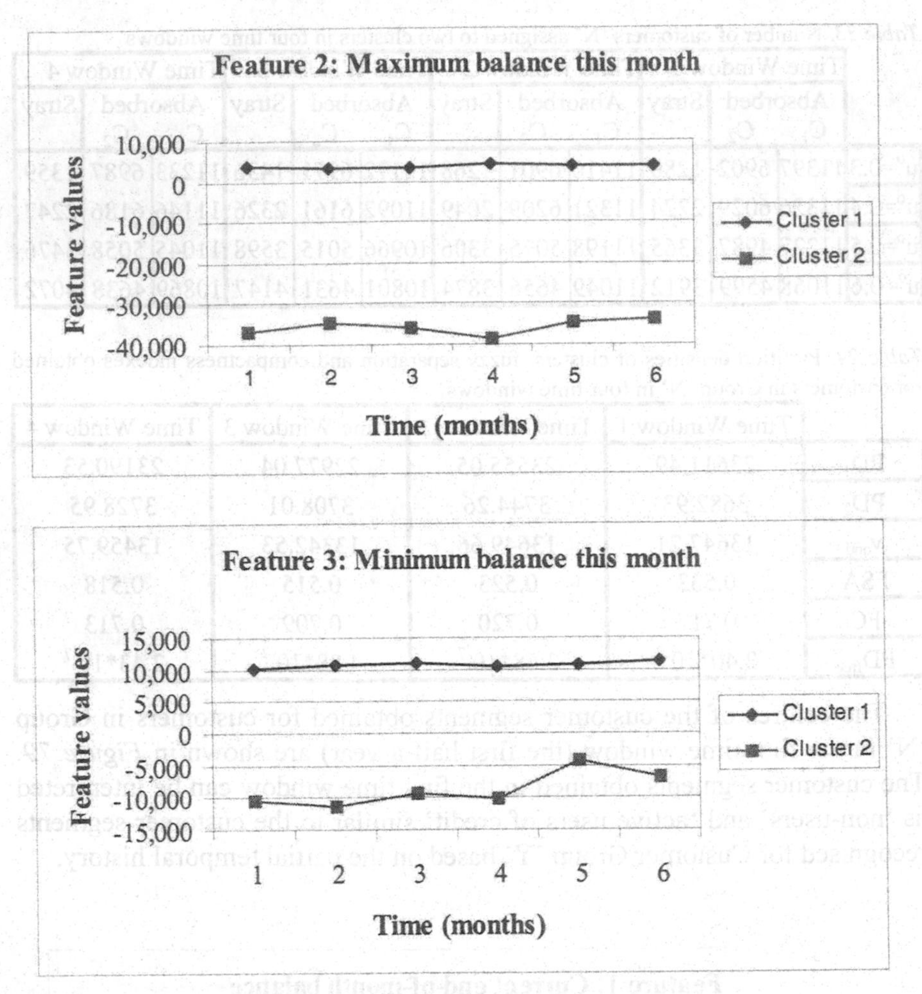

Feature 2: Maximum balance this month
Feature values
10,000
0
-10,000
-20,000
-30,000
-40,000
1
2
3
4
5
6
Time (months)
Cluster 1
Cluster 2
Feature 3: Minimum balance this month
Feature values
15,000
10,000
5,000
0
-5,000
-10,000
-15,000
Time (months)
Cluster 1
Cluster 2

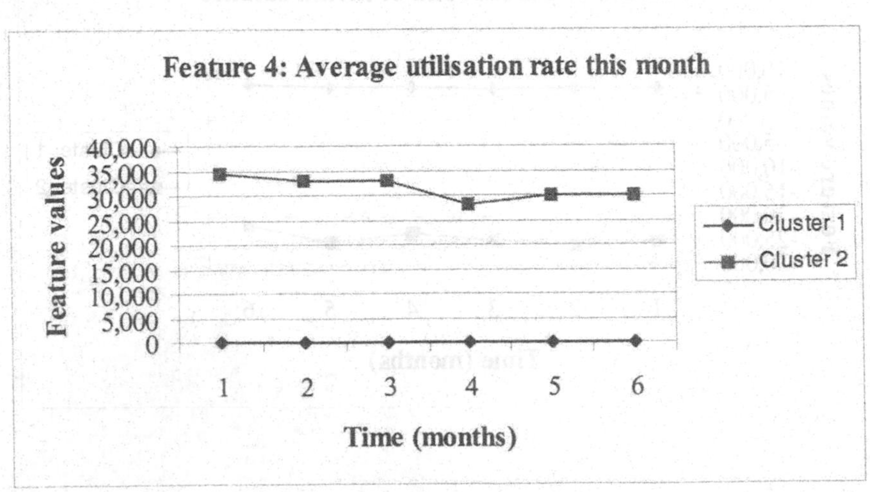

Feature 4: Average utilisation rate this month
Feature values
40,000
35,000
30,000
25,000
20,000
15,000
10,000
5,000
0
1
2
3
4
5
6
Time (months)
Cluster 1
Cluster 2

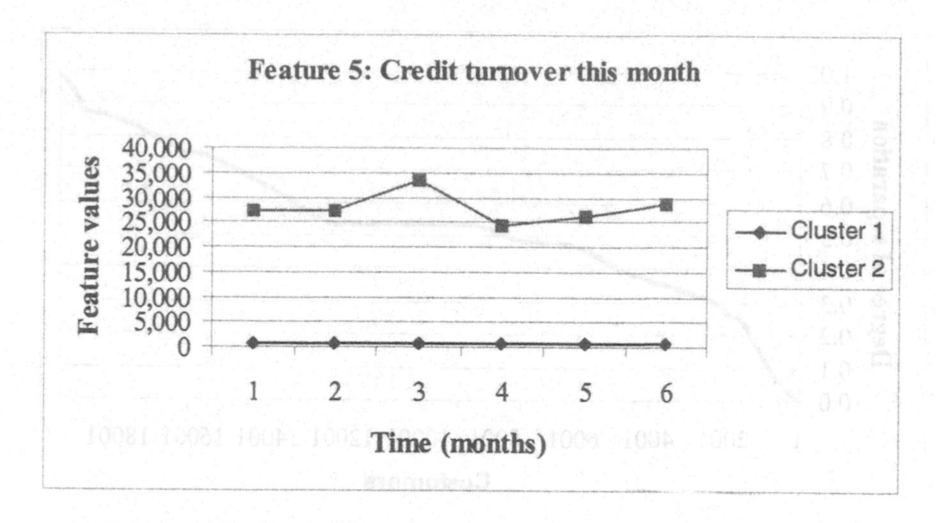

Figure 79. Cluster centres with respect to each feature obtained for customers in Group 'N' in the first time window and based on pointwise similarity between trajectories

Customers in the first segment, 'non-users,' are characterised by always positive end-of-month balances of about 7,000 DM. The account balance of these customers varies between 5,000 DM and 10,000 DM during a given month and the credit turnover constitutes about 1,000 DM per month. Since the average credit utilisation by these customers is very low, it can be stated that this type of customers does not use a credit from the bank. The account seems to be used in a way rather similar to a regular checking account.

Customers in the second segment, 'active users,' are described by a significant variation of their account balance of between –40,000 DM and -10,000 DM, and show a negative end-of-month balance of about –25,000 DM. These customers have a relatively high credit turnover varying between 25,000 DM and 35,000 DM and a high average credit utilisation, which reaches 35,000 DM per month. Obviously, this customer segment uses an available credit from the bank and represents a profitable group of customers.

As can be seen, the description and properties of these two segments are very similar to the ones of segments obtained for customers in Group 'N,' based on the whole temporal history, with the exception that the value ranges of the features in the first time window is slightly different, i.e. the account balance is somewhat smaller in absolute values, as well as the credit turnover and the average credit utilisation. This result confirms the fact that the two customer segments are rather stable over time and their properties change only insignificantly. This can also be seen in *Table 23* presenting the number of customers absorbed into the two segments in the four time windows.

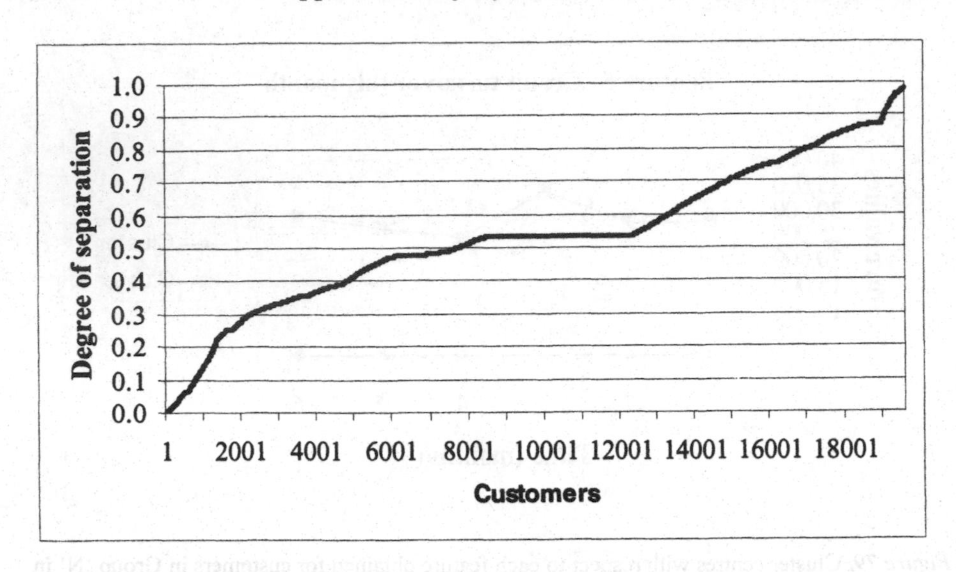

Figure 80. Degrees of separation between clusters obtained for customers in Group 'N' in the first time window

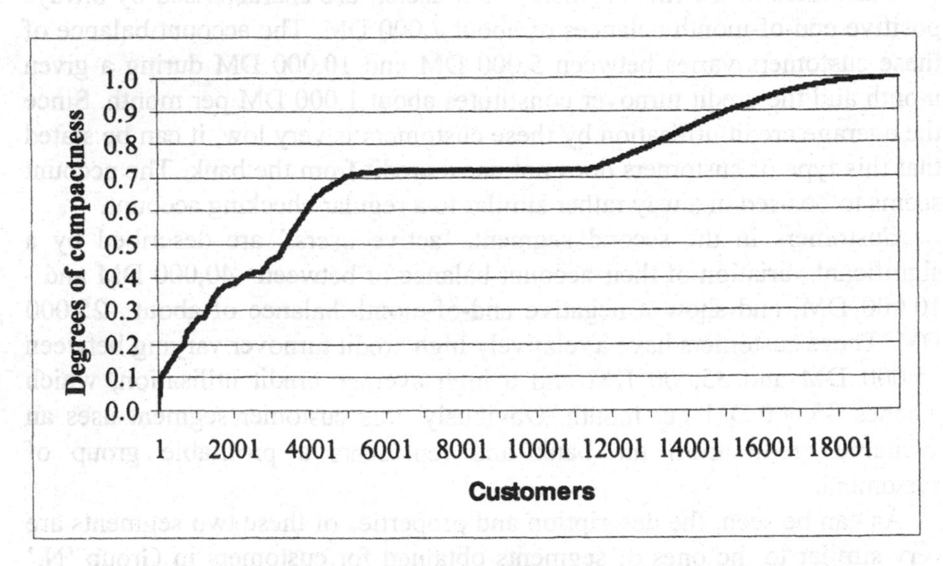

Figure 81. Degree of compactness of clusters calculated for customers in Group 'N' in the first time window

The quality of the fuzzy partition for customers in Group 'N' can be evaluated by considering the degrees of separation between the clusters corresponding to the degrees of ambiguity of object assignment and the degrees of compactness of the clusters expressing the maximum degrees of membership of objects to the clusters. Both validity measures calculated for each customer are shown in *Figure 80* and *Figure 81*, respectively. For the sake of a better visualisation, 19,579 customers are sorted by their degrees of separation or compactness in ascending order.

The analysis of both sequences characterising the validity of the fuzzy partition shows that 64.4% of customers in Group 'N' possess a maximum degree of membership larger that 0.5 and a degree of separation larger than 0.4. Considering more strict thresholds, it can be stated that 54.5% of customers have degrees of membership exceeding 0.7 but only 42.3% of them also have degrees of separation above 0.5. Thus, the fuzzy partition into two clusters in the first time window is similar in its characteristics to the one obtained for objects in Group 'N,' based on the whole time history, and allows a clear and unambiguous assignment of about 50% of customers.

In the next step of analysis, temporal changes of customer assignment to the clusters achieved using the absorption threshold u^o=0.5 are considered (the number of customers assigned to each cluster is given in *Table 23*). The analysis is carried out analogously to the case of customers in Group 'Y' with the purpose of detecting whether customers have remained in the same cluster to which they were assigned in the previous time window for each pair of subsequent time windows or whether they have moved from one cluster to another. The results of temporal changes of customer assignment are summarised in *Table 25*, where the i-th time window is denoted by tw_i, i=1,..., 4. These results show that most of the customers assigned to Segment C_1 or C_2 in the first time window remain in these segments in the following time windows, and just a small percentage of customers move from one cluster to the other between two time windows.

Table 25. Temporal changes of assignment of customers in Group 'N' to clusters

	From tw_1 to tw_2		From tw_2 to tw_3		From tw_3 to tw_4	
Number of customers	C_1	C_2	C_1	C_2	C_1	C_2
Remained in C_i	10466	4323	9951	4045	10250	4321
Moved from C_1 into C_2	108		336		104	
Moved from C_2 into C_1	163		235		123	
Dropped out of C_i	653	501	911	795	612	571
Appeared in C_i	569	644	780	634	672	633

Analogously to the case of customers in Group 'Y,' temporal changes of assignment of customers in Group 'N' between two time windows are also characterised by a small number of customers who were assigned to a certain cluster in one of the time windows but who do not appear in any of the two clusters in the next time window, i.e. these customers are dropped from a cluster due to their decreased degrees of membership in the next time window. Besides this, there are customers who appear in a certain cluster in the subsequent time window due to their increased degrees of membership compared to the previous time window. The number of these customers is represented in *Table 25* as well.

In brief, it can be stated that the generated fuzzy partition with two clusters provides a good segmentation of customers in Group 'N' and fits the natural data structure well.

6.1.8 Comparison of clustering results for customers in Groups 'Y' and 'N'

After conducting four types of analysis for different customer groups and for different lengths of the temporal history it is necessary to compare the results obtained. It has already been stated that the customer segments recognised based on the whole temporal history and in the first time window are very similar, however the feature values characterising cluster centres in the first case are somewhat larger in the absolute values compared to those in the second case.

Comparing the results for customers in Group 'Y' and 'N,' it can be seen that the values of the end-of-month balance of customers in Group 'Y exceed the corresponding values of customers in Group 'N,' and vary in the larger value range. The credit turnover of the first customer group is approximately 10,000-20,000 DM larger than the values of the other customer group, whereas the credit utilisation of the active users is 20,000-30,000 DM lower. Therefore, customers in Group 'Y' belonging to the segment of 'active users' have more entries in their accounts, higher monthly account statements and use bank credit less actively than customers in Group 'N'. Customers in the second segment, 'non-users,' are similar in their behaviour for both groups of customers.

The results of analysis conducted in this section can help a bank to better understand the customer portfolio, to distinguish between different groups of active users and non-users in order to be able to develop particular marketing strategy which may be, for instance, offering special favourable services to a group of the most active users.

Bank customer segmentation was carried out in this book based on the dynamic data representing customers' temporal behaviour and by applying the dynamic fuzzy clustering algorithm. The dynamic analysis allows to take into consideration the payment behaviour of customers over a period of time which characterises customers much better than a single observation. Until now in most applications related to customer segmentation and described in the literature the static analysis of customers was performed based on measurements at a certain moment of time. These analysis results are not obviously very reliable, since clusters, or customer segments, obtained from such analysis can often change due to significant fluctuations of account feature values that requires periodic re-clustering. In contrary the dynamic fuzzy clustering help to save time and affords and can provide more reliable complete results.

6.2 Computer Network Optimisation based on Dynamic Network Load Classification

Computer communications are one of the most rapidly developing technologies. Their purpose is to provide a possibility to transmit data between user's computers, terminals, and application programmes and in this way to support information exchange. The introduction of the techniques and equipment necessary for high-speed data transmission in the late 1960s led to the development of computer networks. In the 1970s government and corporate computer networks became popular as organisations realised the advantages offered in efficiency and profitability. In the mid 1970s extensive computer networks were adopted by financial institutions, such as banks, since they recognised the opportunity to increase their profitability and the need to remain efficient and competitive. The use of local area networks to interconnect computers and terminals within a building or group of buildings has grown in popularity since 1980 and made distributed computing a reality due to the high-capacity low-cost communication that they afford. In 1983 the notion of the Internet appeared as a general data communication network and it has soon become the world's largest computer network. Compared with 6,000 users in 1986, five years later in 1991, the Internet had more than 600,000 users. At present it is estimated that almost 200 millions people use the Internet. According to the International Data Corporation (IDC) this number will reach 500 millions people by2003. This explosive growth of the number of Internet users shows the enormous need for network services.

The availability of information supplied by computer networks is now crucial in many organisations, particularly information from many locations which must be co-ordinated both accurately and quickly. Obviously, the high demand on computer communications leads to high utilisation of computer networks which varies over time. All computer networks have certain limits regarding the speed of data transmission. If the volume of data traffic grows, the network utilisation rapidly increases and above a certain critical threshold the throughput rate of data decreases and time delays in data transmission occur. In order to avoid this problem, the analysis of network load over time can be useful in order to identify specific patterns, or typical states, of network load characteristics for certain months, days of the week, or time intervals during the day. The knowledge of the network load distribution over time can be used to optimise the network resources so that at peak times of network utilisation additional resources are allocated and in periods of low utilisation the regular resources are reduced. This kind of network optimisation is technically possible due to modern techniques and equipment which can provide a dynamically changing bandwidth for data transmission according to current demands. The optimisation of network resources means considerable savings for companies and organisations using network services since the bandwidth for data transmission is usually rented from a network provider. On the other hand, the information about the distribution of network load over time can be used to define prices for network use. It must be noted that the analysis of network load is relevant not only to computer networks but also to telecommunication networks.

In this section it will be shown how the dynamic fuzzy clustering algorithm can be applied to recognise, on-line, different states in network load and to follow state changes. This problem requires the design of a dynamic classifier with adaptive capabilities which is applied for clustering and classification of trajectories representing the development of network load. The problem of identifying typical patterns of network load depending on the day of the week and/or seasons of the year is not considered in this application since it is similar to the problem of customer segmentation presented in sections 6.1.4 and 6.1.5, and is, therefore, concerned with the static classifier design.

6.2.1 Data transmission in computer networks

A communication network is defined as a shared resource used to exchange information between users [Freer, 1988, p. 101]. A computer network is a distributed collection of computers which is viewed by the user as one large computer system allocating jobs without user intervention. Hence a computer communication network is viewed as a collection of several computer systems from which the user can select the service required and communicate with any computer as a local user. According to the definition of the International Standards Organisation (ISO) a network is an interconnected group of nodes or stations. Nodes are determined as the intersection of two or more transmission paths and perform traffic switching, whereas stations may attach to a single transmission path in a network.

The logical manner in which nodes are linked together by channels to form a network is denoted as network topology (for example, star networks, ring networks, bus networks). The way in which cables, which provide channels between stations, are positioned is called topography. Most of the topologies generally broadcast all signals to all the connected stations. The receiving station, or its communication equipment, must select only the signals addressed to the station from all the transmissions on the network.

In order to guarantee the communication between different makes of computers and the co-existence of different application programmes on the network a set of rules is needed to govern the connection and interaction of the network components. This set of rules is called the network architecture which includes the data formats, protocols and logical structures for the functions providing effective communication between data processing systems connected to the network [Freer, 1988, p. 133]. In the late 1970s a reference model for Open System Interconnection (OSI) was developed by ISO to describe the structure of communication protocols. This model forms a basis for the co-ordination of developments in layered network communication standards.

The OSI reference model consists of 7 layers as shown in *Figure 82* [adapted from Black, 1989, p. 285].

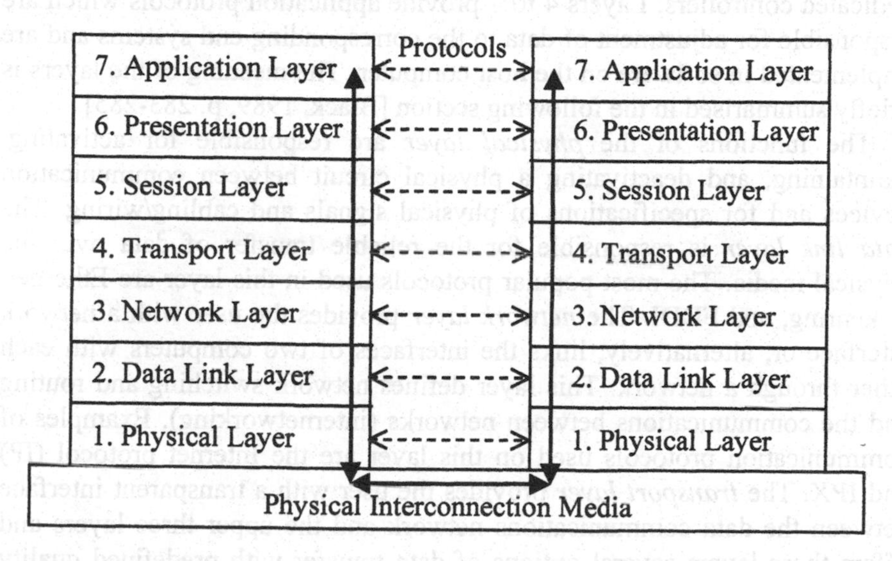

Figure 82. The OSI basic reference model

Each of these layers defines certain functions required for data transfer between computers so that the communication process is represented by a hierarchy of function layers which stack on each other ([Freer, 1988, p. 134-135], [Hunt, 1995, p. 5-7]). Each layer is based on services of the lower layer and offers services to the higher layer. When data are transmitted between application programs on two different computers, the data passes from layer 7 to layer 1 at the sending system and are transmitted over the interconnection media to the receiving system where data travels through layers 1 to 7. The data path through the layers is illustrated by continuous arrows in *Figure 82*. Data does not pass directly between layers, except the physical layer, but layer interfaces are provided instead to define the services of corresponding layers.

The communication of each layer with the same layer of another system requires the definition of communication rules and data formats. Such rules, which ensure the orderly exchange of information between two or more parties, are called *protocols*. The implementation of protocols for each layer is independent from other layers. This allows one to minimise the influence of single technical changes in one layer on the whole protocol family. During the data transmission each layer adds its own protocol control information to data passing from the higher layer. This protocol information is ignored by the lower layers and is interpreted only by the same layer of the receiving system.

Layers 1 to 3 are network related and provide low-level protocols for physical data transfer which are mostly implemented in hardware or by dedicated controllers. Layers 4 to 7 provide application protocols which are responsible for adjustment of data to the corresponding end systems and are implemented in software on the host computer. The meaning of the layers is briefly summarised in the following section [Black, 1989, p. 283-285].

The functions of the *physical layer* are responsible for activating, maintaining, and deactivating a physical circuit between communication devices and for specifications of physical signals and cabling/wiring. The *data link layer* is responsible for the reliable transfer of data over the physical media. The most popular protocols used in this layer are Ethernet, Tokenring, and FDDI. The *network layer* provides the user with a network interface or, alternatively, links the interfaces of two computers with each other through a network. This layer defines network switching and routing and the communications between networks (internetworking). Examples of communication protocols used on this layer are the Internet protocol (IP) and IPX. The *transport layer* provides the user with a transparent interface between the data communications network and the upper three layers and offers these layers several options of data transfer with predefined quality features. The best representatives of communication protocols used in this layer are the transmission control protocol (TCP) and the user datagram protocol (UDP). The *session layer* manages the interactions (sessions) between application programmes (users) providing different services to co-ordinate data exchange between applications. The *presentation layer* is responsible for the description of the data structure and representation and negotiation with the corresponding layer of receiving applications concerning the data format. The *application layer* is concerned with the support of end-user applications.

For a better understanding of the data used in the application example appearing in the following sections, the functions of the network layer will be discussed in more detail below. This layer allows two network users to exchange data with each other through one network connection or multiple network connections. Switching techniques define the method of data transmission between nodes. The task of the routing function is to find the optimal path (route) for data blocks through the network from the sending to the receiving station. For this purpose the addresses of the network layer contain the information concerning the location of stations in the network. The route is chosen based on several criteria such as bandwidth, utilisation rate or least-cost.

There are three primary methods of switching data from one node to another over a network [Freer, 1988, p. 110]. *Circuit switching* provides a direct connection between two nodes and permits an end-to-end transmission between the two end users by utilising the facility within bandwidth and tariff limitations. Many telephone networks are circuit switching systems. *Message switching* is a store-and-forward technology. The switch is used to receive and acknowledge the message from the sending station and to store it temporarily until appropriate circuits are available to forward the message to the receiving station. Private and military teleprinter networks usually employ this technology. *Packet switching* is derived from message switching but messages are broken down into smaller pieces known as *packets* which are interleaved with packets from other virtual channels during transmission through the network. Packets are provided with protocol control information (headers) which enable the complete message to be reassembled by the receiving station and routed through the network as independent entities. The default user data field length in a data packet is 128 bytes but other packet lengths (sizes) are also available: 16, 32, 64, 256, 512, 1024, 2048, and 4096 bytes. Due to a number of advantages such as a fast response time for all users of this facility, a high availability of the network to all users, a distribution of risks of failure to more than one switch, and a sharing of resources, packet switching has become the prevalent technology for switched data networks.

Data transmission over the network requires a method of controlling the access of individual users to the transmission medium. The most popular local area network, Ethernet, represents the shared-medium-concept, i.e. the transmission medium is shared by all stations. If all stations work simultaneously, then the task of the method of access control is to decide what station may transmit data and when. The most common method used by Ethernet is the Carrier Sense Multiple Access with Collision Detection (CSMA/CD). All stations listen into the network and know whether the medium is free or data are being transmitted. If the medium is free, each station can transmit data to another station. If two stations attempt to transmit data simultaneously, a collision appears which is recognised by other stations. After a short pause the network is free again for data transmission. In order to avoid a new collision, stations wait a pseudo-random time interval before attempting retransmission. If, nevertheless, a new collision appears, the waiting interval is doubled. After 16 mistrials an error is reported.

This method of access control and collision detection determines a certain relation between the number of collisions and the data traffic volume [Kiel, 1996, p. 60]. As long as the network load is not too large, there are no problems with data transmission. If the data traffic (the number of transmitted packets) increases, the number of simultaneous accesses to the network and the number of collisions subsequently increase. The situation is getting more and more critical due to the need of retransmission. Above a certain threshold of traffic volume the bandwidth provided for each station, as well as the total data throughput rate, are drastically decreased as is illustrated in *Figure 83* [adapted from Kiel, 1996, p. 60]. Therefore the theoretical throughput rate of Ethernet equal to 10Mbytes/s is rarely reached; effectively it is usually about one or two thirds of this rate if many stations are working on the network.

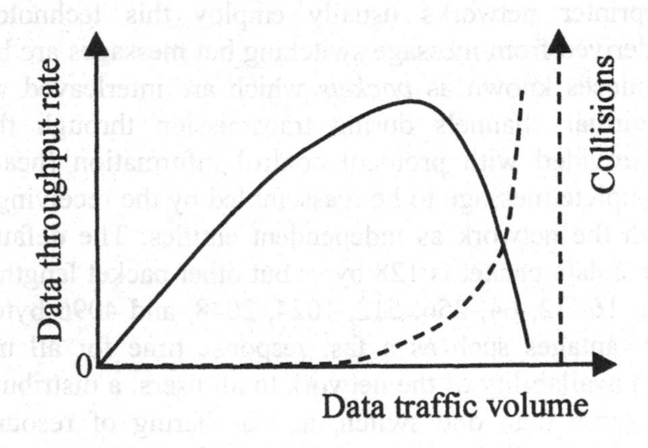

Figure 83. Dependence between the number of collisions and the network load

If the collision rate on more then one computer in the network permanently exceeds 5%, it can be reasonable either to divide the network to reduce the load or to optimise dynamically the network resources by providing some stations with a greater bandwidth if the traffic volume is too high.

6.2.2 Data acquisition and pre-processing for the network analysis

Monitoring is an ideal way to manage overall network performance, as well as analyse the influence of new applications and technologies on successful application operation. There is a big number of software tools called traffic, or network, analysers which enable one to capture and simultaneously view an instant snapshot of network activity for real-time assessment. These data snapshots can be displayed in graphic and numerical formats and stored for future impact analysis and historical reporting. Network analysers are commonly used by companies to increase management productivity by delivering a real-time information that reduces the number of dedicated resources required to evaluate network performance.

In order to acquire data needed for network load analysis the software tool NetXRay was used in the application under consideration. The NetXRay® Protocol Analyser and Network Monitor is a software-based fault and performance management tool that captures data, monitors network traffic and collects key network statistics. The NetXRay Analyser was developed specifically for Windows 95, or NT, and provides a state-of-the-art Graphical User Interface. This 32-bit software-based tool takes full advantage of the Windows operating environment. NetXRay enables network managers to monitor every network segment, to extract and review vital and detailed information needed to effectively troubleshoot, manage and migrate the complex network environments existing today.

NetXRay provides the following features for quick troubleshooting of network problems, real-time network traffic analysis, planning and forecasting of network growth:

- Intuitive, consistent user interface allows for quick capture and display of data to monitor key network performance criteria;
- NetXRay data can be saved in Sniffer® Network Analyser format for automated post-capture expert analysis and enhanced protocol interpretation;
- NetXRay supports all major LAN topologies and decodes over 110 protocols from major protocol suites;
- NetXRay's Traffic Map and traffic matrix show who is talking to whom on the network;
- Protocol distribution information allows to see what protocols are in use, which IP applications are running, and which IPX transport processes are being utilised;
- History reports facilitate an understanding of network usage over long periods of time;
- After establishing the network baseline, day-to-day utilisation can be tracked to predict when changes will need to be made to accommodate increases in network traffic.

During running NetXRay can simultaneously capture data from one, or more, network interfaces, display previously captured data, generate traffic, generate alarms as traffic thresholds are exceeded, automatically discover network hosts, and monitor and store key network statistics. All of this can take place while other Windows applications are running on the same machine, i.e. dedicated hardware is not required. Unfortunately it is not possible to store automatically the history of protocol distribution in the network. The current protocol distribution can only be monitored and some snapshots saved on disk.

The application under consideration takes advantage of the ability of the NetXRay tool to record network activities over a period of time. The programme supports the monitoring of ten network activities, concurrently, which all characterise the data traffic volume, network utilisation and collisions. For the purpose of network load analysis six history statistics concerning the data traffic were selected and collected over a period of 70 days (from mid-April until mid-June 1999) from the computer network of the dormitory of the Technical University of Aachen (Germany). The standard LAN Protocol used in the dormitory is Ethernet with a bandwidth of 10Mbytes/s, and the standard communication protocol is TCP/IP. The six network statistics correspond to six packet sizes used by TCP/IP for data transmission through the network. The number of packets of each category transmitted at the current moment, multiplied with the corresponding packet size, constitutes the current volume of the data traffic and represents a certain degree of network utilisation. The larger the packet size, the larger the influence of the number of these packets on the network utilisation. These six packet sizes characterising the data transmission in the network are considered as dynamic features for the pattern recognition process and are given in *Table 26*.

Table 26. Dynamic features describing data transmission in computer network

Feature	Description
1	Packet size under 64 B/s
2	Packet size between 65 and 127 B/s
3	Packet size between 128 and 255 B/s
4	Packet size between 256 and 511 B/s
5	Packet size between 512 and 1023 B/s
6	Packet size between 1024 and 1518 B/s

Network statistics were monitored each day during 15 hours coinciding with peak network utilisation: from 9am until 24pm. The measurements were recorded with a sampling rate of 30 s. The data traffic and network utilisation during the night are always very low, hence these data are not representative. For the analysis of network activities only the data collected over 14 days in May were used to demonstrate the abilities of the dynamic fuzzy clustering algorithm applied to network load analysis. This results in the temporal sequence of 25200 measurements for each feature, or 6-dimensional trajectory of length 25200 time instances. The main statistical characteristics of this data with respect to each feature are summarised in *Table 27* and *Table 28*. Comparing the value ranges and the main quantiles with a limit of $\mu+3\sigma$, where μ is the mean value and σ is a standard deviation, it can be seen that feature values contain a low percentage of outliers. Due to this fact features are characterised by rather skewed distributions. As an example, two density distributions of the mean values of trajectories of features 1 and 6 are illustrated in *Figure 84*.

In order to be able to use the recorded temporal data for the analysis, they have to be pre-processed to smooth the strong fluctuating behaviour of trajectories and to normalise all features to the same interval [0, 1]. Trajectories of features are smoothed over 200 values leading to trajectories with clear temporal behaviour but filtered from random fluctuations.

Table 27. The value ranges and main quantiles of each feature characterising network data

Features	Value range	s_α (α=0.25)	s_α (α=0.5)	s_α (α=0.75)
1	[0,1374]	38	58	113
2	[0, 390]	9	23	118
3	[0, 1851]	2	6	33
4	[0, 53]	3	5	17
5	[0, 187]	6	12	27
6	[0, 651]	20	69	158

Table 28. Main statistics of each feature of the network data

Features	Mean value μ	Standard deviation σ	$\mu-3\sigma$	$\mu+3\sigma$
1	29.184	33.920	-72.576	130.944
2	27.676	45.249	-108.072	163.424
3	5.497	11.544	-29.135	40.129
4	2.090	3.142	-7.336	11.517
5	4.506	8.641	-21.418	30.430
6	34.913	58.252	-139.843	209.670

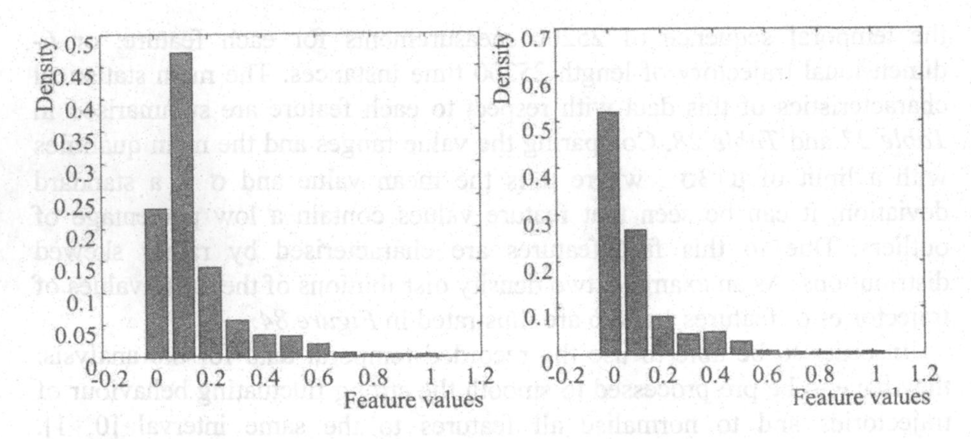

Figure 84. Density distributions of features 1 (left) and 6 (right)

6.2.3 Goals of the analysis of load in a computer network

The primary goal of the computer network analysis conducted in this application is to recognise typical load states in the computer network based on the data traffic volume in order to be able to optimise network resources and to avoid collisions in the network at peak operating times. The load states, or data traffic volumes, are supposed to be time-dependent, low or high data traffic volume is related to certain time intervals during the day. It is possible that time intervals of typical network states change over time, therefore there is a need to follow temporal changes of load states. The consideration of different packet sizes as features allows one to recognise the distribution of packets which provides additional information about the number of users on the network. If there are a lot of large packets transmitted through the network, then it can be assumed that the number of users working simultaneously is not large so that they can use the bandwidth to a full degree. If the number of small packets is very high, it can be assumed that many users are working on the network simultaneously, and in order to minimise the time for transmission and to avoid collisions messages are sent in small packets. In this case the bandwidth allotted to each user is limited. Although in most networks messages are automatically broken into packets of a certain size, there is also equipment which allows an operator to manage this process to adapt it to the current situation.

From the viewpoint of technical analysis, the goal of network analysis consists in the on-line detection of typical states in a 6-dimensional trajectory representing the temporal development of data traffic in the network. This goal can be achieved by applying the algorithm for dynamic fuzzy classifier design and classification which allows one to adapt the classifier to changing network states.

The results of such network analysis can be used to adjust the network resources allotted to the user. As soon as typical patterns of network activities characteristic of specific days are learned, they can be used to derive the basic optimisation strategy of network resources. It should be noted that it is possible to obtain equivalent results by clustering all trajectories representing data traffic during one day which are collected over a long period of time (for instance, over six months). This kind of analysis requires, however, that the corresponding data base would be available in advance. The application under consideration shows that using the dynamic fuzzy clustering algorithm it is possible to start with a small amount of data and to learn typical patterns of data traffic during the adaptation process.

6.2.4 Parameter settings for dynamic classifier design and classification of network traffic

Clustering of multidimensional trajectories describing data traffic in a computer network will be performed based on pointwise, as well as structural, similarity. For the definition of the pointwise similarity the quadratic membership function given by equation (5.11) is chosen to model the fuzzy set 'approximately zero', which is defined with respect to each feature f_r, $r=1, ..., M=6$. Parameter $a(r)$ of each membership function is determined by equation (5.14) and based on the consideration presented in section 6.1.3. The value of parameter $\beta(r)$ is evaluated by the mean value of maximal values of trajectories of the corresponding feature according to equation (6.1) and parameter $a(r)$ is then calculated according to equation (6.2).

For the definition of the structural similarity using the Algorithm 7a parameter a must be defined with respect to each temporal characteristic as well as each feature. Suppose that a set of characteristics chosen to describe the temporal behaviour of trajectories is given by $\{K_1, ..., K_L\}$ and denote the value of characteristic K_i obtained from the j-th trajectory of feature r $x_{jr}(t)$ by $K_i(x_{jr})$, $i=1,..., L$, $j=1, ...,N$. The values of parameter $\beta(i, r)$ for characteristic K_i with respect to feature r and the corresponding value of parameter $a(i,r)$ can be calculated in the following way:

$$\beta(i,r) = \frac{1}{N}\sum_{j=1}^{N} K_i(x_{jr}), \quad i = 1,...,L, r = 1,...,M \tag{6.3}$$

$$a(i,r) = \frac{1}{\beta(i,r)^2}. \tag{6.4}$$

The aggregation of partial similarity measures over all features and characteristics is carried out according to Step 5.2 of Algorithm 7a: partial similarities are transformed to distance measures which are then aggregated to an overall distance in two steps using the Euclidean norm.

The distance measure obtained from the pointwise or structural similarity is used instead of the Euclidean distance measure in the calculation schemes of the functional fuzzy c-means and possibilistic c-means algorithms, where the later is used only for the possibilistic classification of dynamic objects. In both algorithms the distance between objects and cluster centres is involved in the calculation of degrees of membership of objects to clusters (see equation (4.5)). In this application the FFCM algorithm is used instead of the Gath-Geva algorithm, since if the classifier is designed based on a small number of objects (for example 30 objects) some statistical characteristics such as the fuzzy covariance matrixes of clusters used in the Gath-Geva algorithm cannot be considered as representative. A more significant number of objects is required for their evaluation.

For dynamic classifier design it is necessary to define a set of thresholds which are used within the monitoring procedure. The following parameter settings are chosen within the algorithms for detecting new clusters, similar clusters to be merged and heterogeneous clusters to be split:

Table 29. **Parameter settings for three algorithms of the monitoring procedure used during the network analysis**

Parameter	Value
Absorption threshold	$u^o = 0.5$
Share of the average cluster size	$\alpha^{cs} = 0.6$
Share of the average cluster density	$\alpha^{dens} = 0.1$
Threshold for the choice of 'good' free objects	$\alpha^{good} = u^o = 0.5$
Merging threshold	$\lambda = 0.6$
Number of bars in the density histogram	$N_{bars}=10$
Density threshold for splitting	$r^{dens}=10$
Size threshold for dense groups of objects	$r^{diam}=10$

6.2.5 Recognition of typical load states in a computer network using the pointwise similarity measure

For recognising typical states in a 6-dimensional trajectory representing data traffic let it be assumed that new dynamic objects arrive every 200 time instants which correspond to 100 minutes due to a sampling rate of 30 s. Each dynamic object is observed during a time window length of 200 and represented by a 6-dimensional trajectory (temporal history) with a length of 200. As stated above, the data traffic was monitored during 14 days, that is 14*15=210 hours, and 25000 measurements were made. Using the chosen time window length, the total time interval of observation [1, 25000] is broken down into 125 time windows, which are denoted as tw.

For the definition of the membership function 'approximately zero' used to calculate the pointwise similarity between trajectories the following values for parameter a are obtained:

	Feature 1	Feature 2	Feature 3	Feature 4	Feature 5	Feature 6
a	25.6597	30.7279	20.6437	33.1056	11.5258	16.5873

A dynamic fuzzy classifier with the number of clusters c=2 is designed after 31 time windows, i.e. based on 31 dynamic objects obtained during 3 days 6 hours and 40 minutes of observation, by applying the functional fuzzy c-means algorithm based on the pointwise similarity measure for trajectories. After possibilistic classification and assignment of dynamic objects to clusters according to the absorption threshold, the sizes of the clusters are equal to $N_1^{0.5} = 2$ and $N_2^{0.5} = 18$ and the number of free objects is n_{free}=11. The algorithm for detecting new clusters leads to the declaration of a new cluster (both criteria of size and compactness of free objects are satisfied). Therefore, objects are re-clustered with a new number of clusters c=3. The absorption procedure results in cluster sizes $N_1^{0.5} = 4$, $N_2^{0.5} = 14$ and $N_3^{0.5} = 10$, and the number of free objects is equal to n_{free}=3. The quality of a new partition is evaluated during the monitoring procedure. According to the algorithm for detecting heterogeneous clusters, the second cluster can be split. Hence, objects are re-clustered with a new number of clusters c=4. The new classifier is characterised by cluster sizes $N_1^{0.5} = 8$, $N_2^{0.5} = 11$, $N_3^{0.5} = 3$ and $N_4^{0.5} = 3$, and n_{free}=6 objects are left free. Since this fuzzy partition is better than the previous one with 3 clusters according to validity measures, and no more changes can be detected by the monitoring procedure, the classifier is accepted.

In the next time windows new objects are classified and some of them are assigned to existing clusters. In time window tw=34 (after 3 days 11 hours and 40 minutes of observation), when the number of free objects is increased to n_{free}=8, two new clusters are detected during the monitoring procedure. The classifier is designed anew with the number of clusters c=6. After absorption of objects into the new clusters the following cluster sizes are obtained: $N_1^{0.5}=9$, $N_2^{0.5}=15$, $N_3^{0.5}=3$, $N_4^{0.5}=1$, $N_5^{0.5}=2$ and $N_6^{0.5}=3$ whereas only $n_{free}=1$ object cannot be assigned because of its low degree of membership to all clusters. This classifier is accepted and applied to the classification of new objects in subsequent time windows.

In time window tw=64, that is after 7 days and 100 minutes of observation, the number of free objects has increased to $n_{free}=13$ and a new cluster is detected by the monitoring procedure. Thereby, three clusters C_4, C_5, and C_6 have remained unchanged in size and remain very small since the classifier has initially been designed and three other clusters have increased to $N_1^{0.5}=18$, $N_2^{0.5}=23$, $N_3^{0.5}=4$. It can be assumed that the cluster structure learned after the first 3.5 days does not fit the data structure very well after 7 days, which can be explained by the fact that data traffic during days 6 and 7, corresponding to the weekend, differs considerably from data traffic registered on a week day. Thus the classifier is re-learned with a new number of clusters, c=7, resulting in cluster sizes $N_1^{0.5}=17$, $N_2^{0.5}=19$, $N_3^{0.5}=10$, $N_4^{0.5}=2$, $N_5^{0.5}=3$, $N_6^{0.5}=4$ and $N_7^{0.5}=9$, and no free objects. According to validity measures, the new partition represents an improvement compared to the previous one with 6 clusters, however the monitoring procedure is able to detect two similar clusters, C_1 and C_2. Since their similarity measure $s(C_1, C_2)=0.756$ exceeds the merging threshold, these clusters can be merged. Objects are re-clustered with the number of clusters c=6 resulting in the following cluster partition: $N_1^{0.5}=27$, $N_2^{0.5}=14$, $N_3^{0.5}=3$, $N_4^{0.5}=3$, $N_5^{0.5}=4$, and $N_6^{0.5}=12$ and $n_{free}=2$. The new improved classifier is accepted and applied for classification of new objects in the following time windows. Until the last time window tw=125 no further changes in the cluster structure have been detected and the clusters have grown to $N_1^{0.5}=53$, $N_2^{0.5}=27$, $N_3^{0.5}=5$, $N_4^{0.5}=12$, $N_5^{0.5}=8$, and $N_6^{0.5}=14$ and $n_{free}=6$.

It can be assumed that the classifier has learned all typical patterns in the data traffic during 14 days of observation. The results of dynamic clustering and classification over this time period of 125 time windows are summarised in *Table 30*, where v_{apd} stands for average partition density, FSA denotes the fuzzy separation index with respect to ambiguity, and FC is the fuzzy compactness index. The last column contains the result of the monitoring procedure corresponding to changes detected in the cluster structure in the current time window. Time windows where objects were re-clustered due to detected temporal changes in the cluster structure and the corresponding re-clustering results are marked in the table by a light grey shade.

Table 30. Results of dynamic clustering and classification of the network data traffic based on the pointwise similarity measure

	$N(C_i)$	n_{free}	V_{apd}	FSA	FC	Changes
TW=31, c=2	2, 18	11	249466	0.61	0.69	C_{new}=1
TW=31, c=3	4, 14, 10	3	4409678	0.299	0.65	Split C_2
TW=31, c=4	8, 11, 3, 3	6	14833096	0.293	0.68	N_o
TW=34, c=4	8, 12, 3, 3	8	13796752	0.312	0.71	C_{new}=2
TW=34,c=6	9, 15, 3, 1, 2, 3	1	16106656	0.332	0.72	N_o
TW=64, c=6	18, 23, 4, 1, 2, 3	13	9843481	0.303	0.69	C_{new}=1
TW=64, c=7	17, 19, 10, 2, 3, 4, 9	0	14713282	0.297	0.77	Merge C_1 and C_2
TW=64, c=6	27, 14, 3, 3, 4, 11	2	3864317	0.334	0.72	N_o

The resulting cluster centres representing typical states of data traffic are shown in *Figure 85*. The typical states recognised in the data traffic can be interpreted as follows.

Cluster 1 represents a relatively low data traffic with a decreasing tendency: the number of large packets (1024-1518 B/s) decreases from a constant level of about 17 to 10 while the number of small packets (under 64 B/s) increases slowly to 17. The number of other packet types is rather small varying in the interval [6, 8] for packet sizes 65-127 B/s, and below 2 for three other packets sizes. This state characterises a rather small network load.

Cluster 2 describes an average data traffic with a clearly decreasing tendency. The number of small packet sizes under 64 B/s and between 65 and 127 B/s falls from about 60 to 15 whereas the number of large packets (1024-1518 B/s) remains steady at an average level of about 20. Three other packet types also decrease but in a much smaller value range.

Cluster 3 corresponds to a rise from an average to large data traffic. The number of large packets between 1024 and 1518 B/s grows drastically from 30 to 100, along with the small packets under 64 B/s whose number increase from 10 to 40. The number of medium-sized packets (65-127, 128-255 B/s) varies ranging from 20 to 60.

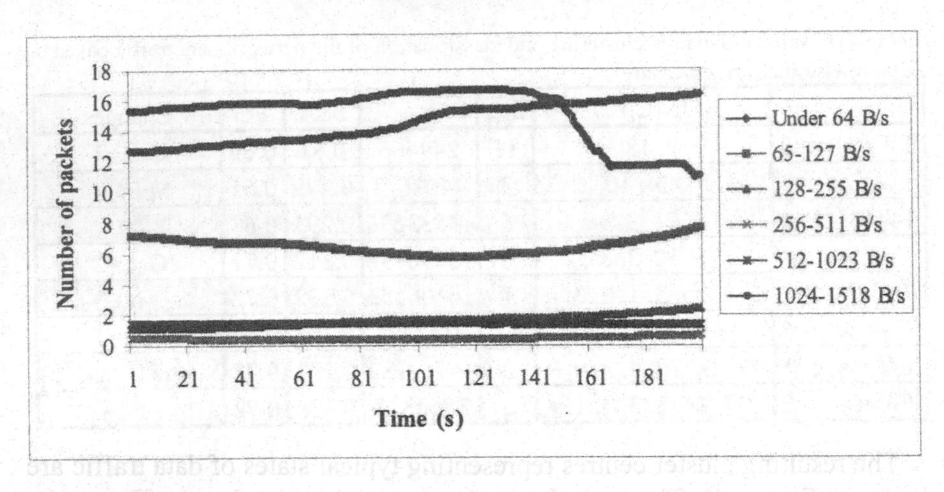
18
16
14
12
10
8
6
4
2
0
Number of packets
1 21 41 61 81 101 121 141 161 181
Time (s)
Under 64 B/s
65-127 B/s
128-255 B/s
256-511 B/s
512-1023 B/s
1024-1518 B/s

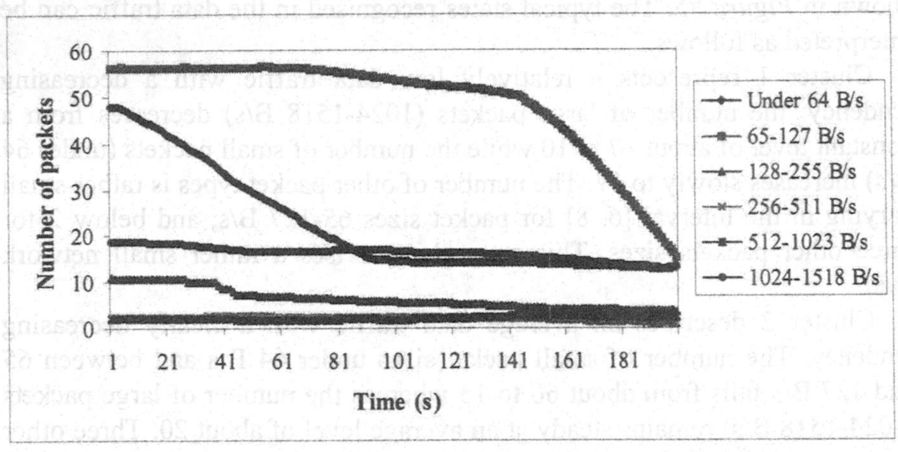
60
50
40
30
20
10
0
Number of packets
1 21 41 61 81 101 121 141 161 181
Time (s)
Under 64 B/s
65-127 B/s
128-255 B/s
256-511 B/s
512-1023 B/s
1024-1518 B/s

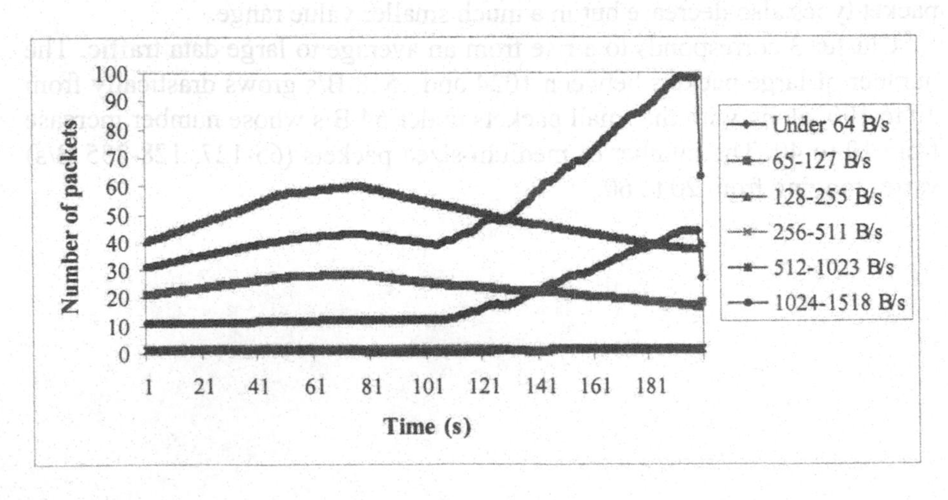
100
90
80
70
60
50
40
30
20
10
0
Number of packets
1 21 41 61 81 101 121 141 161 181
Time (s)
Under 64 B/s
65-127 B/s
128-255 B/s
256-511 B/s
512-1023 B/s
1024-1518 B/s

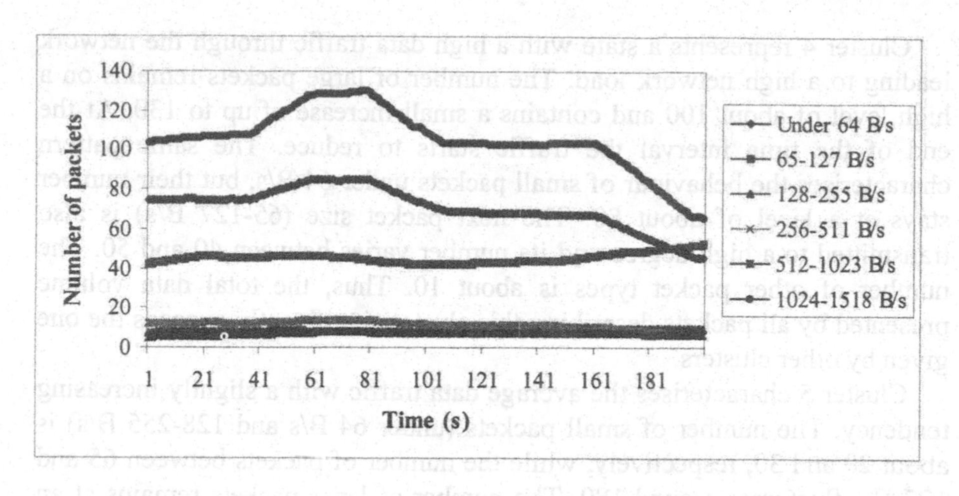

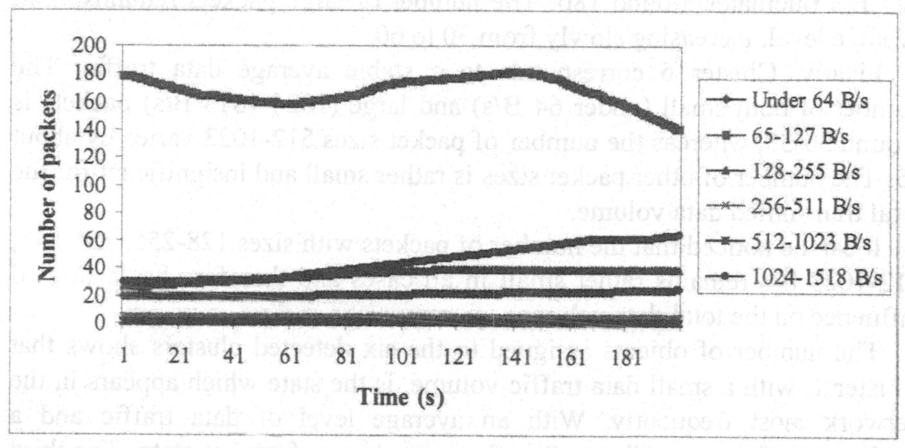

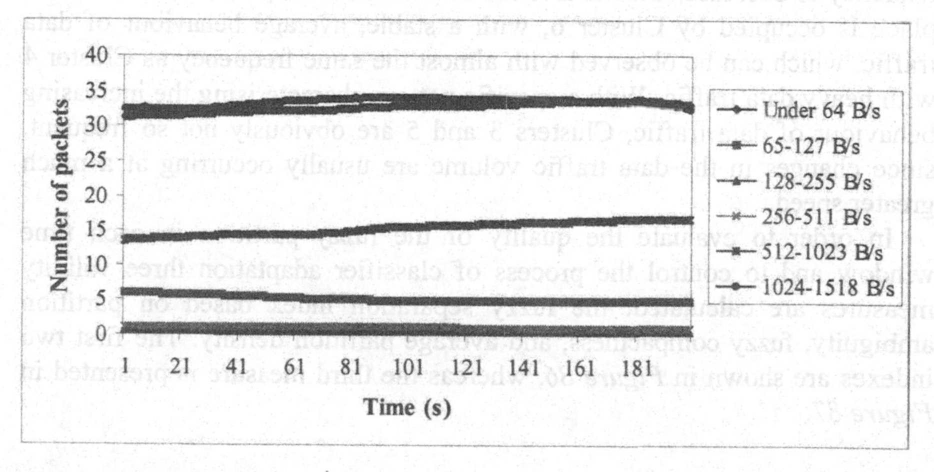

Figure 85. **Six typical states of the data traffic described by six packet sizes and obtained using the pointwise similarity measure**

Cluster 4 represents a state with a high data traffic through the network leading to a high network load. The number of large packets remains on a high level of about 100 and contains a small increase of up to 130. At the end of the time interval the traffic starts to reduce. The same pattern characterises the behaviour of small packets under 64 B/s, but their number stays at a level of about 80. The next packet size (65-127 B/s) is also transmitted to a high degree and its number varies between 40 and 50. The number of other packet types is about 10. Thus, the total data volume presented by all packets describing this cluster significantly exceeds the one given by other clusters.

Cluster 5 characterises the average data traffic with a slightly increasing tendency. The number of small packets (under 64 B/s and 128-255 B/s) is about 20 and 30, respectively, while the number of packets between 65 and 127 B/s fluctuates around 180. The number of large packets remains at an average level, increasing slowly from 30 to 60.

Finally, Cluster 6 corresponds to a stable average data traffic. The number of both small (under 64 B/s) and large (1024-1518 B/s) packets is around 30-35, whereas the number of packet sizes 512-1023 varies by about 15. The number of other packet sizes is rather small and insignificant for the total transmitted data volume.

It can be noticed that the number of packets with sizes 128-255, 256-511, 512-1023 B/s remains rather small in all cases and therefore has a limited influence on the total data volume.

The number of objects assigned to the six detected clusters shows that Cluster 1, with a small data traffic volume, is the state which appears in the network most frequently. With an average level of data traffic and a tendency to decrease, Cluster 2 is the second most frequent state. The third place is occupied by Cluster 6, with a stable, average behaviour of data traffic, which can be observed with almost the same frequency as Cluster 4 with heavy data traffic. With a specific pattern characterising the increasing behaviour of data traffic, Clusters 3 and 5 are obviously not so frequent, since changes in the data traffic volume are usually occurring at a much greater speed.

In order to evaluate the quality of the fuzzy partition in each time window and to control the process of classifier adaptation three validity measures are calculated: the fuzzy separation index based on partition ambiguity, fuzzy compactness, and average partition density. The first two indexes are shown in *Figure 86*, whereas the third measure is presented in *Figure 87*.

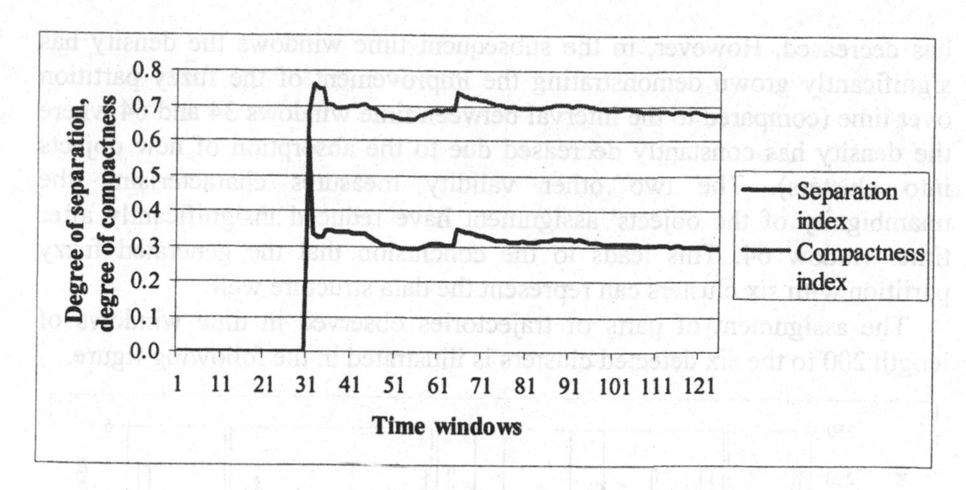

Figure 86. Temporal development of fuzzy separation and fuzzy compactness indexes of fuzzy partitions obtained using the pointwise similarity measure

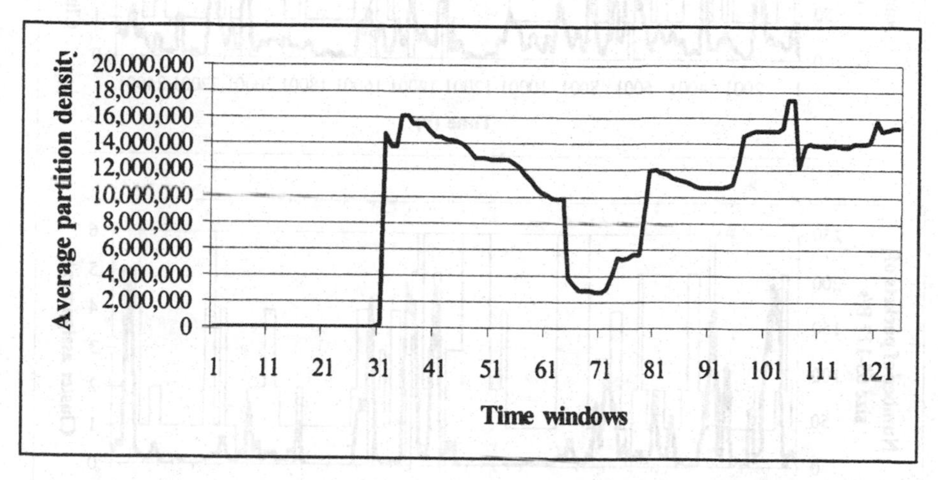

Figure 87. Temporal development of average partition density obtained using the pointwise similarity measure

As can be seen, after the adaptation of the classifier in time window 31 (the classifier with 2 clusters was re-learned with 3 and then with 4 clusters) the values of the separation and compactness indexes have decreased while the average partition density has increased considerably. After the next adaptation of the classifier in time window 34 (the classifier containing 4 clusters was re-learned with 6 clusters) all validity measures have increased slightly. In time window 64, the values of the separation and compactness indexes have increased due to the adaptation (the classifier was re-learned twice with 7 and then with 6 clusters) whereas the average partition density has decreased. However, in the subsequent time windows the density has significantly grown demonstrating the improvement of the fuzzy partition over time (compared to the interval between time windows 34 and 64 where the density has constantly decreased due to the absorption of new objects into clusters). The two other validity measures characterising the unambiguity of the objects' assignment have reduced insignificantly after time window 64. This leads to the conclusion that the generated fuzzy partition with six clusters can represent the data structure well.

The assignment of parts of trajectories observed in time windows of length 200 to the six detected clusters is illustrated in the following figure.

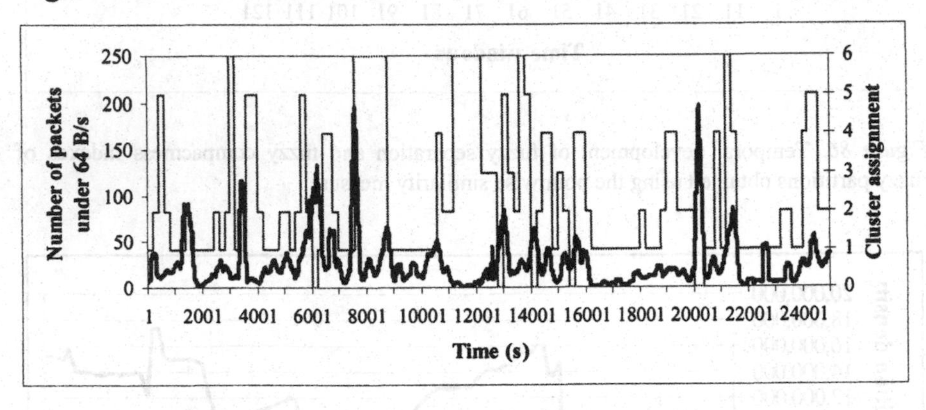

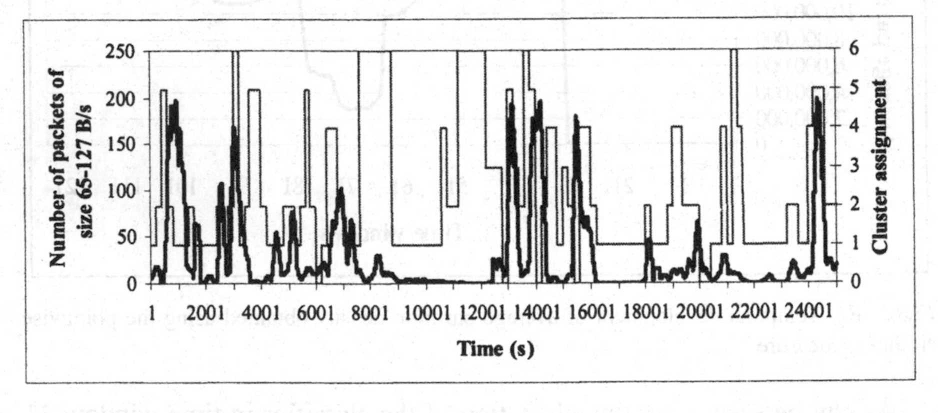

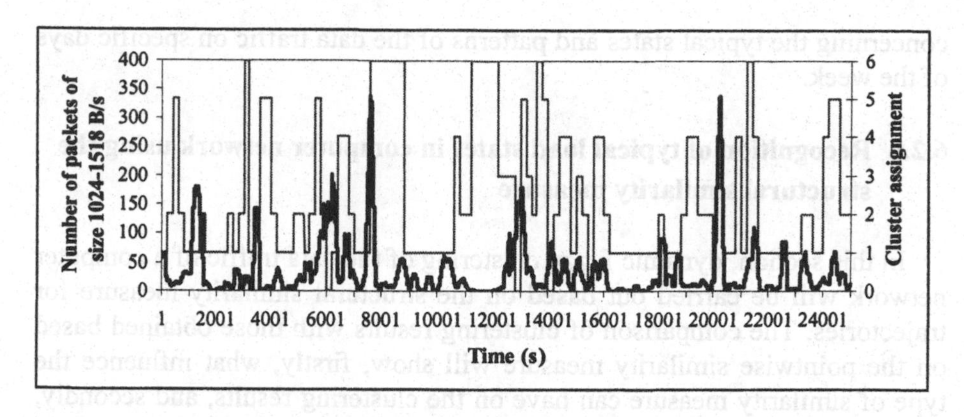

Figure 88. **Assignment of parts of trajectories to six clusters representing data traffic states and obtained based on the pointwise similarity measure**

Figure 88 illustrates only the three trajectories of Features 1, 2 and 6, since these packet sizes are the most frequently transmitted and their number has the greatest influence on the total data traffic as stated in the description of the clusters. Based on this figure, it is possible to determine at what points of time different states of data traffic (clusters) usually appear. Consider some time periods characterised by low data traffic (Cluster 1), an average data traffic with a decreasing tendency (Cluster 2), and a high data traffic (Cluster 4). Choose, for instance, the intervals [8800, 10600], [16200, 18000], [21600, 23200] where the trajectory is assigned to Cluster 1. After the transformation into time of observation, it can be concluded that during the first week data traffic was low from 22:20 on Friday to 22:20 on Saturday, and during the second week all day Wednesday, and on Saturday from 9:00 until 22:20. Notice that the data traffic at night was not observed.

Analogously, one can determine that the level of data traffic was, for example, average on Sunday from 9:00 until 12:20 during the first week, and from 20:40 on Thursday until 10:40 on Friday, and from 22:20 on Saturday until 10:40 on Sunday during the second week. The data traffic was heavy on Thursday from 17:20 until 20:40 during the first and second week, as well as from 19:00 until 22:20 on Tuesday during the second week. Thus, it can be assumed that during the week the data traffic is often heavy in the evening.

The clustering results presented in this section show that the dynamic fuzzy clustering algorithm based on the pointwise similarity measure between trajectories is able to recognise on-line typical states in the behaviour of data traffic in a computer network. The application of this algorithm to the trajectory of the data traffic observed over a much longer time period can make it possible to obtain more reliable statements concerning the typical states and patterns of the data traffic on specific days of the week.

6.2.6 Recognition of typical load states in computer network using the structural similarity measure

In this section, dynamic fuzzy clustering of the data traffic of a computer network will be carried out based on the structural similarity measure for trajectories. The comparison of clustering results with those obtained based on the pointwise similarity measure will show, firstly, what influence the type of similarity measure can have on the clustering results, and secondly, what type of similarity measure is best suited for network load analysis.

For the definition of the structural similarity measure the set of temporal characteristics describing the behaviour of trajectories of features must first be defined. A preliminary analysis of the trajectories under consideration shows that they are generally characterised by a fluctuating behaviour with hills of different amplitudes and duration and a base line at different levels. In order to describe such a behaviour of parts of trajectories in time windows of length 200, the following characteristics were implemented: the mean value; the standard deviation; the range of values of a trajectory in the current time window; the maximum value of a trajectory and its position (time instant of appearance); the minimum value and its position; the temporal trend; the degree of smoothness; the maximum length of the interval with a positive slope (i.e. increasing behaviour) and the start time of this interval; the maximum length of the interval with a negative slope (i.e. decreasing behaviour) and the start time of this interval; the maximum length of the interval with a zero derivative (i.e. constant behaviour) and the start time of this interval; the two largest and the two smallest local extreme values of a trajectory and the time instants of their appearance. Calculating the structural similarity measure between trajectories based on different subsets of these characteristic features and clustering trajectories based on the generated similarity measure, it was observed that some subsets of these characteristics possess very low discriminating ability, leading to similar cluster centres. After investigating different subsets and their contribution to the recognition process, the following 11 temporal characteristics have been chosen to describe the temporal behaviour of the network load trajectories: the mean value; the range of values; the degree of smoothness of the trajectories; the values of four local extreme values and the time instants of their occurrence.

In order to define the structural similarity measure, parameter a for the fuzzy set 'admissible difference for characteristic K_i' determining the admissible deviation of characteristic's values from the desired value is calculated according to (6.4) with respect to each temporal characteristic and each feature and yields the following values:

Table 31. Values of parameter a with respect to temporal characteristics and features

Characteristics	Feature 1	Feature 2	Feature 3	Feature 4	Feature 5	Feature 6
1	26.673	30.357	19.636	29.763	9.926	14.203
2	18.502	23.215	17.450	32.230	12.273	17.526
3	65.483	69.813	52.719	70.045	113.837	93.275
4	83.040	87.418	66.412	113.663	61.972	67.625
5	15.550	21.273	11.388	25.409	8.076	9.253
6	151.080	144.593	68.481	142.675	134.099	149.952
7	11.221	12.221	5.879	12.822	5.948	5.646
8	247.001	207.070	126.554	185.283	244.075	240.799
9	8.173	6.409	4.686	7.825	3.002	4.590
10	214.297	145.235	101.703	163.038	273.188	301.105
11	4.131	1.027	1.671	5.637	2.115	2.908

The procedure of classifier design and classification over 125 time windows is carried out in the same manner as in the case of the pointwise similarity measure. A dynamic fuzzy classifier with the number of clusters c=2 is designed after 31 time windows by applying the functional fuzzy c-means algorithm based on the structural similarity measure for trajectories. After possibilistic classification and absorption of dynamic objects into the clusters, the sizes of the clusters are equal to $N_1^{0.5}=6$ and $N_2^{0.5}=10$ and the number of free objects is n_{free}=15. According to the monitoring procedure two new clusters can be formed. Therefore, the objects are re-clustered with a new number of clusters c=4. The absorption procedure provides clusters of sizes $N_1^{0.5}=7$, $N_2^{0.5}=12$, $N_3^{0.5}=3$ and $N_4^{0.5}=3$, whereas the number of free objects is equal to n_{free}=6. Since this fuzzy partition is better than the previous one with 2 clusters according to the validity measures (the average partition density and the fuzzy compactness index have increased) and no more changes can be detected by the monitoring procedure, this classifier is accepted.

In the next time windows new objects are classified by this classifier and some of them are absorbed into existing clusters. In time window tw=34 (after 3 days 11 hours and 40 minutes of observation) the number of free objects has increased to n_{free}=8, and the monitoring procedure declares a new cluster. The classifier is designed anew with a number of clusters c=5, leading to clusters of sizes $N_1^{0.5}=10$, $N_2^{0.5}=11$, $N_3^{0.5}=4$, $N_4^{0.5}=5$, $N_5^{0.5}=2$ and n_{free}=1 free object which cannot be assigned to clusters because of its low degree of membership to all clusters. This classifier is accepted as the fuzzy separation and the fuzzy compactness indexes have increased and is used for classifying new objects in the subsequent time windows.

Temporal changes in the cluster structure are then detected in time window tw=62, that is after 7 days and 100 minutes of observation. Since the number of free objects n_{free}=8 and the density of this group of objects are sufficiently high compared to the existing clusters, a new cluster is declared by the monitoring procedure. As a result, the classifier has to be re-learned using the existing cluster centres and an estimated new centre for initialisation in order to fit a new cluster structure representing the data traffic after 7 days. The new classifier designed with c=6 clusters is characterised by clusters of sizes $N_1^{0.5}=2$, $N_2^{0.5}=16$, $N_3^{0.5}=14$, $N_4^{0.5}=6$, $N_5^{0.5}=9$ and $N_6^{0.5}=5$ and no free objects. According to the validity measures the new partition represents an improvement compared to the previous one, with 5 clusters (average partition density, fuzzy separation and fuzzy compactness indexes have increased). Therefore, the new improved classifier is accepted and used for classifying new objects in the following time windows. Until the last time window tw=125 no further changes in the cluster structure have been detected and the clusters have grown to $N_1^{0.5}=23$, $N_2^{0.5}=28$, $N_3^{0.5}=31$, $N_4^{0.5}=13$, $N_5^{0.5}=16$, and $N_6^{0.5}=14$ and n_{free}=0. Thus, it can be assumed that the number of typical states of data traffic during 14 days of observation is equal to six and the classifier fits the cluster structure nicely. The results of dynamic clustering and classification over this time period of 125 time windows are summarised in *Table 32*. Time windows where the classifier was re-learned due to detected temporal changes in the cluster structure and the corresponding re-clustering results are shaded light grey.

Table 32. Results of dynamic clustering and classification of the network data traffic based on the structural similarity measure

	$N(C_i)$	N_{free}	V_{apd}	FSA	FC	Changes
TW=31, c=2	6, 10	15	$1.52\cdot10^{-10}$	0.242	0.608	c_{new}=2
TW=31, c=4	7, 12, 3, 3	6	$2.16\cdot10^{-10}$	0.238	0.621	No
TW=34, c=4	8, 12, 3, 3	8	$1.94\cdot10^{-10}$	0.233	0.615	c_{new}=1
TW=34,c=5	10, 11, 4, 5, 2	1	$6.05\cdot10^{-11}$	0.247	0.618	No
TW=62, c=5	11, 19, 13, 5, 6	8	$4.05\cdot10^{-11}$	0.185	0.558	c_{new}=1
TW=62, c=6	12, 16, 14, 6, 9, 5	0	$4.76\cdot10^{-11}$	0.225	0.657	No

Cluster centres generated by the dynamic fuzzy clustering algorithm based on the structural similarity measure are not represented by trajectories as in the case of pointwise similarity but by 11 temporal characteristics of trajectories of features. These characteristics describe the temporal behaviour of trajectories, or the structure of the typical patterns of data traffic. The six cluster centres detected are shown in *Table 33* with respect to each feature, where sm denotes the degree of smoothness of a trajectory, T_i, i=1, ..., 4, is the time instant of the occurrence of the i-th extreme value, and X_i, i=1, ..., 4, is the i-th extreme value of a trajectory.

Table 33. Cluster centres representing data traffic states obtained based on the structural similarity measure

Feature 1											
	Mean	Range	Sm	T_1	X_1	T_2	X_2	T_3	X_3	T_4	X_4
C_1	23.13	13.26	9.70	42.69	22.28	106.65	21.21	140.45	25.07	152.91	23.23
C_2	24.27	26.90	3.19	29.53	19.19	45.22	19.02	38.34	14.32	9.94	1.40
C_3	17.15	10.70	6.48	60.23	16.22	85.11	16.86	112.67	16.06	143.66	14.90
C_4	26.16	14.22	5.93	37.91	29.92	56.28	29.92	77.20	28.82	127.15	18.30
C_5	32.52	47.55	1.13	31.84	6.55	11.90	3.05	19.71	2.62	17.46	0.89
C_6	54.50	64.13	1.56	71.49	22.29	53.70	19.90	13.57	3.38	6.63	1.30
Feature 2											
	Mean	Range	Sm	T_1	X_1	T_2	X_2	T_3	X_3	T_4	X_4
C_1	43.58	24.81	9.31	32.42	48.17	54.82	47.10	103.85	45.52	101.67	7.28
C_2	102.48	45.60	2.48	62.07	90.85	104.45	105.91	15.49	4.86	12.00	1.33
C_3	20.77	17.21	6.42	58.71	20.94	73.89	15.24	92.73	5.20	71.30	1.32
C_4	45.16	78.74	1.87	108.01	23.18	18.46	14.84	10.69	2.94	14.70	2.04
C_5	22.45	25.12	3.27	65.37	19.50	52.78	18.46	72.86	18.23	30.44	0.54
C_6	15.37	11.78	6.48	42.41	13.13	87.46	13.67	138.50	11.98	162.24	12.73
Feature 3											
	Mean	Range	Sm	T_1	X_1	T_2	X_2	T_3	X_3	T_4	X_4
C_1	5.28	5.11	8.40	28.81	5.93	47.85	5.78	125.57	1.97	131.34	1.52
C_2	19.25	12.56	8.68	83.05	20.59	110.12	21.44	130.63	21.82	146.87	20.60
C_3	5.33	6.13	3.60	37.44	4.32	45.24	4.24	46.24	3.18	32.17	2.64
C_4	4.68	8.35	1.49	76.89	1.98	8.92	0.95	13.66	1.28	13.14	0.28
C_5	3.90	5.27	2.38	67.45	2.68	46.03	2.73	22.94	2.53	27.53	2.39
C_6	3.47	3.98	5.97	33.82	2.94	44.54	3.04	70.98	2.86	108.67	1.81

Feature 4											
	Mean	Range	Sm	T_1	X_1	T_2	X_2	T_3	X_3	T_4	X_4
C_1	2.41	1.70	6.15	33.80	2.34	78.89	2.48	115.56	2.22	66.42	2.37
C_2	3.54	1.86	4.91	50.21	3.56	73.64	3.03	107.28	3.28	115.48	3.25
C_3	1.05	1.03	6.00	83.42	0.91	76.53	0.75	94.70	0.65	92.76	0.63
C_4	1.83	1.72	6.26	47.80	1.63	85.73	1.67	133.26	1.52	139.16	1.57
C_5	1.91	2.17	2.62	84.50	1.71	13.63	0.55	9.57	0.14	11.25	0.10
C_6	1.45	1.27	7.42	34.33	1.49	63.21	1.39	90.09	1.10	113.84	1.15
Feature 5											
	Mean	Range	Sm	T_1	X_1	T_2	X_2	T_3	X_3	T_4	X_4
C_1	2.04	2.53	7.58	48.17	2.02	79.74	2.05	112.59	1.61	143.99	1.79
C_2	2.13	2.10	9.50	41.92	1.92	57.56	1.88	128.33	2.27	142.36	2.29
C_3	5.68	5.56	3.68	76.08	5.12	32.41	4.03	28.98	2.67	13.36	0.40
C_4	1.57	1.30	11.68	40.08	1.28	80.15	1.37	157.30	1.41	171.31	1.40
C_5	4.33	6.41	2.41	39.36	3.82	66.95	3.72	20.56	0.90	6.17	0.12
C_6	8.05	5.31	9.50	22.76	9.06	69.98	8.96	154.75	6.45	169.66	6.63
Feature 6											
	Mean	Range	Sm	T_1	X_1	T_2	X_2	T_3	X_3	T_4	X_4
C_1	32.67	22.56	8.13	39.02	32.92	77.56	30.52	110.36	36.92	146.89	32.51
C_2	30.40	33.34	6.50	25.54	26.56	54.96	25.36	76.09	23.08	118.10	21.61
C_3	16.64	18.23	6.16	48.54	12.79	76.01	12.63	110.27	19.18	143.40	18.87
C_4	28.17	30.58	4.02	41.52	18.61	73.19	33.62	146.19	28.98	13.83	3.11
C_5	34.07	45.24	2.52	32.37	33.23	45.40	7.19	9.56	1.59	8.43	1.00
C_6	78.71	118.92	1.75	57.08	22.25	36.55	20.88	7.89	1.85	6.50	1.51

Considering the values of the characteristics, these cluster centres can be interpreted as follows.

Cluster 1 represents an average data traffic with a high degree of fluctuations (degree of smoothness is about 9) in a small value range, so that it can be considered as being stable. The number of small packets varies by 23 (under 64 B/s) and 43 (65-127 B/s) whereas the number of large packets is about 32. The number of other packet sizes is rather small.

Cluster 2 describes an average level of data traffic with an increasing tendency for small packets and decreasing tendency for large packets. The mean value of the number of packets under 64 B/s is similar to that of Cluster 1 (around 24) and packets of size 65-127 B/s are around 102 with a larger value range in both cases compared to Cluster 1. This part of data traffic is characterised by a rather small degree of fluctuations. The number of large packets fluctuates more and is around 30.

Cluster 3 corresponds to a stable state of the data traffic at a low level with strong fluctuations (degree of smoothness is about 6) in a small value range. The number of both small and large packets is under 20, whereas the number of other packet types is very low.

Cluster 4 represents data traffic increasing from a small to an average level, and characterised by a low degree of fluctuations for most types of packets. The numbers of small packets under 64 B/s and between 65 and 127 B/s vary by 26 and 45 and rise until 30 and 80, respectively, whereas the number of large packets grows from 3 to 33. Other packet sizes remain at low level.

Cluster 5 is characterised by an almost monotonously increasing tendency from a small to an average level, where the number of small packets is lower (around 22 for packet sizes 65 and 127 B/s), and the number of large packets is somewhat larger (around 34, rising to 45 for packet sizes 1024-1518 B/s) compared to those of Cluster 4.

Cluster 6 describes heavy data traffic with an increasing tendency. The number of small packets (under 64 B/s) increases almost monotonously and is characterised by a mean value of about 54, whereas the number of large packets exhibits the same behaviour and varies by around 78, with a value range of about 118. The number of other packet sizes exceeds those of the other clusters as well.

In order to evaluate the quality of the fuzzy partition in each time window and to control the process of classifier adaptation three validity measures such as the fuzzy separation index based on ambiguity of partition, the fuzzy compactness, and the average partition density are calculated. Their temporal development is shown in *Figure 89* and *Figure 90*, respectively. As can be seen after the adaptation of the classifier in time window 31 (the classifier with 2 clusters was re-learned with 4 clusters), the values of the compactness index and the average partition density have increased while the separation index has decreased. After the next adaptation of the classifier in time window 34 (the classifier containing 4 clusters was re-learned with 5 clusters), the separation and compactness indexes have increased, however the average partition density has decreased. During classification and absorption of new objects into the 5 existing clusters in the following time windows, all validity measures have slowly decreased indicating a deterioration of the classifier over time. In time window 62 the values of all validity measures have considerably increased due to the adaptation (the classifier was re-learned with 6 clusters). In the subsequent time windows the density has first slightly decreased and then started to grow, demonstrating the improvement of the fuzzy partition over time. The two other validity measures characterising the unambiguity of the objects' assignment have remained partly on the same level, or decreased insignificantly, after time window 64. Thus it can be concluded that the generated fuzzy partition with six clusters can provide a good representation of the data structure.

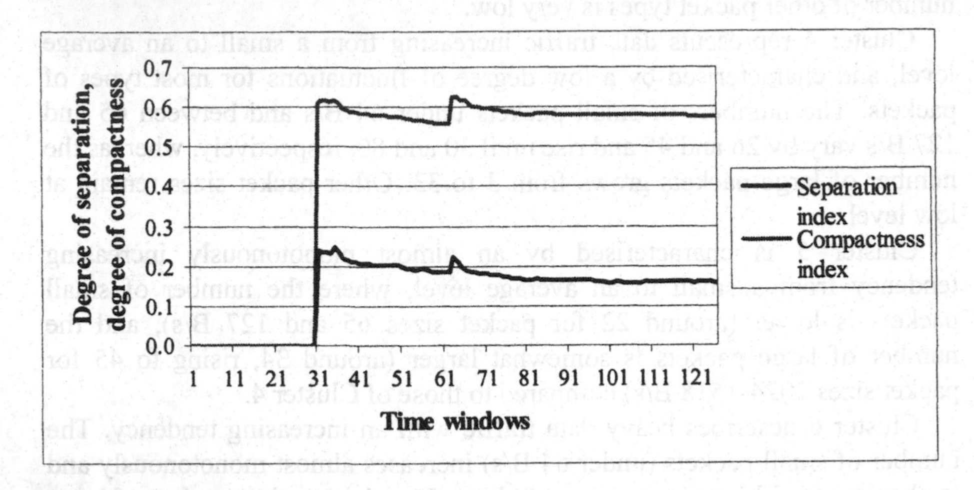

Figure 89. Temporal development of fuzzy separation and fuzzy compactness indexes of fuzzy partitions obtained using the structural similarity measure

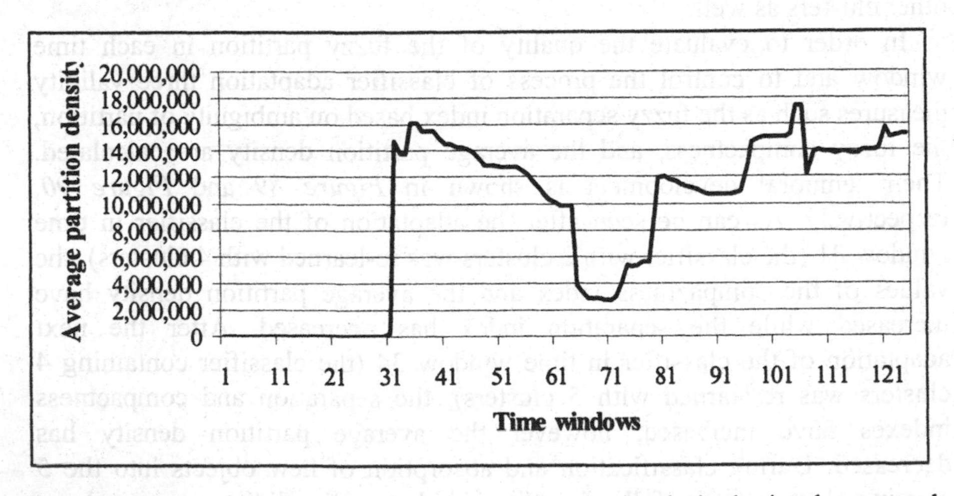

Figure 90. Temporal development of average partition density obtained using the structural similarity measure

The assignment of parts of trajectories observed in time windows of length 200 to six clusters detected is illustrated in the following figure, where only trajectories of Features 1, 2 and 6 are shown.

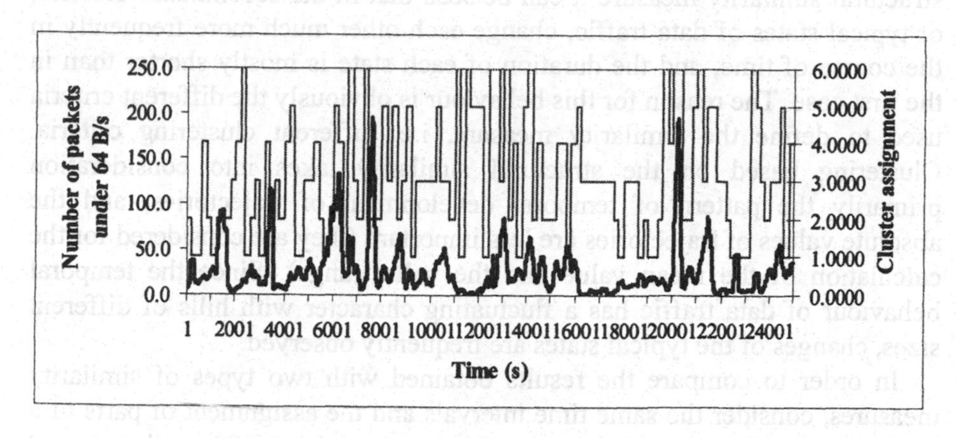

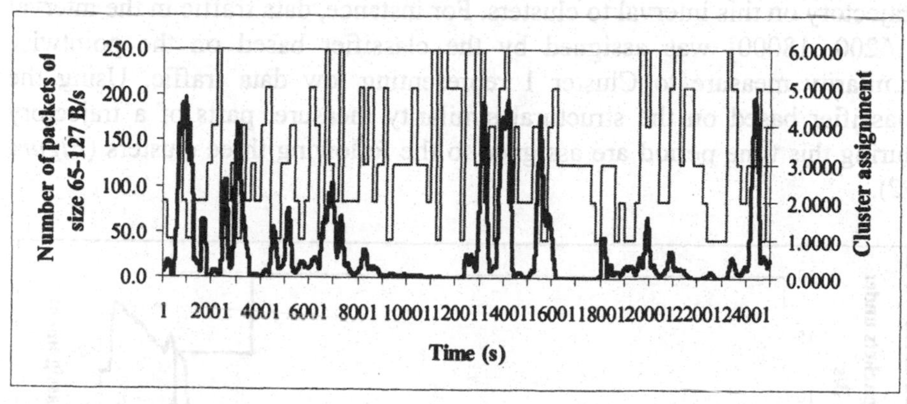

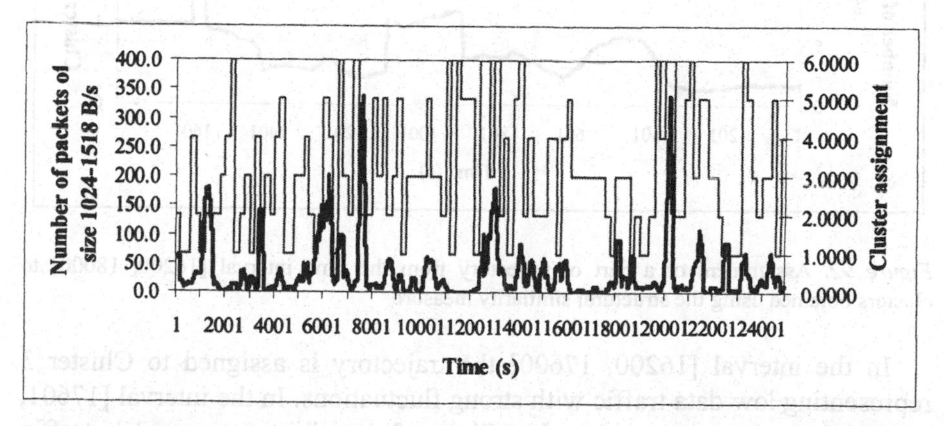

Figure 91. Assignment of parts of trajectories to six clusters obtained based on the structural similarity

Comparing the assignment of parts of trajectories to clusters obtained using the pointwise similarity measure and to clusters generated using the structural similarity measure it can be seen that in the second case clusters, or typical states of data traffic, change each other much more frequently in the course of time, and the duration of each state is mostly shorter than in the first case. The reason for this behaviour is obviously the different criteria used to define the similarity measure, i.e. different clustering criteria. Clustering based on the structural similarity takes into consideration primarily the pattern of temporal development of trajectories, and the absolute values of trajectories are less important (they are considered for the calculation of the mean value and the value range). Since the temporal behaviour of data traffic has a fluctuating character with hills of different sizes, changes of the typical states are frequently observed.

In order to compare the results obtained with two types of similarity measures, consider the same time intervals and the assignment of parts of a trajectory on this interval to clusters. For instance, data traffic in the interval [16200, 18000] was assigned by the classifier based on the pointwise similarity measure to Cluster 1 representing low data traffic. Using the classifier based on the structural similarity measure, parts of a trajectory during this time period are assigned to the following three clusters (*Figure 92*).

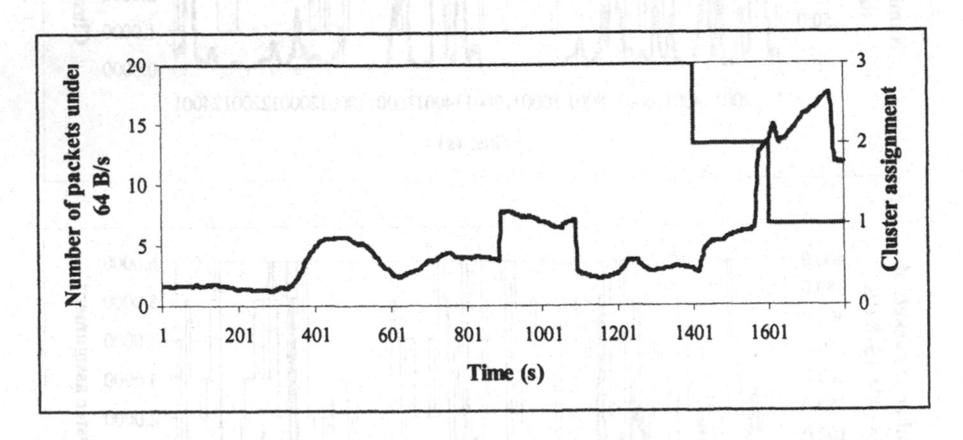

Figure 92. Assignment of a part of trajectory from the time interval [16200, 18000] to clusters obtained using the structural similarity measure

In the interval [16200, 17600] the trajectory is assigned to Cluster 3 representing low data traffic with strong fluctuations. In the interval [17601, 17800] the trajectory is assigned to Cluster 2 describing average data traffic with an increasing behaviour of small packets, decreasing behaviour of large packets and small degree of fluctuations. Finally, in the interval [17801, 18000] the trajectory is assigned to Cluster 1 where data traffic is at an average level with strong fluctuations in a small range. Thus, the typical states of data traffic obtained using the structural similarity measure have a much stronger relation to the behaviour of a trajectory than to its absolute values.

In general, it can be seen that the descriptions of clusters obtained based on the pointwise and structural similarity measures are comparable and allow a clear identification of low, average and high levels of data traffic with increasing or decreasing tendencies. In case of structural similarity the distinction is, however, more comprehensive. It can be concluded that the choice of a similarity measure which is the best suited for the analysis of load in computer networks requires the consideration of the technical aspects of network optimisation and the available equipment. It should be decided what is more important for optimisation: to react to changes in data traffic behaviour which can be used for short-term forecasting or to react to changes in data traffic volume representing the current network load. In both cases a company can gain considerable advantages due to the network optimisation using on-line monitoring and dynamic fuzzy pattern recognition techniques.

7 CONCLUSIONS

The research area of dynamic fuzzy pattern recognition is new and challenging from both the theoretical and the practical point of view. There is a huge number of applications where for a correct recognition of structure in data the consideration of the temporal development of objects over time is required, for instance state-dependent machine maintenance and diagnosis, the analysis of bank customers' behaviour for the evaluation of their creditworthiness, recognition of typical scenarios for strategic planning or monitoring of patients in medicine. So far only a limited number of algorithms for clustering and classification in dynamic environments has been developed. They can be separated into two main groups. The first group is represented by algorithms which consider static objects at a fixed moment in time. During the dynamic process of pattern recognition these algorithms try to recognise typical (static) states in the behaviour of a system/process given as points in the feature space at a certain time instant, and follow their temporal changes as time passes using updating techniques. Most of these algorithms are able to detect only gradual changes in the data structure by monitoring characteristics of the classifier performance. Another group includes algorithms which are capable of processing dynamic objects represented by temporal sequences of observations, or trajectories, in the feature space. These algorithms result in cluster prototypes given by trajectories and describing typical dynamic states of a system over a certain period of time . In this case, the order of states of an object, or the history of its temporal development, determine the membership of an object to a certain pattern, or cluster. Most of the algorithms in this area use statistical properties of trajectories to determine their similarity.

This book has suggested a new algorithm for dynamic classifier design and classification which can be applied to static as well as dynamic objects and combined with different types of fuzzy classifiers. The main property of this algorithm is its ability to automatically recognise changes in the cluster structure in the course of time and to adapt the classifier to these changes. The algorithm takes advantage of fuzzy set theory which provides a unique mechanism for gradual assignment of objects to clusters and allows to detect a gradual temporal transition of objects between clusters. The algorithm proposed consists of four algorithms constituting the monitoring procedure and based primarily on properties of the membership function and two algorithms for the adaptation procedure. This algorithm is unique in so much as that it has been developed for the unsupervised dynamic design of point-prototype-based fuzzy classifiers and is capable of recognising gradual as well as abrupt changes in the cluster structure possessing a flexible mechanism for an adaptation of the classifier.

In order to be able to apply the new algorithm for clustering and classification of dynamic objects, a number of similarity measures for trajectories were proposed which take into consideration either the pointwise closeness of trajectories in the feature space or the best match of trajectories with respect to their shape. The definition of structural similarity measures is of primary importance for the comparison of trajectories based on their temporal behaviour. It depends on the meaning of similarity in the given application and is formulated with respect to specific mathematical properties and characteristics of trajectories. Different definitions of fuzzy similarity between trajectories introduced in this book allow to model a gradual representation of similarity according to human judgement which provides the most plausible tool for the recognition process.

The adaptive fuzzy clustering algorithm for dynamic objects developed in this book should help the practitioner to design systems for on-line monitoring of system states and for the recognition of the typical behaviour of a system under consideration, taking into account the history of temporal development of objects. This kind of problems arises in many application areas such as medicine, marketing, finance, or technical diagnosis.

In order to demonstrate the practical relevance of the algorithm developed here, it was applied to two different problems. The first economic problem is concerned with bank customer segmentation based on the analysis of the customers' behaviour. By contrast to many similar tasks solved by companies until now, the analysis is based on the consideration of the temporal development of customers over a long period of time, or in time windows. Clustering and classification in time windows allow a bank to recognise typical customer segments and follow their possible temporal changes which may appear due to changing bank services or economic circumstances, and to detect changes in customer assignment to typical segments. The results of this analysis can help a bank to increase its profits by preparing special offers for the best users of bank services and to adapt their products and services to recognised changes in the customer portfolio.

The second technical problem is related to the recognition of typical states in computer network load. The knowledge of time specific network states with a particularly high or low load provides companies with the opportunity to optimise their network resources by purchasing different band widths for data transmission depending on time of day, thus producing savings. The problem of load optimisation is also relevant for the area of telecommunication where the prices for the use of telecommunication services can be determined depending on the different load states.

The performance of the algorithm for dynamic fuzzy classifier design and classification depends on the clustering algorithm chosen for classifier design and on the correct choice of several threshold values used within the monitoring procedure. Improvements of the algorithm can be attained by introducing more powerful clustering algorithms capable of recognising clusters of different forms, sizes and densities. Although appropriate values of thresholds can be determined by an expert depending on the specific application, it seems reasonable to try to develop automatic procedures, or to find dependencies between these values and data properties, in order to simplify the choice of the correct threshold values. Further enhancement of the performance of dynamic pattern recognition systems can be expected by applying new methods for time-dependent feature selection / generation which is also an important aspect in static pattern recognition. On the one hand, a variable set of features can make the analysis of the clustering results more difficult, but on the other hand it can improve clustering results by representing only the most relevant information as time passes.

The design of an adaptive fuzzy classifier relies to a large degree on the validity measure used to control the adaptation process. Most of the existing validity measures depend on the number of clusters chosen for the fuzzy partition of objects, and none of them can guarantee the correct evaluation of the partition quality in all cases. Thus, there is a need for a definition of new improved validity measures which can provide a more reliable statement regarding the partition quality achieved with a dynamic fuzzy classifier.

Considering the modification of static fuzzy clustering algorithms by using a similarity measure for trajectories, extensions of the list of proposed similarity measures are possible. In particular, structural similarity measures should be a subject of further investigations since they can describe a context-dependent similarity of trajectories with respect to their form and temporal development. And the temporal characteristics of trajectories should be studied to discover which of them seem to be the most relevant for certain classes of applications or for clustering problems based on fuzzy numbers as a special case of trajectories.

One can conclude that dynamic fuzzy pattern recognition represents a new promising research area with a wide field of potential applications. The results presented in this book show that the adaptive fuzzy clustering algorithm for dynamic objects can be successfully used to solve the pattern recognition problem in a dynamic environment. The main power of the algorithm lies in the combination of a new method for dynamic classifier design based on the principles of adaptation with a set of fuzzy similarity measures for trajectories. Further modifications of the algorithm for different types of classifiers and an extension of the set of similarity measures will lead to an expansion of this new class of methods for dynamic pattern recognition.

REFERENCES

Abrantes, A.J., Marques, J.S. (1998)
A Method for Dynamic Clustering of Data. *Proceedings of the 9th British Conference*, University of Southampton, UK, 1998, p. 154-163, http://peipa.essex.ac.uk/BMVC98/bmvc/papers/d038/h038.htm

Agrawal, R., Lin, K.-I., Sawhney, H.S., Shim, K. (1995)
Fast Similarity Search in the Presence of Noise, Scaling and Translation in Time-Series Databases. In: *Proceedings of the 21st International Conference on Very Large Data Bases (VLDB'95)*, Zurich, Switzerland, 1995, p. 490-501

Angstenberger, J. (1997)
Data Mining in Business Anwendungen. *Anwendersymposium zu Fuzzy Technologien und Neuronalen Netzen*, 19. - 20. November 1997, Dortmund, Transferstelle Mikroelektrtonik / Fuzzy-Technologien, c/o MIT GmbH, Promenade 9, D-52076 Aachen, 1997, p. 131-138

Angstenberger, J., Nelke, M., Schrötter, T. (1998)
Data Mining for Process Analysis and Optimization. CE Expo, Houston, Texas, 1998

Arneodo, A., Bacry, E., Muzy, J.F. (1995)
The Thermodynamics of Fractals Revisited with Wavelets. *Physika A*, Vol. 213, 1995, p. 232-275

Åström, K. J., Wittenmark, B. (1995)
Adaptive Control. Addison-Wesley Publishing Company, Inc., 1995

Bakshi, B.R., Locher, G., Stephanopoulos, G., Stephanopoulos, G. (1994)
Analysis of Operating data for Evaluation, diagnosis and Control of Batch Operations. *Journal of Process Control*, Vol. 4 (4), 1994, Butterworth-Heinemann, p. 179-194

Bandemer, H., Näther, W. (1992)
Fuzzy Data Analysis. Kluwer Academic Publishers, Dordrecht, Boston, London, 1992

Bensaid, A.M., Hall, L.O., Bezdek, J.C., Clarke, L.P., Silbiger, M.L., Arrington, J.A., Murtagh, R.F. (1996)
Validity-Guided (Re)Clustering with Applications to Image Segmentation. *IEEE Transactions on Fuzzy Systems*, Vol. 4 (2), 1996, p. 112-123

Bezdek, J.C. (1981)
Pattern Recognition with Fuzzy Objective Function Algorithms. New York, Plenum, 1981

Binaghi, E., Della Ventura, A., Rampini, A., Schettini, R. (1993)
Fuzzy Reasoning Approach to Similarity Evaluation in Image Analysis. *International Journal of Intelligent Systems*, Vol. 8, 1993, p. 749-769

Black, U. (1989)
Data Networks. Concepts, Theory, and Practice. Prentice Hall, Englewood Cliffs, New Jersey 07632, 1989

Bock, H. H. (1974)
Automatische Klassifikation: theoretische und praktische Methoden zur Gruppierung und Strukturierung von Daten (Cluster-Analyse). Vandenhoek & Ruprecht, Göttingen, 1974

Bocklisch, S. (1981)
Experimentelle Prozeßanalyse mit unscharfer Klassifikation. Dissertation, Technische Hochschule Karl-Marx-Stadt, 1981

Bocklisch, S. (1987)
Hybrid Methods in Pattern Recognition. In: Devijver, P., Kittler, J. *Pattern Recognition Theory and Applications*. Springer-Verlag, Berlin, Heidelberg, 1987, p. 367-382

Boose, J.H. (1989)
A Survey of Knowledge Acquisition Techniques and Tools. *Knowledge Acquisition*, 1, 1989, p. 3-37

Boutleux, E., Dubuisson, B. (1996)
Fuzzy Pattern Recognition to Characterize Evolutionary Complex Systems. Application to the French Telephone Network. *Proc. IEEE Int. Conference on Fuzzy Systems*, New Orleans, LA, September 1996

Brachman, R.J., Anand, T. (1996)
The process of Knowledge Discovery in Databases. In: Fayyad, U.M., Piatetsky-Shapiro, G., Smyth, P., Uthurusamy, R. *Advances in Knowledge Discovery and Data Mining*. AAAI Press / The MIT Press, Menlo Park, California, 1996, p. 37-57

Bunke, H. (1987)
Hybrid Methods in Pattern Recognition. In: Devijver, P., Kittler, J. *Pattern Recognition Theory and Applications*. Springer-Verlag, Berlin, Heidelberg, 1987, p. 367-382

Cayrol, M., Farreny, H., Prade, H. (1980)
Possibility and Necessity in a Pattern-Matching Process. *Proceedings of IXth International Congress on Cybernetics*, Namur, Belgium, September, 1980, p. 53-65

Cayrol, M., Farreny, H., Prade, H. (1982)
Fuzzy Pattern Matching. *Kybernetes,* 11, 1982, p. 103-116

Chow, C.K. (1970)
On Optimum Recognition Error and Reject Tradeoff. *IEEE Transactions on Information Theory*, IT-16, 1970, p. 41-46

Das, G., Gunopulos, D., Mannila, H. (1997)
Finding Similar Time Series. In: Komorowski, J., Zytkow, J. (Eds.) *Principles of Data Mining and Knowledge Discovery. Proceedings of the First European Symposium PKDD'97,* Trondheim, Norway, June 1997, Springer, 1997, p. 88-100

Davies, D.L., Bouldin, D.W., (1979)
A Cluster Separation Measure. *IEEE Transactions on Pattern Analysis and Machine Intelligence*, Vol. PAMI-1, 1979, p. 224-227

Denoeux, T., Masson, M., Dubuisson, B. (1997)
Advanced Pattern Recognition Techniques for System Monitoring and Diagnosis: a Survey. *Journal Europeen des Systemes Automates*, Vol. 31, Nr. 9/10, 1997, p. 1509-1540

Devijver, P., Kittler, J. (1982)
Pattern Recognition: A Statistical Approach. Prentice Hall, 1982

Dilly, R. (1995)
Data Mining: An Introduction. Queens University Belfast, December 1995, http://www.pcc.qub.ac.uk/tec/courses/datamining/stu_notes/dm_book_1.html

Dorf, R.C. (1992)
Modern Control Systems. Addison-Wesley, Reading, Mass., 1992

Dubois, D., Prade, H. (1988)
Possibility Theory - An Approach to computerized Processing of Uncertainty. Plenum Press, New York, 1988

Dubois, D., Prade, H., Testemale, C. (1988)
Weighted Fuzzy Pattern Matching. *Fuzzy Sets and Systems,* 28, 1988, p. 313-331

Duda, R.O., Hart P.E. (1973)
Pattern Classification and Scene Analysis. Wiley, New York, 1973

Dunn, J.C. (1974)
Well Separated Clusters and Optimal Fuzzy Partitions. *Journal of Cybernetics*, Vol. 4 (3), 1974, p. 95-104

Eckardt, H., Haupt, J., Wernstedt, J. (1995)
Fuzzygestützte Substanzerkennung in der Umweltanalytik. *Anwendersymposium zu industriellen Anwendungen der Neuro-Fuzzy technologien*, Lutherstadt Wittenberg, March 14-15, 1995, p. 105-111

Engels, J., Chadenas, C. (1997)
Learning Historic Traffic Patterns as Basis for Traffic Prediction. *Proceedings of the 5th European Conference on Intelligent Techniques and Soft Computing (EUFIT'97)*, Aachen, Germany, September 8-11, 1997, p. 2021-2025

Falconer, K. (1990)
Fractal Geometry – Mathematical Foundations and Applications, John Wiley, 1990

Famili, A., Shen, W.-M., Weber, R., Simoudis, E. (1997)
Data Pre-processing and Intelligent Data Analysis. *Intelligent Data Analysis*, Vol. 1 (1), 1997, http://www.elsevier.com/locate/ida

Fayyad, U.M., Piatetsky-Shapiro, G., Smyth, P. (1996)
From Data Mining to Knowledge Discovery: An Overview. In: Fayyad, U.M., Piatetsky-Shapiro, G., Smyth, P., Uthurusamy, R. *Advances in Knowledge Discovery and Data Mining*. AAAI Press / The MIT Press, Menlo Park, California, 1996, p. 1-34

Föllinger, O., Franke, D. (1982)
Einführung in die Zustandsbeschreibung dynamischer Systeme. Oldenbourg Verlag, München, 1982

Frawley, W.J., Piatetsky-Shapiro, G., Matheus, C. J. (1991)
Knowledge Discovery in Databases: An Overview. In: Piatetsky-Shapiro, G., Frawley, W.J. (Ed.) *Knowledge Discovery in Databases*. Cambridge, Mass: AAAI Press / The MIT Press, 1991, p. 1-27

Frawley, W.J., Piatetsky-Shapiro, G., Matheus, C. J. (1992)
Knowledge Discovery in Databases: An Overview. *AI Magazine*, Vol. 14 (3), 1992, p. 53-70

Freer, J. R. (1988)
Computer Communications and Networks. Pitman Publishing, Computer Systems Series, London, 1988

Frélicot, C. (1992)
Un système adaptatif de diagnostic prédictif par reconnaissance des formes floues. PhD thesis, Université de Technologie de Compiégne, Compiégne, 1992

Frélicot, C., Masson, M.H., Dubuisson, B. (1995)
Reject Options in Fuzzy Pattern Classification Rules. *Proceedings of the 3th European Conference on Intelligent Techniques and Soft Computing (EUFIT'95)*, Aachen, Germany, August 28-31, 1995, p. 1459-1464

Frigui, H., Krishnapuram, R. (1997)
Clustering by Competitive Agglomeration. *Pattern Recognition*, Vol. 30 (7), 1997, p. 1109-1119

Fu, K.S. (1974)
Syntactic Methods in Pattern Recognition. Academic Press, New York, 1974

Fu, K.S. (1982a)
Syntactic Pattern Recognition with Applications. Prentice-Hall, Englewood Cliffs, NJ, 1982

Fu, K.S. (Ed.) (1982b)
Applications of Pattern Recognition. CRC Press, Boca Raton, FL, 1982

Gath, I., Geva, A.B. (1989)
Unsupervised optimal Fuzzy Clustering. *IEEE Transactions on Pattern Analysis and Machine Intelligence*, 11, 1989, p. 773-781

Geisser, S. (1975)
The Predictive Sample Reuse Method with Applications. *Journal Of American Statistical Association*, 70, 1975, p. 320-328

Geva, A.B., Kerem, D.H. (1998)
Forecasting Generalized Epileptic Seizures from the EEG Signal by Wavelet Analysis and Dynamic Unsupervised Fuzzy Clustering. *IEEE Transactions on Biomedical Engineering*, 45 (10), 1998, p. 1205-1216

Gibb, W.J., Auslander, D.M., Griffin, J.C. (1994)
Adaptive classification of myocardial electrogram waveforms. *IEEE Transactions on Biomedical Engineering*, 41, 1994, p. 804-808

Grabisch, M., Bienvenu, G., Grandin, J.F., Lemer, A., Moruzzi, M. (1997)
A Formal Comparison of Probabilistic and Possibilistic Frameworks for Classification. *Proceedings of 7th IFSA World Congress,* Prague 1997, p. 117-122

Graham, I., Jones, P.L. (1988)
Expert Systems, Knowledge Uncertainty and Decision. London Chapman and Hall Computing, 1988

Grenier, D. (1984)
Méthode de détection d'évolution. Application a l'instrumentation nucléaire. Thèse de docteur -ingénier. Université de Technologie de Compiègne, 1984

Gunderson, R. (1978)
Application of fuzzy ISODATA Algorithms to Star Tracker Printing Systems. *Proceedings of the 7th Triannual World IFAC Congress*, Helsinki, Finland, 1978, p. 1319-1323

Gustafson, D., Kessel, W. (1979)
Fuzzy Clustering with a Fuzzy Covariance Matrix. *Proceedings of IEEE CDC*, San Diego, California. IEEE Press, Piscataway, New Jersey, 1979, p. 761-766

Hofmeister, P. (1999)
Evolutionäre Szenarien: dynamische Konstruktion alternativer Zukunftsbilder mit unscharfen Regelbasen. Verlag Dr. Kovac, Hamburg, 2000

Hofmeister, P., Joentgen, A., Mikenina, L., Weber, R., Zimmermann, H.-J. (1999)
Reduction of Complexity in Scenario Analysis by Means of Dynamic Fuzzy Data Analysis. *OR Spektrum*, (to appear)

Hogg, R.V., Ledolter, J. (1992)
Applied Statistics for Engineers and Physical Scientists. Macmillan, New York, 1992

Höppner, F., Klawonn, F., Kruse, R. (1996)
Fuzzy Clusteranalyse: Verfahren für die Bilderkennung, Klassifikation und Datenanalyse. Vieweg, Braunschweig, 1996

Hornik, K., Stinchcombe, M., White, H. (1989)
Multilayer Feedforward Networks are Universal Approximators. *Neural Networks,* 2 (5), 1989, p. 359-366

Hunt, C. (1995)
TCP/IP Netzwerk Administration. O'Reilly International Thomson Verlag, Bonn, 1995

Huntsberger, T.L., Ajjmarangsee, P. (1990)
Parallel self-organizing feature maps for unsupervised pattern recognition. *International Journal on Genetic Systems*, 16, 1990, p. 357-372

Jain, A.K. (1987)
Advances in statistical pattern recognition. In: Devijver, P., Kittler, J. *Pattern Recognition Theory and Applications*. Springer-Verlag, Berlin, Heidelberg, 1987, p. 1-19

Joentgen, A., Mikenina, L., Weber, R., Zimmermann, H.-J. (1998)
Dynamic Fuzzy Data Analysis: Similarity between Trajectories. In: Brauer, W. (Ed.) *Fuzzy-Neuro Systems'98 – Computational Intelligence*. Infix, Sankt Augustin, p. 98-105

Joentgen, A., Mikenina, L., Weber, R., Zimmermann, H.-J. (1999a)
Theorie und Methodologie der dynamischen unscharfen Datenanalyse. *Report DFG-Zi 104/27-1*, 1999

Joentgen, A., Mikenina, L., Weber, R., Zimmermann, H.-J. (1999b)
Dynamic Fuzzy Data Analysis based on Similarity between Functions. *Fuzzy Sets and Systems*, 105 (1), 1999, p. 81-90

Joentgen, A., Mikenina, L., Weber, R., Zeugner, A., Zimmermann, H.-J. (1999)
Automatic Fault Detection in Gearboxes by Dynamic Fuzzy Data Analysis. *Fuzzy Sets and Systems,* 105 (1), 1999, p. 123-132

Kastner, J.K., Hong, S.J. (1984)
A Review of Expert Systems. *European Journal of Operations Research,* 18, 1984, p. 285-292

Kiel, H.-U. (1996)
Integration studentischer Wohnanlagen in die Datenkommunikationsstruktur einer Hochschule. Diplomarbeit, Institut für Informatik und Rechenzentrum, Technische Universität Clausthal, Germany, 1996

Kohavi, Z. (1978)
Switching and Finite Automata Theory. 2nd edition, McGraw-Hill, New York, 1978

Kohonen, T. (1988)
Self-Organization and Associative Memory. Springer-Verlag, New-York, 1988

Kosko, B. (1992)
Neural Networks and Fuzzy Systems. Prentice-Hall, Englewood Cliffs, NJ, 1992

Krishnapuram, R., Nasraoui, O., Frigui, H. (1992)
The Fuzzy C Spherical Shells Algorithms: A new Approach. *IEEE Transactions on Neural Networks*, 3, 1992, p. 663-671

Krishnapuram, R., Keller, J. (1993)
A Possibilistic Approach to Clustering. *IEEE Transactions on Fuzzy Systems*, 1, 1993, p. 98-110

Kunisch, G. (1996)
Anpassung und Evaluierung statistischer Lernverfahren zur Behandlung dynamischer Aspekte im Data Mining. Master's Thesis, Fakultät für Mathematik und Wirtschaftswissenschaften, Universität Ulm, Germany, 1996

Lanquillon, C. (1997)
Dynamic Neural Classification. Diplomarbeit, Institut für Betriebssysteme und Rechnerverbund, Abteilung Fuzzy-Systeme und Soft-Computing, Universität Braunschweig (TU), Germany, 1997

Lee, S.C. (1975)
Fuzzy Neural Networks. *Mathematical Bioscience*, 23, 1975, p. 151-177

Looney, C.G. (1997)
Pattern Recognition Using Neural Networks: Theory and Algorithms for Engineers and Scientists. Oxford University Press, Oxford, New York, 1997

Mallat, S.G., Zhong, S. (1992)
Complete Signal Representation with Multiscale Edges, *IEEE Transactions PAMI,* Vol. 14, 1992, p. 710-732

Mann, S. (1983)
Ein Lernverfahren zur Modellierung zeitvarianter Systeme mittels unscharfer Klassifikation. Dissertation, Technische Hochschule Karl-Marx-Stadt, Germany, 1983

Marsili-Libelli, S. (1998)
Adaptive Fuzzy Monitoring and Fault Detection. *International Journal of COMADEM,* 1(3), 1998, p. 31-37

Minsky, M.L., Papert, S.A. (1988)
Perceptrons. Expanded Edition, MIT Press, Cambridge, MA, 1988

Nakhaeizadeh, G., Taylor, C., Kunisch, G. (1997)
Dynamic Supervised Learning. Some Basic Issues and Application Aspects. In R. Klar, O. Opitz (Eds.) *Classification and Knowledge Organization.* Springer Verlag, Berlin, Heidelberg, 1997, p. 123-135

Nauck, D., Klawonn, F., Kruse, R. (1996)
Neuronale Netze und Fuzzy Systeme. (2. Aufl.), Vieweg, Braunschweig, 1996

Nemirko, A.P., Manilo, L.A., Kalinichenko, A.N. (1995)
Waveform Classification for Dynamic Analysis of ECG. *Pattern recognition and Image Analysis,* Vol. 5 (1), 1995, p. 131-134

OMRON ELECTRONICS GmbH (1991)
OMRON: Fuzzy Logic – Ideen für die Industrieautomation. Hannover-Messe, 1991

Pal, S.K. (1977)
Fuzzy Sets and Decision making Approaches in Vowel and Speaker Recognition. *IEEE Transactions on Man, Machine, and Cybernetics*, 1977, p. 625-629

Pal, R.N., Bezdek, J.C. (1994)
Measuring Fuzzy Uncertainty. *IEEE Transactions on Fuzzy Systems*, Vol. 2, 1994, p. 107-118

Pedrycz, W. (1990)
Fuzzy Sets in Pattern Recognition: Methodology and Methods. *Pattern Recognition*, Vol. 23, 1990, p. 121-146

Pedrycz, W. (1990)
Fuzzy Sets in Pattern Recognition: Accomplishments and Challenges. *Fuzzy Sets and Systems,* 90, 1997, p. 171-176

Peltier, M.A., Dubuisson, B. (1993)
A Human State Detection System Based on a Fuzzy Approach. *Tooldiag'93, International Conference on Fault Diagnosis*, Toulouse, April, 1993, p. 645-652

Peschel, M., Mende, W. (1983)
Leben wir in einem Volterra-Welt? Akademie-Verlag, Berlin, 1983

Petrak, J. (1997)
Data Mining – Methoden und Anwendungen. Austrian Research Institute for Artificial Intelligence (ÖFAI), Vienna, TR-97-15, 1997, http://www.ai.univie.ac.at/cgi-bin/tr-online?number+97-15]

Pokropp, F. (1996)
Stichproben:Theorie und Verfahren. R. Oldenbourg Verlag, München, Wien, 1996

Rosenberg, J.M. (1986)
Dictionary of Artificial Intelligence and Robotics, 1986

Rosenblatt, F. (1958)
The Perceptron: A Probabilistic Model for Information Storage and Organization in the brain. *Psychological Review,* 654, 1958, p. 386-408

Rüger, B. (1989)
Induktive Statistik: Einführung für Wirtschafts- und Sozialwissenschaftler. R. Oldenbourg Verlag, München, Wien, 1989

Sato, M., Sato, Y., Jain, L.C. (1997)
Fuzzy Clustering Models and Applications, Physica-Verlag, 1997

Schleicher, U. (1994)
Vergleich und Erweiterung von c-Means Verfahren und Kohonen Netzten zur Fuzzy Clusterung: eine Untersuchung unter dem Aspekt der Anwendung in der Qualitätskontrolle. Magisterarbeit, Lehrstuhl für Unternehmensforschung, Rheinisch-Westfälische Technische Hochschule (RWTH) Aachen, Germany, 1994

Schreiber, T., Schmitz, A. (1997)
Classification of Time Series Data with Nonlinear Similarity Measures. *Physical Review Letters*, Vol. 79 (8), 1997, p. 1475-1478

Setnes, M., Kaymak, U. (1998)
Extended Fuzzy c-Means with Volume Prototypes and Cluster Merging. *Proceedings of the 6th European Conference on Intelligent Techniques and Soft Computing (EUFIT'98)*, Aachen, Germany, September 7-10, 1998, p. 1360-1364

Shewhart, W. A. (1931)
Economic Control of Quality Manufactured Product. Princeton, NJ, D. Van Nostrand Reinhold, 1931

Smith, A. E. (1994)
X-bar and r-control Chart Interpretation using Neural Computing. *International Journal of Production Research*, 32, 1994, p. 309-320

Stone, M. (1974)
Cross-validatiry Choice and assessment of statistical predictions. *Journal of Royal Statistical Society B*, 36, 1974, p. 111-147

Stöppler, S. (Hrsg.) (1980)
Dynamische Ökonomische Systeme: Analyse und Steuerung. Gabler GmbH, Wiesbaden, 1980

Struzik, Z.R. (1995)
The Wavelet Transform in the Solution to the Inverse Fractal Problem, *Fractals,* Vol. 3 (2), 1995, p. 329-350

Struzik, Z.R., Siebes, A. (1998)
Wavelet Transformation in Similarity Paradigm. In: Wu, X., Kotagiri, R., Korb, K.B. (Eds.) *Research and Development in Knowledge Discovery and Data Mining. Proceedings of the Second Pacific-Asia Conference PAKDD-98,* Melbourne, Australia, April 1998, Springer, 1998, p. 295-309

Stutz, C. (1998)
Partially Supervised Fuzzy c-Means Clustering with Cluster Merging. *Proceedings of the 6th European Conference on Intelligent Techniques and Soft Computing (EUFIT'98)*, Aachen, Germany, September 7-10, 1998, p. 1725-1729

Taylor, C. Nakhaeizadeh, G. (1997)
Learning in Dynamically Changing Domains: Theory Revision and Context Dependence Issues. In M. Someren, G. Widmer (Eds.) *Machine Learning: 9th European Conference on Machine Learning (ECML'97)*, Prague, Czech Republic, p. 353-360

Taylor, C. Nakhaeizadeh, G., Lanquillon, C. (1997)
Structural Change and Classification. In G. Nakhaeizadeh, I. Bruha, C. Taylor (Eds.) *Workshop Notes on Dynamically Changing Domains: Theory Revision and Context*

Dependence Issues, 9th European Conference on Machine Learning (ECML'97), Prague, Czech Republic, p. 67-78

Therrien, C.W. (1989)
Decision Estimation and Classification: An Introduction to Pattern Recognition and Related Topics. John Wiley & Sons, New York, 1989

Thole, U., Zimmermann, H.-J., Zysno, P. (1979)
On the suitability of minimum and product operators for the intersection of fuzzy sets. *Fuzzy Sets and Systems,* Vol. 2, 1979, p. 167-180

Vesterinen, A., Särelä, A., Vuorimaa, P. (1997)
Pre-processing for Fuzzy Anaesthesia Machine Fault Diagnosis. *Proceedings of the 5th European Conference on Intelligent Techniques and Soft Computing (EUFIT'97),* Aachen, Germany, September 8-11, 1997, p. 1799-1803

Welcker, J. (1991)
Technische Aktienanalyse: die Methoden der technischen Analyse mit Chart-Übungen. Verlag Moderne Industrie, Zürich, 1991

Widmer, G., Kubat, M. (1993)
Effective Learning in Dynamic Environments by explicit context tracking. In P. Brazdil (Ed.) *Machine Learning: Proceedings of the 6th European Conference on Machine Learning (ECML'93),* Vienna, Austria, April 1993. Lecture Notes in Artificial Intelligence (667), Springer-Verlag, Berlin, Germany, p. 227-243

Widmer, G., Kubat, M. (1996)
Learning in the Presence of Context Drift and Hidden Contexts. *Machine Learning*, 23, 1996, p. 69-101

Windham, M.P. (1981)
Cluster Validity for Fuzzy Clustering Algorithms. *Fuzzy Sets and Systems*, Vol. 5, 1981, p. 177-185

Xie, L.X., Beni, G. (1991)
A Validity Measure for Fuzzy Clustering: *IEEE Transactions on Pattern Recognition,* Annals Machine Intelligence, Vol: 13, 1991, p. 841-847

Yager, R.R., (1988)
On Ordered Weighted Averaging Aggregation Operators in Multicriteria Decision Making. *IEEE Transactions on Systems, Man & Cybernetics*, Vol. 18, 1988, p. 183-190

Yazdani, N., Ozsoyoglu, Z.M. (1996)
Sequence Matching of Images. In: *Proceedings of the 8th International Conference on Scientific and Statistical Database Management,* Stockholm, 1996, p. 53-62

Zadeh, L.A. (1977)
Fuzzy Sets and their Application to Classification and Clustering. In J. Van Ryzin (Ed.) *Classification and Clustering.* Academic Press, New York, 1977

Zahid, N., Abouelala, O., Limouri, M., Essaid, A. (1999)
Unsupervised Fuzzy Clustering. *Pattern Recognition Letters*, 20, 1999, p. 123-129

Zieba, S., Dubuisson, B. (1994)
Tool Wear Monitoring and Diagnosis in Milling using Vibration Signals. *SafeProcess'94,* Espoo, Finland, June 1994, p. 696-701

Zimmermann, H.-J. (1992)
Methoden und Modelle des Operations Research. Vieweg, Braunschweig, 1992

Zimmermann, H.-J. (Hrsg.) (1993)
Fuzzy Technologien: Prinzipien, Werkzeuge, Potentiale. VDI Verlag, Düsseldorf, 1993

Zimmermann, H.-J. (Hrsg.) (1995)
Datenanalyse. Düsseldorf, 1995

Zimmermann, H.-J. (1996)
Fuzzy Set Theory - and its Applications. Third edition, Boston, Dordrecht, London, 1996
Zimmermann, H.-J. (1997)
A fresh perspective on uncertainty modeling: Uncertainty vs. Uncertainty modeling. In M. Ayyub, M.M. Gupta (Hrsg.) *Uncertainty Analysis in Engineering and Sciences:Fuzzy Logic, Statistics, and Neural Approach.* Kluwer Academic Publishers, Boston, 1997, p. 353-364
Zimmermann, H.-J., Zysno, P. (1980)
Latent Connectives in Human Decision Making. *Fuzzy Sets and Systems,* Vol. 4, 1980, p. 37-51.
Zimmermann, H.-J., Zysno, P. (1985)
Quantifying Vagueness in Decision Models. *European Journal of Operational Research,* Vol. 22, 1985, p. 148-158.

APPENDIX

Unsupervised Optimal Fuzzy Clustering Algorithm of Gath and Geva

The advantage of the UOFC algorithm is the unsupervised initialisation of cluster prototypes and the ability to detect automatically the correct number of clusters based on the performance measures for cluster validity which use fuzzy hypervolume and density functions. The idea of the algorithm consists in iteration of the basic calculation scheme of the UOFC with an increasing number of clusters, evaluation of the cluster validity measure in each iteration and choice of the optimal number of clusters based on the validity criterion.

The calculation scheme of the UOFC algorithm is closely related to probability theory. Objects are interpreted as observations of an M-dimensional normally distributed random variable. Assignments of N observations x_j, j=1, ..., N, to c normal distributions f_i, i=1, .., c correspond to hard memberships $u_{ij} \in \{0, 1\}$. Expected values of normal distributions are interpreted as cluster prototypes. The distance function in the algorithm is chosen reverse proportional to the a posteriori probability (likelihood) to which a certain object x_j is an observation of a normal distribution f_i. This choice is motivated by the fact that a smaller distance of an object to a cluster prototype corresponds to a larger probability of a membership of an object into this cluster and vice versa. Generalising these results for the case of fuzzy memberships of objects to clusters (observations to normal distributions), the final calculation scheme of the UOFC algorithm is obtained.

The main steps of the UOFC algorithm are summarised in the following [Gath, Geva, 1989].

Algorithm 8: The unsupervised optimal fuzzy clustering (UOFC) algorithm.

1. Choose the initial cluster prototype as the mean location of all objects.

2. Calculate a new partition of objects by performing the following two phases:

2.1 apply the FCM algorithm with an Euclidean distance function;

$$d^2(\mathbf{x}_j, \mathbf{v}_i) = (\mathbf{x}_j - \mathbf{v}_i)^T \cdot (\mathbf{x}_j - \mathbf{v}_i) \qquad (7.1)$$

where $\mathbf{x}_j$ *is the j-th M-dimensional feature vector, j=1,...,N,* $\mathbf{v}_i$ *is the centre of the i-th cluster, i=1,...,c.*

2.2 apply the FCM algorithm with an exponential distance function (a fuzzy modification of the MLE algorithm):

$$d^2(\mathbf{x}_j, \mathbf{v}_i) = \frac{1}{P_i} \cdot \sqrt{\det(\mathbf{F}_i)} \cdot \exp\left[\frac{1}{2} \cdot (\mathbf{x}_j - \mathbf{v}_i) \cdot \mathbf{F}_i^{-1} \cdot (\mathbf{x}_j - \mathbf{v}_i)^T\right] \quad (7.2)$$

where P_i *is the a priory probability of selecting cluster i, which is determined as the normalised sum of membership degrees of objects to cluster i:*

$$P_i = \frac{1}{N}\sum_{j=1}^{N} u_{ij}, \quad (7.3)$$

F_i *is the fuzzy covariance matrix of cluster i given by:*

$$\mathbf{F}_i = \frac{\sum_{j=1}^{N} u_{ij}(\mathbf{x}_j - \mathbf{v}_i)(\mathbf{x}_j - \mathbf{v}_i)^T}{\sum_{j=1}^{N} u_{ij}} \quad (7.4)$$

3. *Calculate the following cluster validity measures:*

3.3 The fuzzy hypervolume criterion is calculated by:

$$v_{HV}(c) = \sum_{i=1}^{c} h_i \quad (7.5)$$

where the hypervolume of the i-th cluster is defined as $h_i = \sqrt{\det(\mathbf{F}_i)}$.

3.4 The partition density is calculated by:

$$v_{PD}(c) = \frac{\sum_{i=1}^{c} S_i}{\sum_{i=1}^{c} h_i} \quad (7.6)$$

where S_i *is the sum of 'good objects' in cluster i whose distance to the cluster centre does not exceed the standard deviation of features for this cluster:*

$$S_i = \sum_{j=1}^{N} u_{ij} \quad \forall x_j \in \left\{ x_j \left| (x_j - v_i)^T F_i^{-1} (x_j - v_i) < 1 \right. \right\}$$

3.5 The average partition density is calculated by:

$$v_{APD}(c) = \frac{1}{c} \sum_{i=1}^{c} \frac{S_i}{h_i} \qquad (7.7)$$

4. ***Add another cluster prototype equidistant (with large values of standard deviations) from all objects.***

5. ***If the current number of clusters is smaller than a predefined maximum number, then go to step 2.***

6. ***Else, stop and choose the optimal partition using the validity measure criteria.***

In contrast to FCM, PCM and the algorithm of Gustafson and Kessel, equations to determine cluster prototypes are obtained not from the objective function with a modified distance function, but by fuzzification of statistic equations of the expected value and the covariance matrix. Thus, the proposed equations represent a suitable heuristic which can not guarantee a global optimal solution. Due to the exponential distance function, the UOFC algorithm searches for an optimum in a narrow local region. Therefore it requires 'good' cluster prototypes for the initialisation which are calculated by the FCM in the first phase. These initial prototypes are refined in the second phase by the use of fuzzy modification of the MLE algorithm and the partition is adapted according to different cluster shapes, sizes and densities.

Description of Implemented Software

All algorithms proposed in this book and clustering algorithms used for classifier design and classification were implemented using software package MATLAB 5.1.

Adp_fpcm_traject	Algorithm for adaptive fuzzy classifier design and classification of dynamic objects represented by multidimensional trajectories, which uses the functional fuzzy c-means algorithm for classifier design and functional possibilistic c-means for classification and is based on the pointwise similarity measure between trajectories
Adp_ggpos_traject	Algorithm for adaptive fuzzy classifier design and classification of dynamic objects represented by multidimensional trajectories, which uses the modified Gath-Geva algorithm for classifier design and its possibilistic version for classification and is based on the pointwise similarity measure between trajectories
FCM_traject	Functional fuzzy c-means algorithm for clustering multidimensional trajectories based on the pointwise similarity measure
PCM_class_traject	Possibilistic c-means algorithm used for classification of multidimensional trajectories based on the pointwise similarity measure
GG_traject	The modified Gath-Geva algorithm for clustering multidimensional trajectories based on the pointwise similarity measure
GG_possclass_traject	The modified Gath-Geva algorithm for possibilistic classification of multidimensional trajectories based on the pointwise similarity measure
Density_clusters_tra	Calculation of densities of existing clusters and the average partition density, where the pointwise similarity measure between trajectories is used to obtain the covariance matrixes of clusters
Density_free_tra	This program partitions new free objects (trajectories) into cnew clusters using functional fuzzy c-means and possibilistic c-means based on the pointwise similarity measure for trajectories. Densities of new clusters and the average partition density of free objects are calculated using the localisation procedure for detecting compact dense groups within free objects.
Point_sim	Calculation of the distance measure based on the pointwise similarity measure between trajectories; the program includes the choice between different shapes of the membership function and different types of aggregation operators
Adp_fpcm_traject_whole	Algorithm for adaptive fuzzy classifier design and classification of dynamic objects applied to the whole temporal history of objects. The classifier design is static and the monitoring and adaptation procedures of the

	algorithm are used to determine the correct number of clusters. The algorithm uses the functional fuzzy c-means algorithm for classifier design and functional possibilistic c-means algorithm for classification, which are based on the pointwise similarity measure between trajectories
Adp_ggpos_traject_whole	Algorithm for adaptive fuzzy classifier design and classification of dynamic objects applied to the whole temporal history of objects. The classifier design is static and the monitoring and adaptation procedures of the algorithm are used to determine the correct number of clusters. The algorithm uses the modified Gath-Geva algorithm for classifier design and its possibilistic version for classification and is based on the pointwise similarity measure between trajectories
Adp_fpcm_traject_str	Algorithm for adaptive fuzzy classifier design and classification of dynamic objects represented by multidimensional trajectories, which uses the functional fuzzy c-means algorithm for classifier design and functional possibilistic c-means for classification and is based on the structural similarity measure between trajectories
Adp_ggpos_traject_str	Algorithm for adaptive fuzzy classifier design and classification of dynamic objects represented by multidimensional trajectories, which uses the modified Gath-Geva algorithm for classifier design and its possibilistic version for classification and is based on the structural similarity measure between trajectories
FCM_traject_str	Functional fuzzy c-means algorithm for clustering multidimensional trajectories based on the structural similarity measure
PCM_class_traject_str	Possibilistic c-means algorithm used for classification of multidimensional trajectories based on the structural similarity measure
GG_traject_str	The modified Gath-Geva algorithm for clustering multidimensional trajectories based on the structural similarity measure
GG_possclass_traject_str	The modified Gath-Geva algorithm for possibilistic classification of multidimensional trajectories based on the structural similarity measure
Density_clusters_tra_str	Calculation of densities of existing clusters and the average partition density, where the structural similarity measure between trajectories is used to obtain the covariance matrixes of clusters.
Density_free_tra_str	This program partitions new free objects (trajectories) into cnew clusters using functional fuzzy c-means and possibilistic c-means based on the structural similarity measure for trajectories. Densities of new clusters and the average partition density of free objects are calculated using the localisation procedure for detecting compact dense groups within free objects.

Struct_sim	**Calculation of the distance measure based on the structural similarity measure between trajectories; the program includes the choice between different shapes of the membership function and different ways of aggregation of partial similarity measures**
Temp_characteristics	**Calculation of 18 temporal characteristics of trajectories, which can be chosen in different combinations depending from the application**
Absorp_traject	**Absorption procedure and calculation of the number of new clusters within free objects**
Merging_proced	**Algorithm for detection of similar clusters and merging procedure**
Splitting_proced	**Algorithm for detection of heterogeneous clusters and splitting procedure**

INDEX